Basics of Laser Material Processing

Advances
in
Science
and
Technology

Technology Series

ADVANCES IN SCIENCE AND TECHNOLOGY

Basics of Laser Material Processing

Alexander G. Grigoryants

Translated from the Russian by P.S. Ivanov

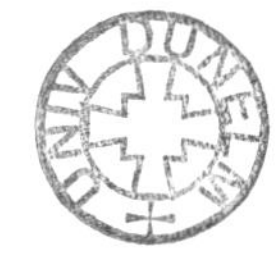

Mir Publishers
Moscow

CRC Press
Boca Raton Ann Arbor Tokyo London

Library of Congress Cataloging-in-Publication Data

Grigor'iants, A.G. (Aleksandr Grigor'evich)
Basics of laser material processing / A.G. Grigoryants; translated from the Russian by P.S. Ivanov.
p. cm. -- (Advances in science and technology)
Includes bibliographical references and index.
ISBN 0-8493-7534-7
1. Lasers--Industrial applications. 2. Manufacturing processes. I. Title. II. Series: Advances in science and technology in the USSR.
TA1677.G75 1994
621.36'6--dc20 94-6353
CIP

This edition published by CRC Press, Inc., 2000 Corporate Blvd. N.W., Boca Raton, Florida.
Direct all inquiries to CRC Press, Inc., 2000 Corporate Blvd., N.W., Boca Raton, Florida 33431.

International Standard Book Number 0-8493-7534-7
Library of Congress Card Number 94-6353
Printed in the United States of America 1 2 3 4 5 6 7 8 9 0
Printed on acid-free paper

To my wife Nataliya and children Grigorii and Natasha

CONTENTS

Part I
BASIC THEORY OF LASER PROCESSING

1 Types of Industrial Lasers

The laser is a stimulated emission device that generates a very narrow, intense beam of coherent light in the wavelength range covering the near ultraviolet, visible, and infrared regions. This explains why laser radiation is easy to focus to an extremely small spot comparable in area with the square of the wavelength [1, 2, 3]. Present-day lasers can provide very high levels of power per unit area. For comparison, Fig. 1.1 presents the power densities q, W/cm^2, for various energy sources.

The first industrial lasers came into being in the late 1960s. Since they emitted low-power radiation, the early lasers had limited uses, mainly in instrument making for welding, cutting, and other kinds of treatment of thin workpieces [4, 5].

Further advances in science and engineering have spurred the development of continuous wave and repetitively pulsed high-power lasers for processing structural materials employed in machine building [6, 7, 8].

The modern lasers most extensively applied in industry for material processing are solid-state neodymium-doped yttrium-aluminum-garnet (Nd-YAG) lasers, Nd-glass lasers, and molecular or gas carbon dioxide (CO_2) lasers.

They have high output powers, good efficiencies, and stable radiation parameters. Solid-state lasers employed in engineering industry can yield an average power of 50 to 500 W, and commercial models of CO_2 laser can generate over 15 kW [9].

Powerful solid-state lasers use either a crystalline host medium, such as YAG crystals, or an amorphous host medium, such as glass, which acts as a matrix to hold dopant Nd^{3+} ions implanted into the host. These ions form the lasing medium, called the lasant, which actually emits the light when excited by the working host material [10].

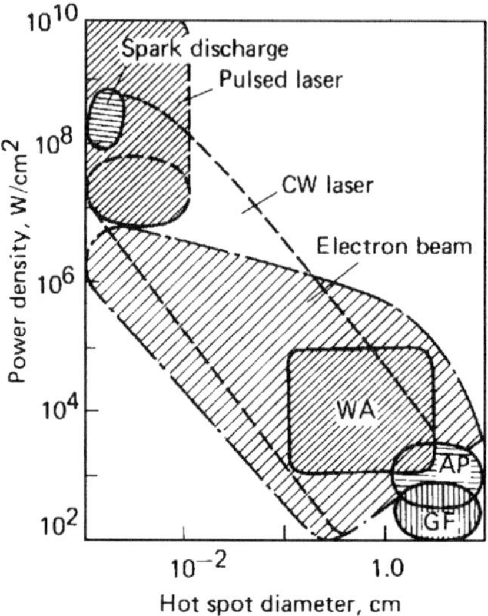

Fig. 1.1. Power densities of energy sources. Welding arc (WA); arc plasma (AP); gas flame (GF)

The industrial application of solid-state Nd-doped lasers attests that they are reliable, safe in operation, and simple to control. They can emit as high power in a pulse as 10^7 W or more. Solid-state neodymium lasers can process materials at an extremely high rate. A typical repetition rate varies from 0.05 to 50.0 kHz and the average power reaches 20 to 50 W with a high lasing efficiency of 4 to 7% for solid-state lasers. At low frequencies of 0.1 to 1.0 Hz, these lasers can yield up to 10 J in a 100-μs pulse and emit 10^5 W of peak power [11]. However, even the best solid-state lasers generate beams of comparatively low powers because of the small linear dimensions of synthetic crystals and their low thermal conductivity, which causes difficulties in cooling lasing elements.

Progress in crystal growth technology for solid-state lasers has offered the ways of increasing the lasing efficiency and power with an attendant increase in the quality and capacity of material processing [12, 13]. Yet crystal lasers today have a large beam divergence because of inhomogeneous crystal structure or irregular doping of the host material.

Laser technology harnessed to material processing entered a new development stage with the advent of gas lasers, modern versions of which can emit at a very high average power in continuous-wave and repetitive-pulsed modes.

The working medium of a gas laser is a pure gas, a mixture of gases, or a mixture of a gas and metal vapor, the excitation of which relies on an electric discharge, chemical reactions, and adiabatic outflow of the heated gas through a supersonic nozzle. The excitation methods categorize three types of gas lasers: gas-discharge, chemical, and gas-dynamic lasers. Gas-discharge lasers are the most widely applied in industry.

Various models of gas-discharge lasers are treated in numerous works. Here we look at some typical commercial gas-discharge lasers [9, 11].

Laser equipment extensively employs CO_2 lasers in which the lower vibrational levels of excited CO_2 molecules provide the mechanism for generating infrared radiation at a wavelength of 10.6 μm. Most CO_2 lasers use a mixture of carbon dioxide, CO_2, nitrogen, N_2, and helium in different ratios to increase the power output. Adding nitrogen intensifies the lasing action. Helium features a high heat capacity and increased heat conduction, so adding it speeds up heat removal, thereby reducing the mixture temperature.

CO_2 lasers largely operate on the principle of a self-sustained discharge that combines the functions of laser mixture pumping with ionization. This type of laser is the simplest in design and is able to provide up to 10 kW of power.

The laser with a nonself-sustained discharge depends for its operation on an external ionization and electric-discharge pumping of the gas mixture. Lasers are available in which electron beams and repetitive pulses perform the function of external ionization. Commercial small-size lasers with power outputs of 10 kW and above use fast electron beams for the purpose. Examples of this type of industrial laser which can provide up to 15 kW of power are the CO_2-ETL [14, 15] and the HPL-15 models [11]. The Tsiklon laser setup of Russian manufacture is quite simple in design and uses a 6-kW laser with repetitive pulse ionization and high beam quality [16].

An example of a fast gas transport laser that offers much promise for industrial systems is the Lantan model which uses short capacitive discharges for preionization [9, 17]. This laser can produce 2.0 kW power in CW mode and an average of 1.2 kW in repetitive pulse mode. What is important from the practical viewpoint is that this laser can be converted from CW lasing to pulse lasing by switching over the power supply circuits and changing the gas mixture composition.

In modern CO_2 lasers provision is made to maintain an optimal temperature of the gas mixture and thus utilize the laser more efficiently. For this, laser designs envisage either diffusion cooling of the gas mixture on the principle of heat removal from the discharge tube or convective cooling through circulation of the gas mixture.

Diffusion-cooled CW CO_2 lasers, also called slow-flow lasers, generally come in water-cooled discharge tube configurations, which yield high output powers per unit tube length. However, they display an increased beam divergence because of a large number of turning mirrors and the multimode character of emission. For this reason, the power density of a focused beam does not exceed 10^4-10^5 W/cm^2. Commercial versions of this type of laser include the Iglan-3 (3 kW) [18]. MKTL-1 (1.0 kW) [19], and Yupiter (1.0 kW) [20], all of Russian manufacture, and the M-400 of Ferranti (0.4 kW) [11] and Photon Soures Inc. (1.0 kW) [11].

Diffusion-cooled low-energy density lasers are only suitable for surface hardening but are not effective for welding, cutting, and precision material working. Convectively cooled CW CO_2 lasers also called fast-flow lasers, offer much higher emitting powers than diffusion-cooled ones. Both axial flow and transverse flow lasers exist, depending on the direction of gas flow relative to the optical axis of the resonator. In an axial fast flow laser, the gas mixture is blown through the tubes at high speed and recycled through a heat exchanger to cool it before delivering the mixture to the discharge zone.

A developed laser of this type is the Karat model [21]. A good sealed-off configuration does not require the laser mixture of CO_2, N_2, and He to be refreshed in the closed-loop cycle of gas flow, so lasing occurs with a minimum of gas mixture consumption. The Karat ensures stability of power output from 1 to 2 kW over time and has a long lifetime. It is used in commercial laser setups such as the Kometa and Latus-31 for various kinds of material treatment. A fast axial flow laser of United Technology Co. (USA) can deliver 6 kW of power. However, specialists believe axial fast flow lasers for power levels in excess of 5 kW will be too complex in design and may yield an inadequate beam [11].

The transverse fast flow lasers used in present-day industrial laser equipment can provide considerable power outputs. This type of laser is available in a number of designs which sustain a dc-excited transverse discharge. Typical commercial versions are the GTE-971 (1.2 kW), GTE-820 (2.5 kW), and GTE-975 (5 kW) of Spectra Physics Co. (USA). The gas discharge chamber consists of a hollow water-cooled cathode and segmented anode with each segment arranged transverse to the gas flow. This configuration ensures power levels from 1 to 5 kW [11].

A laser with cathode pins segmented in both transverse and axial directions to the gas flow can generate more power than a laser with a gas discharge chamber of the same dimensions. For example, the LT-1 model based on this cathode configuration can provide 5-kW power output [22] and the LSU welding setup can deliver 7 kW of power [11]. The TL-10 setup whose discharge chamber is arranged on the same principle can emit at a power of 10 kW [23].

The development of a series of lasers based on a 5-kW laser setup using a fast transverse gas flow with a self-sustained discharge transverse to the flow is of interest [9]. Examples are the 1.2-kW LOK-2 and 2-kW LOK-3M models of CW lasing. But building such lasers with powers exceeding 10 kW does not seem expedient [11].

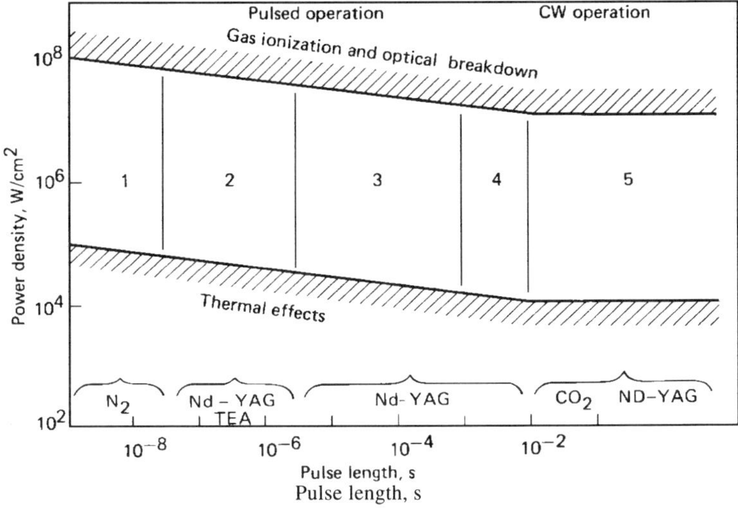

Pulse length, s

Fig. 1.2. Use of lasers for treatment processes: (*1*) thin film vaporization; (*2*) scribing and film vaporization; (*3*) hole drilling and punching; (*4*) spot welding and surface heat treatment; (*5*) deep melting, gas-assisted cutting, surface hardening, and splitting. TEA, transversely excited atmospheric pressure laser

CO_2 lasers can operate both in the continuous wave and in the repetitively pulsed mode on the basis of either self-sustained discharge or nonself-sustained discharge controlled by a pulsed beam of electrons. The laser mixture used in both modes of operation consists of the same components [9, 25].

Pulsed lasers with a small pulse length and high pulse repetition rate are suitable for discrimination technology. They can also be used for heat treatments such as amorphization, welding, cutting, and hole piercing. Pulsed CO_2 lasers such as the Lantan [9, 17], Dyatel [26], and IPL-1 [27] have the following main parameters: pulse energies E_p of 6 J, 5 to 10 J, and 2 J; pulse repetition rates f_p of 500 Hz, 100 to 200 Hz, and 500 Hz; average powers P_{av} of 1.2 kW, 1.0 kW, and 1.0 kW; and pulse lengths t_p of 5×10^{-5} to 15×10^{-5} s, 10^{-7} s, and 1×10^{-5} to 1×10^{-4} s, respectively.

A molecular CO laser can ensure a high conversion efficiency and can operate in various beam generation modes at shorter wavelengths than its CO_2 counterpart. Metal vapor lasers feature a high gain of the laser medium and so they can provide high power levels in the visible and ultraviolet regions at small overall dimensions. Excimer lasers using the excimer molecules (dimers) of some gases and their halogens can give high powers in the ultraviolet region and offer the possibility of smooth adjustment of the radiation frequency over a wide range of emission wavelengths.

Promising lasers for material processing include devices relying on the electrode-free high-frequency excitation of the lasant and compact waveguide CO_2 lasers which are noted for peak values in their power density.

The use of heavy-duty lasers for various kinds of material processing depends on their output characteristics and operating time-varying parameters. A diagram illustrative of the possible applications of lasers for various heat treatment processes is shown in Fig. 1.2.

Most of the processes illustrated in Fig. 1.2 find advantageous use in industry while some processes, such as shock hardening and amorphization, are at the stage of experiments. Material processing has indisputable advantages because lasers can provide a high energy density, do not require a vacuum as electron beams do in order to carry out a heat treatment process, can process elements of any dimensions, provide easy beam energy transfer in space, and allow for simple automation of the material treatment process. This explains why the last years have seen increased interest in the application of lasers for material processing in all developed countries.

Industrial lasers will gain still wider acceptance with the introduction of simple and reliable automatic laser work cells capable of performing an extended range of various treatment processes.

2 Focusing of Laser Radiation

An optical system is the basic building block of any laser setup, which generally consists of an optical resonator that acts as the first beam-shaping element, beam focusing and control optics, and other optical elements intended to transform the laser beam parameters.

2.1. Optical Resonators

An optical resonator consists of a system of mirrors arranged such that the effective length of the laser medium is increased by multiple reflection. A simple resonator used in early solid-state lasers consisted of parallel-plane mirrors. This however incurred high diffraction losses and led to unstable energy parameters.

Modern optical resonators commonly employ spherical mirrors or combinations of spherical and plane mirrors. The main parameters of an optical resonator are the curvature radii r_1 and r_2 of the reflecting surfaces, the distance L between the mirrors, i.e. the true optical path length of the laser, and the diameter D of the aperture stop limiting the beams cross section (Fig. 2.1). The generalized parameters are the form factors defined as

$$g_1 = 1 - L/r_1 \text{ and } g_2 = 1 - L/r_2. \tag{2.1}$$

All stable resonators (Fig. 2.2*g* and *h*) obey the relation

$$0 < g_1 g_2 < 1. \tag{2.2}$$

In a stable resonator the light rays remain near the optical axis as they are multiply reflected inside the resonator.

If condition (2.2) is not met, viz. if

$$g_1 g_2 < 0 \quad \text{or } g_1 g_2 > 1, \tag{2.3}$$

the resonator is said to be unstable (Fig. 2.2*d* and *e*). In an unstable resonator

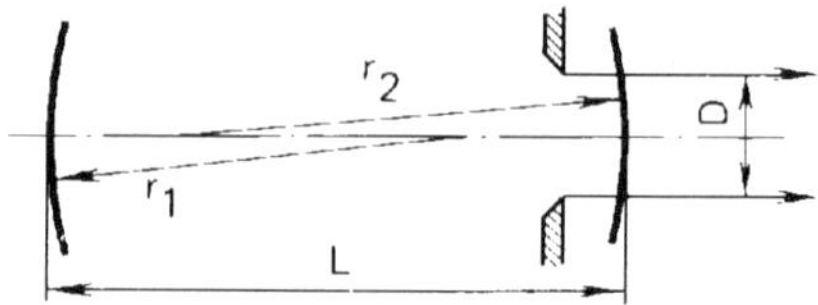

Fig. 2.1. Main parameters of resonators

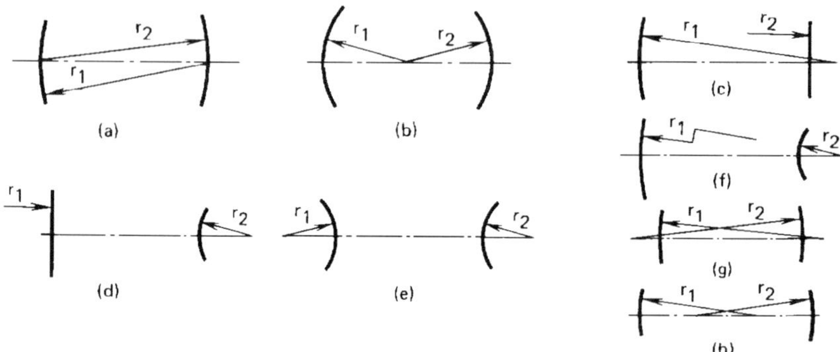

Fig. 2.2. Optical resonators using various combinations of mirrors:
(*a*) confocal resonator, $r_1 = r_2 = L$ and $g_1 = g_2 = 0$; (*b*) concentric resonator, $r_1 = r_2 = L/2$ and $g_1 = g_2 = -1$; (*c*) planoconcave resonator, $r_1 = L$, $r_2 = \infty$, $g_1 = 0$, and $g_2 = 1$; (*d*) planoconvex resonator, $r_1 = \infty$, $r_2 = -L$, $g_1 = 1$, and $g_2 = 2$; (*e*) resonator emitting beams in two opposite directions, $r_1 = r_2 = -L$ and $g_1 = g_2 = 2$; (*f*) telescopic confocal resonator, $r_1 = 2L$, $r_2 = -L$, $g_1 = 0.5$, and $g_2 = 2$; (*g*) resonator with nearly plane mirrors, $r_1 >> L$, $r_2 >> L$, and $0 < g_1 = g_2 < 1$; (*h*) nearly concentric resonator, $r_1 = r_2 > L/2$ and $0 < g_1 g_2 < 1$

the light rays diverge from the axis after repeated reflections. A resonator, such as those illustrated in Fig. 2.2*a*, *b*, *c* and *f* [28], is on the verge of stability if

$$g_1 g_2 = 0 \text{ or } g_1 g_2 = 1. \tag{2.4}$$

Both stable and unstable resonators are applicable. The stable resonator of Fig. 2.2*g* is most popular and offers the highest efficiency. For CO or CO_2 lasers, unstable resonators are preferable as they produce a hollow-ring beam, i.e. yield a ring-shaped beam at the output.

Laser radiation emitted from a resonator displays a sharp directionality, i.e. a small angle of divergence, so it is possible to transfer the beam quite efficiently for a large distance and focus it on a very small area.

A laser beam diverges because of beam diffraction at the output aperture of the resonator, inhomogeneity of the laser medium, mirror deformation, and other factors. Diffraction-limited divergence is amenable to computation, whereas other types of divergence are most often defined experimentally.

It is practical to use the notion of divergence determined by a flat or a solid angle of a cone which encloses a beam with a specified share of energy or power. Proceeding from this concept, we need first to define the share of power in order to estimate the divergence. In [29] the suggestion was to adopt an angle that accounts for 83.8% of the total beam power. This share of power enables us to correlate the divergence of a circular solid beam that is specific to stable resonators with the divergence of an annular (hollow-ring) beam typical of unstable resonators.

The divergence of a circular beam is commonly defined by the angular radius for the first minimum of the diffraction pattern [30]. In accordance with this definition, the divergence angle (the sine of the half-angle of divergence) is given by

$$\theta' = 1.22\ \lambda/D, \tag{2.5}$$

where λ is the emission wavelength and D is the beam diameter. The above formula shows that nearly 83.8% of the beam power falls within the optical cone with such an angle.

Let us define the beam divergence for an unstable telescopic resonator often used in industrial lasers. To estimate the share of beam power P that falls within the angle of diffraction θ, we use the expression derived in [30]:

$$P(\theta') = \frac{2}{\left(1 - 1/m^2\right)^2} \int_0^x \left[\frac{J_1(x)}{x} - \frac{1}{m^2}\frac{J_1(x/m)}{x/m}\right]^2 x\,dx\,, \tag{2.6}$$

where $m = D/d_{in}$ is the magnification of the telescopic resonator (Fig. 2.3); d_{in} is the inner diameter of the output ring; $x = kD\theta/2$; $k = 2\pi/\lambda$ is the wave number; and J_2 is a Bessel function of order one.

Using expression (2.6) for an annular beam, we can calculate the angle of divergence that encloses 83.8% of the total power. We need to solve (2.6) for the divergence angle θ', assuming P to be equal to 0.838. The solution can be found in numerical form on a computer. Figure 2.4 presents the results obtained for a generalized angle of divergence $D\theta'/\lambda$ as a function of reciprocal of magnification $1/m$. Solid curve *1* is the plot of $D\theta'/\lambda$ versus $1/m$ laid off on the x axis with a step length of 0.025. Dashed curve *2* represents approximated results that can be found to a sufficient accuracy for design

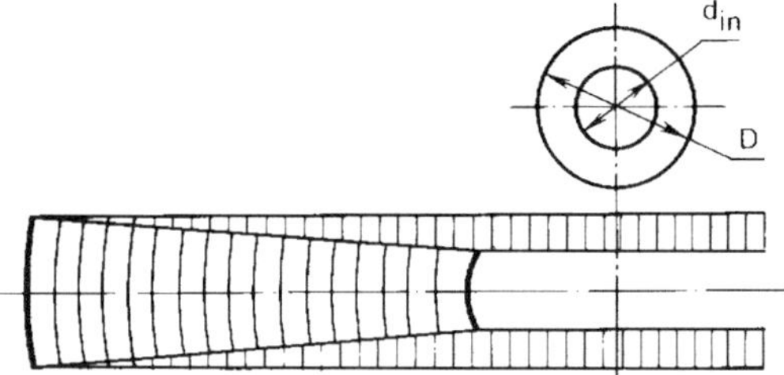

Fig. 2.3. Unstable telescopic resonator ($m = D/d_{in}$)

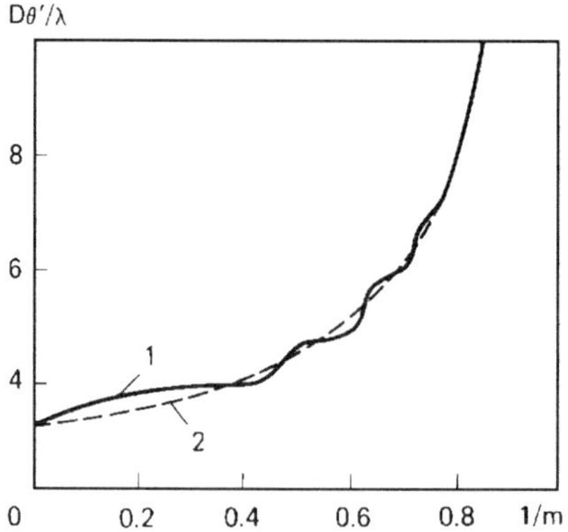

Fig. 2.4. Generalized divergence angle $D\theta'/\lambda$ as a function of $1/m$

calculations from the relation

$$D\theta' / \lambda \cong 2.4\left(\frac{1}{1-1/m}\right). \tag{2.7}$$

We can readily obtain from (2.7) a relation for the divergence angle as a function of the beam diameter and resonator magnification:

$$\theta' = 2.4\frac{\lambda}{D}\left(\frac{1}{1-1/m}\right). \tag{2.8}$$

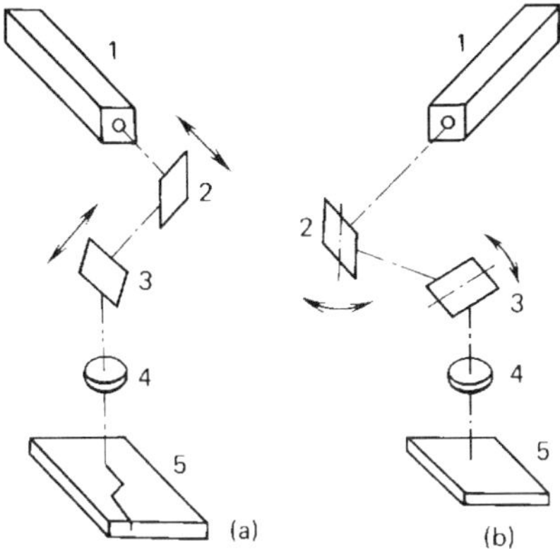

Fig. 2.5. Laser beam guiding systems using (*a*) sliding mirrors and (*b*) rotating mirrors: (*1*) laser head; (*2* and *3*) movable plane mirrors; (*4*) focusing lens; (*5*) target

By transforming (2.8), the final expression for θ' assumes the form

$$\theta' = 1.2\ \lambda/b, \tag{2.9}$$

where $b = (D - d_{in})/2$ is the width of the output ring. Comparing (2.5) with (2.9) reveals that the resulting angles of divergence differ insignificantly. The only difference is that the divergence defined by expression (2.5) for a circular beam depends on the beam diameter, whereas the divergence given by (2.9) depends on the ring width of the ring-shaped image of the beam.

The divergence defined by formulas (2.5) and (2.9) determine the minimum possible diffraction-limited divergence. The actual divergence is always greater than the diffraction-limited divergence. The actual divergences for a circular beam and an annular beam, respectively, can be found from the following expressions:

$$\theta = 1.22\ e\lambda/D, \tag{2.10}$$

$$\theta = 1.2\ e\lambda/b, \tag{2.11}$$

where e is an empirical coefficient accounting for the increase in the divergence due to other influences. The value of this factor is usually preset after measuring the actual divergence for a specific laser setup.

2.2. Laser Optical Systems

A laser beam leaving the resonator cannot as a rule be used directly for heat treatment since its energy density, power density distribution, and other output characteristics do not yet comply with the requirements for a particular process. Laser apparatus employ various optical transformation systems to enhance the efficiency of the heat treatment process and make the operation of the setup easier.

Whatever the purpose of a laser technology setup, its main task is to position the beam with respect to the target either by moving the beam if the workpieces are large and heavy or by moving small workpieces relative to the stationary beam.

Simple arrangements for a laser beam positioning system to guide the beam on a stationary workpiece with two moving mirrors along any desired path appear in Fig. 2.5 [4]. In the system of Fig. 2.5*a*, the flat mirrors *2* and *3* move linearly without changing their angular positions. Mirror *2* moves along the axis of the beam emitted from laser *1* and mirror *3* does so along the axis of the beam reflected from mirror *2*. The focusing lens *4* and mirror *3* are moved in step with mirror *2*. The system of Fig. 2.5*b* guides the beam with two mirrors rotating in two mutually perpendicular planes.

A diagram of a laser beam positioning system using three rotating mirrors to scan the beam across the surface of a large stationary object is shown in Fig. 2.6 [31]. The system has a frame *6* moving along rails *7* above a workpiece *8*. The first mirror *2* guides the beam of laser *1* parallel to the direction of travel of the frame and the second mirror *3* directs the beam along the frame. The third mirror *4* is mounted on a carriage *5* which moves it together with the focusing lens *9* along the frame.

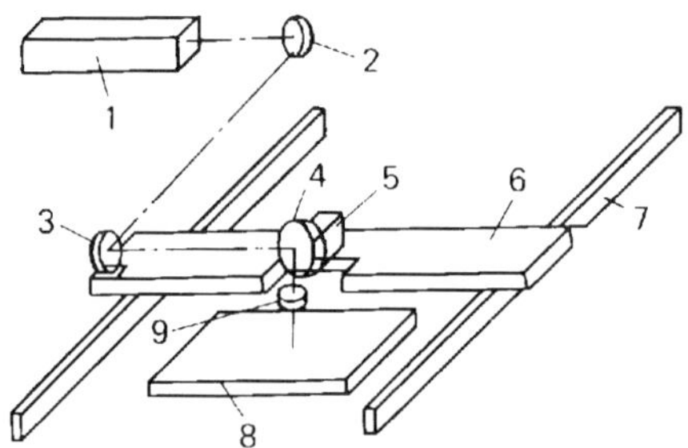

Fig. 2.6. Beam positioning system using three mirrors

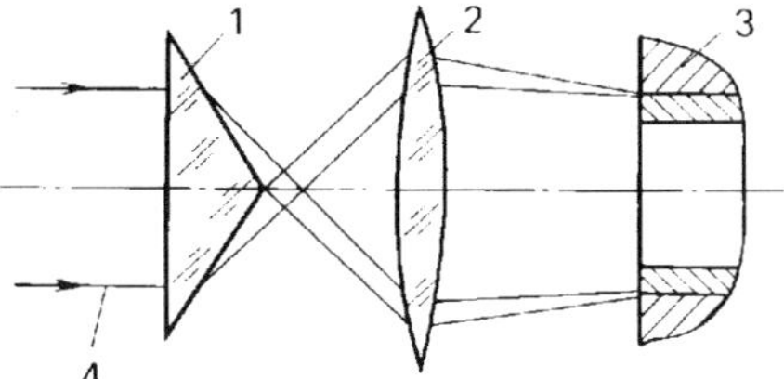

Fig. 2.7. Diagram of an optical system for transformation of a circular beam into an annular beam:
(*1*) axicon; (*2*) lens; (*3*) target; (*4*) laser beam

Some laser processing techniques require beam handling equipment to move the beam in a circular path. It is often sufficient to use rotating flat mirrors installed at an angle of 45° to the beam axis or a set of flat mirrors for the laser beam focusing system [31]. In this respect, a system based on rotating lenses answers the need, which is quite simple in design.

For circular weld seams to be run on a plane, it is best to use an axicon in the form of a transparent cone with its base facing the beam (Fig. 2.7) or a shield with a ring-shaped opening. As illustrated in Fig. 2.7, the hollow beam emerging from the axicon passes through a focusing lens which increases the output fluence (energy density) as it reduces the ring width and provides a ring-shaped distribution of energy on the target. A single pulse can be sufficient to form the weld without moving the beam or parts. This system does not afford a high energy density on the target because of the large cross-sectional area of the focused beam. It can be used for welding or cladding at a small penetration depth or for surface hardening of annular sections.

It should be noted that a laser with an unstable resonator does not require the transformation system such as illustrated in Fig. 2.7 since the resonator emits a ring-shaped beam. The focusing lens alone suffices to direct the beam on the entire ring-shaped surface.

Figure 2.8 illustrates an optical system for the heat treatment of cylindrical surfaces. It consists of three mirrors and a focusing lens. The mirror *2* has an outer reflecting surface and mirrors *6* and *3* have inner reflecting surfaces. The mirrors *2* and *6* transform the original circular or hollow beam *1* into a large-diameter annular beam which passes through the focusing lens *5* and falls on the workpiece *4* after reflection from the mirror 3. The reflected radiation encircles the workpiece and simultaneously heats it over the entire circumference. Moving the lens along its axis makes it possible to adjust the fluence over the target surface. A disadvantage of the system is that it produces a low beam fluence on the work surface.

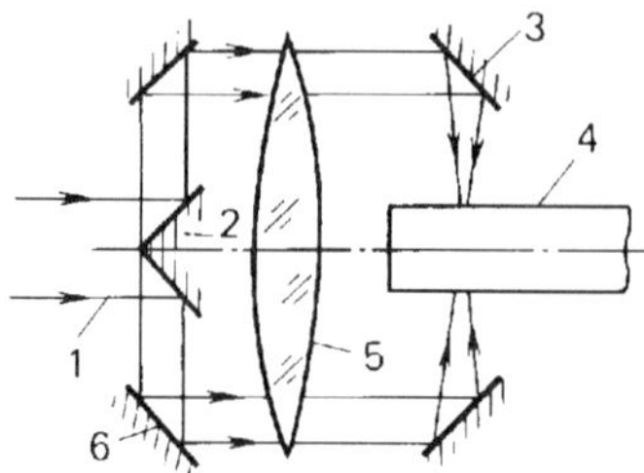

Fig. 2.8. Diagram of an optical system for heat treatment of the surface of a cylindrical part

In a number of cases, an elongated hot spot produced on the target helps increase the efficiency of a laser treating process, especially the process involving the pulse-laser to material interaction. A cylindrical lens can focus the laser beam to a narrow strip. An elongated hot spot can also be produced with a focusing lens whose optical axis is shifted relative to the beam axis in the direction of motion of a workpiece.

Laser radiation is easy to adjust and control by dividing the beam into a few split beams with a simple optical device. This property of laser radiation can be best used by concurrently coupling the beam energy to the heat-affected zone (HAZ) in order to reduce the rate of cooling of the weld and also to heat the bead or deposited metal just after welding to temper the treated area and attain other objectives. In [31] a simple system was put forward for splitting the beam by a focusing lens with a central opening (Fig. 2.9). The central

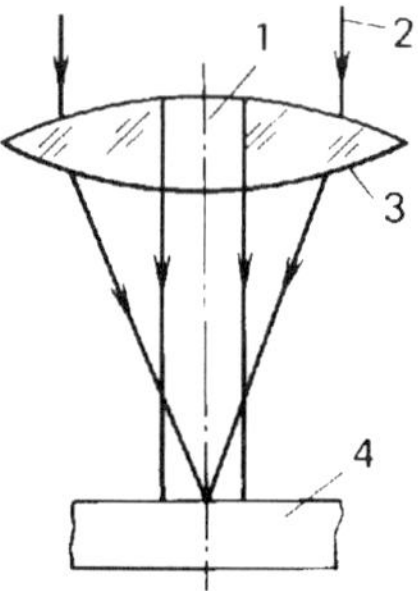

Fig. 2.9. Laser beam splitter for concurrent heating of weld edges along the heat treat track

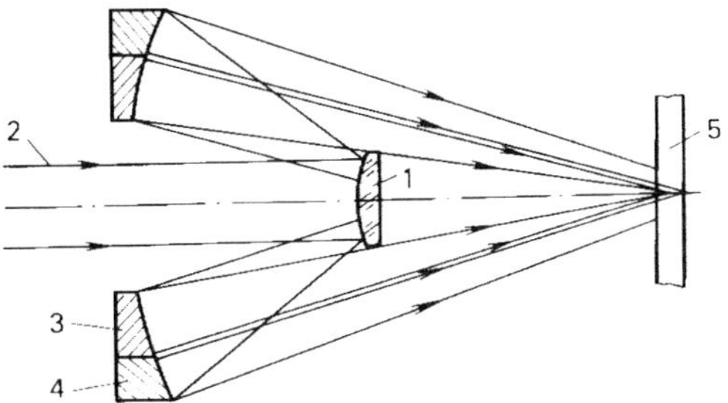

Fig. 2.10. Inverted-type Cassegrain axisymmetric reflector used as a beam splitter to heat up weld edges in the heat-affected zone

portion of the beam *2* freely passes through the hole *1* in the lens *3* and impinges on the part *4* to be welded. This beam portion produces a round hot spot of a low heat power density that is merely sufficient to keep the weld zone hot. The lens focuses the peripheral portion of the beam to a smaller spot of a much higher energy that is enough to weld the edges together. Changing the hole diameter enables the welding unit to adjust the hot spot temperature and thus control the rate of cooling of the weld and the heat-affected zone.

A Casssegrain inverted axisymmetrical reflector is also suitable for beam splitting (Fig. 2.10). The laser beam *2* reflected from a convex mirror *1* falls on two concave mirrors *3* and *4* which differ in focal length. The inner mirror *3* focuses the rays to a small spot of high energy density sufficient to weld the part 5. The outer mirror *4* which has a larger focal length deflects the rays to produce an annular hot spot whose energy density is enough to heat up the edges to be welded and to control the rate of cooling of the HAZ. Moving the mirror *4* in the direction of the beam axis changes the hot spot temperature and adjusts the rate of cooling.

The efficiency of laser processing strongly depends on the absorptivity of a material. All materials reflect some of the incident radiation and reduce the energy efficiency of a heat treatment process. To raise the efficiency a chamber covered with an inner reflecting coating can be used. The chamber has a cover gas inlet and an inlet for the laser beam. The scattered rays are reflected from the inner wall and directed to the target. The cover gas delivered to the chamber provides coverage of the work surface.

An optical system is able to smoothly adjust the irradiance during heat treatment by moving the focusing lens along the beam axis, which changes the depth of lens focus and, hence, the energy density at the hot spot.

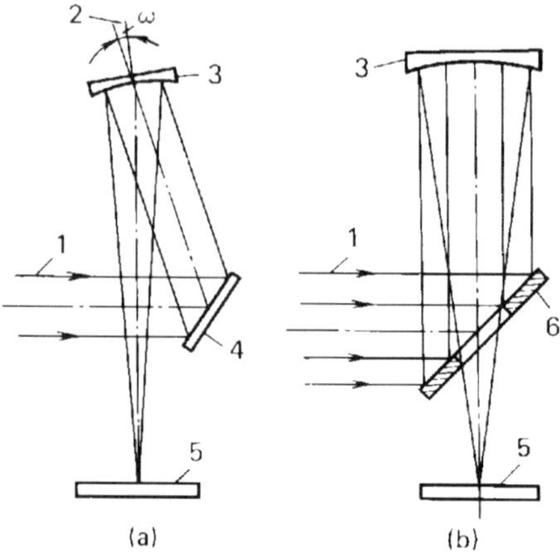

Fig. 2.11. Laser beam focusing systems using single focusing mirrors:
(*1*) laser beam; (*2*) focusing mirror axis; (*3*) focusing spherical mirror; (*4*) flat mirror; (*5*) workpiece; (*6*) plane ring-shaped mirror

Beam handling equipment uses both light-transmission optics (feedthrough lenses) and reflective metal optics (mirrors) for beam focusing. Focusing lenses are applicable for beam powers up to 1 kW. Higher-power lasers use reflective focusing systems of metal optics that feature a much longer service life.

A focusing reflector can be built around a single focusing mirror or any two-mirror Cassegrain reflector. Schematic diagrams of single focusing reflectors are shown in Fig. 2.11. The reflector of Fig. 2.11*a* has a focusing spherical mirror *3* placed at an angle ω to the axis of the incident beam. The reflector can also include a flat reflecting mirror *4* if it is necessary to heat-treat large parts. If laser-treated parts are small, the laser beam can be focused as illustrated in Fig. 2.11*b*. The plane ring-shaped mirror *6* reflects the beam and the spherical mirror *3* coaxial with the reflected beam focuses the beam reflected from its surface on the target *5*. This reflector is used in practice where the laser beam takes the form of a ring in section.

A two-mirror Cassegrain reflector can focus a high-power beam. The original diagram of this reflector was suggested by Cassegrain in the 17th century and is given in Fig. 2.12. The reflector consists of a large parabolic concave mirror *1* and a small hyperbolic convex mirror *2*. A feature of the reflector is that it ensures an ideal image of a point at infinity and can thus focus the beam to a spot with a high power density [30].

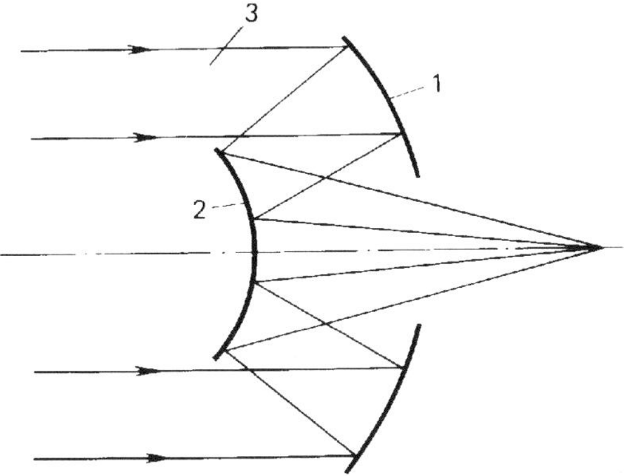

Fig. 2.12. Original configuration of the Cassegrain reflector:
(*1*) parabolic concave mirror; (*2*) hyperbolic convex mirror; (*3*) laser beam

In its original configuration, the Cassegrain reflector if used to focus a laser beam for material processing suffers from several disadvantages: (1) its secondary mirror shields the radiation if the inner diameter of the annular beam is smaller than the diameter of this mirror; (2) the mirror surfaces of the second order involve manufacturing difficulties; and (3) the high irradiance on the secondary mirror overheats the reflecting surface and leads to its geometric distortion, which impairs the focusing characteristics.

An inverted symmetric Cassegrain reflector (Fig. 2.13) consists of two spherical mirrors and eliminates the 2nd and the 3d shortcoming of the classical configuration, but the effect of shielding of the central beam portion remains. This on-axis reflective system is suitable for laser processing only if the inner diameter of the annular beam is smaller than the diameter of the secondary mirror. The system also imposes limits on the distance from the laser head to the target because the output ring quickly becomes irregular with increasing distance from the resonator.

The oblique-angled (off-axis) Cassegrain reflector, such as illustrated in Fig. 2.14, is more appropriate for laser processing. This reflective system does not shield the beam and all the beam energy reaches the target irrespective of its distribution in the cross section.

Mirror optics for powerful industrial lasers are commonly made of pure copper which has high reflectivity and thermal conductivity. Metal mirrors must be rigid enough to withstand high levels of laser radiation and not to undergo appreciable thermal deformations.

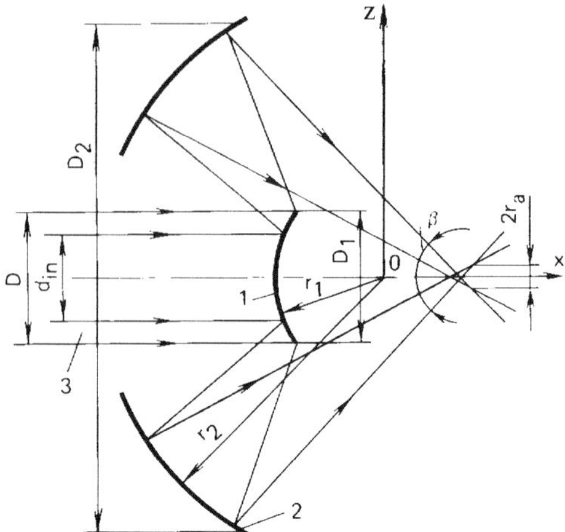

Fig. 2.13. Schematic of the center-fed Cassegrain reflector for focusing the laser beam:
(*1*) secondary convex mirror; (*2*) primary concave ring-shaped mirror; (*3*) laser beam; r_a, radius of aberration-limited focused spot

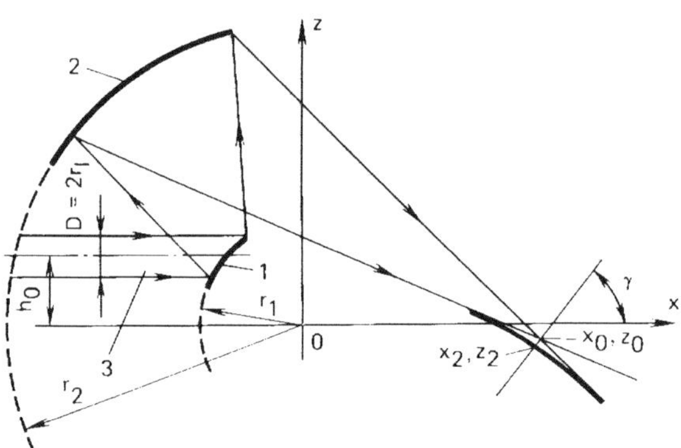

Fig. 2.14. Schematic of the oblique-angled Cassegrain reflector:
(*1*) secondary convex mirror; (*2*) primary concave mirror; (*3*) laser beam, r_l, radius of laser beam; h_0, height of beam axis above reflector axis

Multikilowatt lasers are made complete with forced water cooling systems to reduce the thermal strains of the copper mirrors and to increase their stability. To stabilize the mirrors in storage and operation, it is good practice to cover them with protective, multilayer transparent coatings.

Lasers with powers up to 1 kW generally operate with focusing lenses from optically transparent materials. The most widely used CO_2 lasers emit at a wavelength of 10.6 μm and use lenses produced from artificial optical crystals of sodium chloride, potassium chloride, sodium fluorine, potassium bromide, thallium bromide-iodide of grade KRS-5, thallium bromide-chloride of grade KRS-6, and barium fluoride. Less popular materials for optical lenses include germanium, zinc selenide, cadmium telluride, silicon, and gallium arsenide, which have a higher refractive index and thus increased reflection losses. Thus, lenses from the above materials require transparent covering.

Lenses from KCl and NaCl single crystals are cheap and easy to manufacture. They have a low refractive index, suffer small reflection losses, and do not require coatings. However, KCl and NaCl are hygroscopic, have low strength and poor stability.

Improving optical systems to the level of perfection of laser techniques has been a task for researchers of the field for some time. However, a great deal of effort is still required to investigate thoroughly the characteristics of optical systems and work out adequate recommendations of how to optimize them. Further advances in laser material processing call for detailed studies of the effects of focusing system characteristics on the process parameters. The efficiency of most laser heating processes greatly depends on the amount of laser energy coupled to the hot spot of a minimum size.

2.3. General Concepts of Focusing System Design

Focusing systems are designed on the basis of the laws of physical and geometrical optics [30]. Physical optics considers the wave nature of light and deals with such properties of light as monochromaticity, interference, coherence, and polarization. Geometrical optics considers a light beam as the source of energy propagating in a definite direction. Design involved in geometrical optics is performed based on the laws of rectilinear and independent propagation of a light beam, refraction and reflection.

Let us consider the main concepts of geometrical optics necessary to design an optical system. A light ray can be regarded as the axis of a light tube. The rays emanating from a point or converging at a point are said to be homocentric and the rays parallel and close to the optical axis are said to be paraxial. In order to formulate the relations for the propagation of laser beams in optical systems it is important to evaluate aberration which is understood to be an image error in the optical system due to the beam deflection from an ideal direction. This means that a homocentric beam, having passed through an optical system, will deviate from its homocentric configuration at the exit of the system. In the general case, the deviation occurs after the refraction and reflection of the light on the first surface.

Every optical medium has a refractive index n which is defined as the ratio between the sine of the angle of beam incidence to the sine of the angle of refraction given the beam enters the optical medium from a vacuum whose refractive index is unity. In approximate calculations, the refractive index of air can be taken to be unity.

In optical systems, plane or spherical surfaces mainly act as reflective or refractive surfaces. Surfaces such as ellipsoid, hyperboloid, or paraboloid of revolution can reduce aberrations, but aspheric mirrors are difficult to manufacture. Being expensive, they do not practically find use in industrial lasers.

Refractive and reflective surfaces form an optical system, for example, a center-fed system in which the centers of spherical surfaces lie on the same straight line.

In an ideal case, parallel rays entering a focusing system converge at a point called the focus. In geometrical optics, the refractive surface for parallel rays is changed for an equivalent plane, referred to as the principal plane, which is normal to the optical axis and displays the points of intersection of the rays heading toward the plane or traveling from it. The distance from this plane to the focal point of a lens is its focal length.

Figure 2.15 displays the diagram of travel of a ray through a lens and demonstrates the basic concepts of geometrical optics. The incoming ray travels

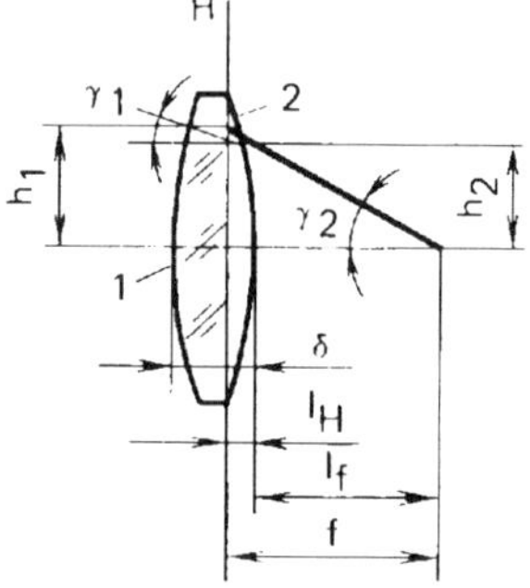

Fig. 2.15. Travel of the light ray through a lens with principal plane H

parallel to the lens axis at a height h_1 above this axis, so the angle of incidence of the ray is zero. The refractive index of the lens in air is, n, the lens thickness is δ, and the radii of curvature of surfaces *1* and *2* are r_1 and r_2 respectively. After refraction of the light ray on the surface *1*, the angle γ_1 of incidence of the ray to the lens axis becomes

$$\gamma_1 = h_1(n - 1)/nr_1. \tag{2.12}$$

The height at which the ray intersects the surface *2* is

$$h_2 = h_1[1 - \delta(n - 1)/nr_1]. \tag{2.13}$$

The angle of incidence of the ray coming from the lens is

$$\gamma_2 = h_1\left[(n-1)\left(\frac{1}{r_1}-\frac{1}{r_2}\right)+\frac{(n-1)^2}{nr_1r_2}\delta\right]. \tag{2.14}$$

As seen from the diagram of Fig. 2.15, the lens focal length

$$f = h_1/\tan \gamma_2. \tag{2.15}$$

Assuming the rays to be paraxial, we may let $\gamma_2 = \tan \gamma_2$ and thus write (2.14) in the form

$$f = h_1/\gamma_2. \tag{2.16}$$

Substituting (2.14) into (2.16), we define the relation for the focal power of the lens which is the reciprocal of the focal length f,

$$1/f = (n-1)\left(\frac{1}{r_1}-\frac{1}{r_2}\right)+\frac{(n-1)^2}{nr_1r_2}\delta. \tag{2.17}$$

The distance from the focal point to the vertex of the second surface of the lens is

$$l_f = f\,[1 - \delta(n - 1)/nr_1]. \tag{2.18}$$

The distance from this vertex to the principal plane H is

$$l_H = f\,\delta(n - 1)/nr_1. \qquad (2.19)$$

For a thin lens in which the distance between the lens vertices, i.e. the lens thickness δ, is much smaller than the curvature radii r_1 and r_2 of the lens surfaces, we may take $\delta = 0$. Formula (2.17) then assumes the form

$$1/f = (n - 1)(1/r_1 - 1/r_2). \qquad (2.20)$$

In order to generalize the calculations for a focusing system's parameters, we can regard reflection from spherical mirrors as a special case of refraction for which the refractive index retains its absolute value, but changes sign.

An important property of laser radiation is its monochromaticity, that is there are no chromatic aberrations when focusing a laser beam. Chromatic aberrations arise from the dependence of the refractive index on wavelengths. The main types of monochromatic aberration are astigmatism, coma, distortion, and spherical aberration. The first three types of aberration are typical of an inclined beam. Since in the focusing system of a laser device the principal plane of the lens is usually normal to the laser beam, the design only has to deal with spherical aberration [32].

To illustrate the mechanism of spherical aberration, consider an example of travel of parallel rays through a lens (Fig. 2.16). For simplicity, the first surface is shown flat, so it does not refract rays.The second surface is spherical. The oncoming rays travel through the second surface at different angles of incidence. The farther the parallel rays are from the axis, the closer the points of ray convergence behind the lens to the lens vertex. The lens thus does not focus the rays at a common point. Instead, the focused beam is tapered.The outer surface of this configuration near its narrow section is called the caustic surface.

The distance b_l along the axis between the points of convergence of inner rays and extreme rays is the longitudinal spherical aberration. The radius b_t of the blurred circle in the focal plane is the transverse (lateral) spherical aberration.

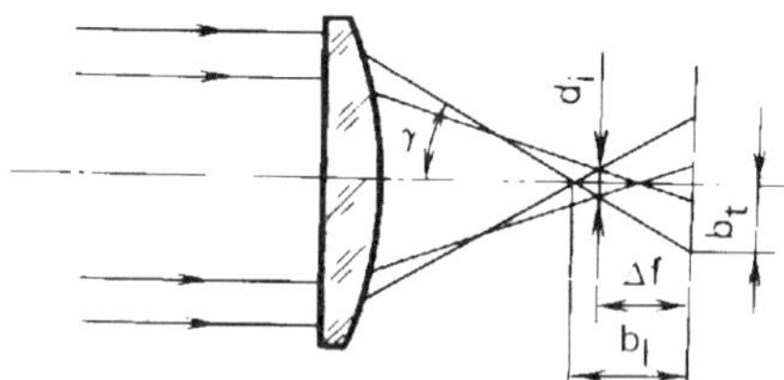

Fig. 2.16. Illustration of spherical aberrations

The smallest blur circle forms a distance Δf from the focal plane in the plane of the best focal setting. The ideal diameter d_i of this blur circle (blur of an ideal focused spot) is about half the transverse spherical aberration:

$$d_i = b_t/2. \tag{2.21}$$

The distance from the ideal focused spot to the focal plane is about three-fourths of b_l:

$$\Delta f \cong 3b_l / 4. \tag{2.22}$$

In the general case, to define the aberrations of an optical system, we need to calculate the paths of the refracted and reflected rays and determine the coordinates of the points where the rays intersect the focal plane. The difference between the coordinates of these points and the point of an ideal paraxial focus will give the aberration.

Calculating the optical system's parameters that satisfy a specified aberration involves a repeated estimation of the ray paths for different values of the parameters. This design approach is time-consuming and can only be done on a computer. All the same, it does help establish a general pattern or a functional relationship between aberrations and a focusing system's parameters. A more preferable approach is to formulate approximate relations between aberrations and system parameters using the theory of aberrations.

The aberrations that are functionally dependent on optical system parameters are generally written in the form of power series where the odd terms starting from the third-order term are left intact.

In practical calculations it is enough to deal with the third-order term proceeding from the third-order theory of aberrations [30]. This theory enables us to obtain a formula for a transverse spherical aberration b_t when the beam of parallel rays is circular in section and the principal plane of a thin lens ($\delta = 0$) is normal to the optical axis:

$$b_t = \sum_{k=1}^{k} p_k (D/2)^3 / 2f^2. \tag{2.23}$$

Here k is the order of the optical surface; D is the diameter of the incoming beam; f is the lens focal length; and p_k is a factor defined as

$$p_k = (\Delta\alpha_k / \Delta\mu_k)^2 \Delta\alpha_k \mu_k, \tag{2.24}$$

where α_k is the slope of the outer refracted ray; $\Delta\alpha_k = \alpha_{k+1} - \alpha_k$; $\mu_k = 1/n_k$ is the reciprocal of the refractive index; and $\Delta\mu_k = \mu_{k+1} - \mu_k$.

The theory of third-order aberrations thus helps us to obtain simple and illustrative expressions for the size of a focal spot, and hence the energy density in the focused beam.

The size of the focal spot, d_f, is the sum of the divergence-limited component d_θ and the aberration-limited component d_a,

$$d_f = d_\theta + d_a, \tag{2.25}$$

where

$$d_\theta = f\theta \tag{2.26}$$

and

$$d_a = b_t/2. \tag{2.27}$$

2.4. Simple Focusing Lenses

Industrial laser devices largely use single-element focusing lenses coaxial with the beam. The principal plane of such a lens is normal to the beam axis. Using the statements presented in Sect. 2.3, we can estimate the aberration-limited spot diameter d_a from the spherical aberration b_t as defined by (2.23). The sum $p' = \sum_{k-1}^{k} p_k$ in (2.23) is found assuming that the lens is in the air whose refractive index is unity,

$$p' = \left(\frac{\alpha_2}{1/n-1}\right)^2 (\alpha_2/n), \tag{2.28}$$

where α_2 is the slope of an outer ray after refraction at the first surface with a radius of curvature of r_1. The value of α_2 can be estimated from the expression

$$\alpha_2 = (n-1)f/r_1 n. \tag{2.29}$$

For a thin lens, expression (2.29) subject to (2.20) can be written as

$$\alpha_2 = \frac{1}{n(1-c)}, \tag{2.30}$$

where the ratio of the lens radii is

$$c = r_1/r_2. \tag{2.31}$$

Formula (2.23) defines the transverse aberration, which is half the aberration-limited focal spot diameter, i.e.,

$$d_a = p'D^3/32\,f^2. \tag{2.32}$$

The least diameters d_a correspond to minimum values of p', which is the case where the condition below holds:

$$c = (2n^2 - n - 4)/(2n^2 + n). \tag{2.33}$$

Formula (2.33) permits us to determine the shape of a lens of a specified refractive index n that leads to the lowest aberration. For example, the widely used KCl lens ($n = 1.455$) suffers the least aberration at $c = -0.215$ and $p' = 2.45$. This is a biconvex lens with different radii of curvature of the surfaces, of which the more convex surface must face the oncoming light.

An analysis of formula (2.32) reveals that aberrations strongly vary with the diameter of the laser beam. Low-power lasers emit narrow beams, and so the aberrations involved in focusing them are insignificant. In high-power lasers, the laser beam diameter is many times larger and can reach several centimeters.

In laser cutting, welding, and cladding, optimal heat treatment requires the highest energy density at the prescribed parameters of laser radiation. Consider relation (2.25) for determining the focal spot size with regard to beam divergence and aberrations. Substituting expression (2.11) for the divergence of the ring mode output, (2.26) for a diffraction-limited ring-shaped spot size, and (2.32) for the aberration-limited spot size into relation (2.25), we obtain

$$d_f = 1.2\,e\lambda\,f/b + p'D^3/32\,f^2. \tag{2.34}$$

From formula (2.34) we can derive an expression for an optimal focal length f_{opt} that ensures a minimum focal spot diameter $d_{f\,min}$. We differentiate equation (2.34) with respect to f and equate the derivative to zero. Solving this equation yields

$$f_{opt} \cong 0.37 D\left(\frac{p'b}{e\lambda}\right)^{1/3}. \tag{2.35}$$

To simplify the procedure of calculating the optimal focal length f_{opt} by expression (2.35) at the specified values of the laser beam diameter and beam divergence defined by equation (2.11), we use the chart in Fig. 2.17, which illustrates how the beam divergence θ varies with the beam diameter D for a KCl lens with $n = 1.455$, which has the lowest aberrations at $c = -0.215$ and $p' = 2.45$. After determining f_{opt} from the nomograph, we can estimate the lens parameters by calculating α_2 from equation (2.30) and the radii of curvatures of the first and the second surface using (2.29) and (2.31) respectively. The lens diameter is chosen such that it is no less than the beam diameter.

If a KCl lens of another shape but with a known ratio c is needed, we can estimate the parameter p' from (2.28) and (2.30) and then calculate the optimal focal length from the expression

$$f_{opt} = (p'/2.45)^{1/3} f'_{opt}, \tag{2.36}$$

where f'_{opt} is the focal length determined from the chart of Fig. 2.17 for a biconvex lens with $c = -0.215$.

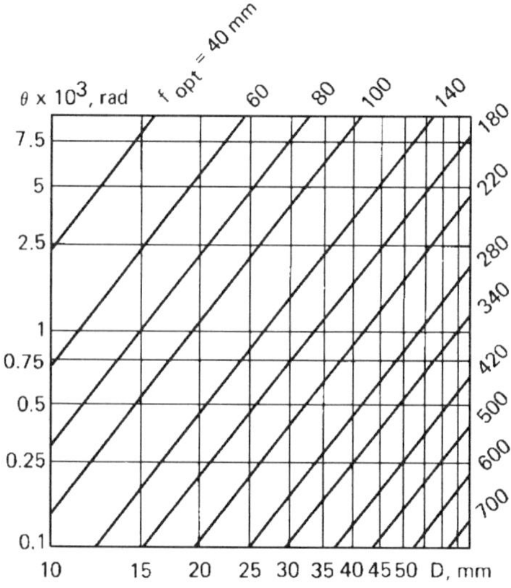

Fig. 2.17. Nomograph of divergence angles versus beam diameter D and divergence θ for estimating f_{opt} of the lens having the lowest aberration at $c = -0.215$ and $p' = 2.45$

Formula (2.36) yields $f_{opt} = 1.05\, f'_{opt}$ for a convexo-plane lens at $c = 0$; $f_{opt} = 1.14f'_{opt}$ for a biconvex lens at $c = -1$; and $f_{opt} = 1.61f'_{opt}$ for a plano-convex lens at $c = -\infty$.

The nomographs in Fig. 2.17 can also be used to optimize the single spherical mirror, for example, in the system of Fig. 2.11*b*. The radius of curvature of a spherical focusing mirror is $r = 0.93f'_{opt}$.

By substituting formula (2.35) into (2.34) we can derive a relation for a minimum possible focal spot size

$$d_{f\,min} \cong 0.67\, D(p')^{1/3}(e\lambda/b)^{2/3}. \qquad (2.37)$$

In the general case, the above statements enable us to optimize the parameters of a focusing system in the following sequence. First, we must know the diameter of a circular beam or the diameter and width of an output ring of the beam emerging from the resonator. Given the experimental values of e for the resonator, we estimate the beam divergence from equation (2.10) or (2.11). Further, after fixing n for the selected lens, we determine the ratio c from (2.33) that offers the lowest aberration.

Next we calculate α, p', and f_{opt} with equations (2.30), (2.28), and (2.35), respectively, and then we obtain the radii of curvature of the first and the second surface from (2.29) and (2.31), respectively. Finally, we calculate a minimum focal spot size with equation (2.37).

Equation (2.37) implies that there is a possibility of further minimizing the spot diameter $d_{f\,min}$. As described above, focusing systems with reduced aberrations has the greatest potential for the purpose. One more approach is to decrease the beam diameter D. An aperture placed across the beam reduces the diameter but entails a reduction in the beam power. A telescopic system is a better choice, but it leads to an increase in the beam divergence in proportion to a decrease in the beam diameter. Thus, a decrease in D by a factor of m causes $d_{f\,min}$ to decrease by a factor of $m^{1/3}$, which is indicative of the weak effect of the energy density increase.

As regards lasers with unstable resonators, conical beam integrators, or beam formers (Fig. 2.18) are preferable. A beam integrator can reduce the beam diameter without as increase in beam divergence [32] because the width of the output ring remains the same after beam diameter reduction, as follows from equation (2.11). A decrease in $d_{f\,min}$ is proportional to a decrease in D, and so this way of increasing the energy density is more effective.

The above design statements for laser beam focusing were checked experimentally by studying the melting efficiency of laser radiation focused with potassium chloride lenses of different focal lengths. The refractive index n of KCl lenses is 1.455. The ratio $c = r_1/r_2$ was zero for all lenses. The parameters α_2 and p' were the same for all lenses and equal to 0.68 and 2.81

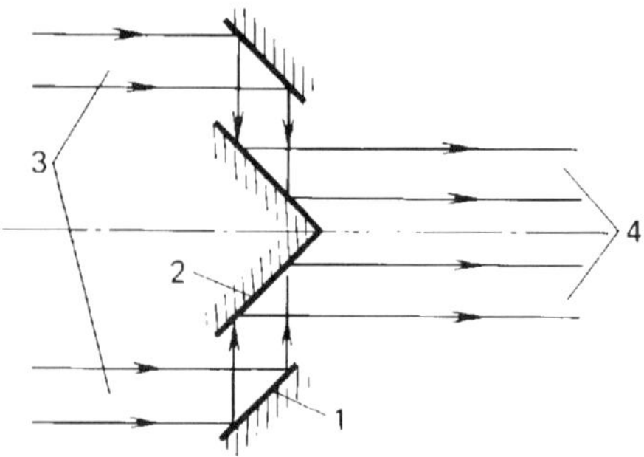

Fig. 2.18. Beam integrator for reducing the diameter of a ring-shaped beam:
(*1*) outer cone; (*2*) inner cone; (*3*) incoming ring-shaped beam; (*4*) outcoming ring-shaped beam

respectively. In the experiments, use was made of a laser setup that provided a ring-shaped beam 35 mm in diameter with a width of the output ring of 7 mm [22].

All things being equal, the melt depth varies with the energy density at the hot spot and is generally independent of the angle of beam divergence for a low beam energy. The changes in the melt depth yield an optimal focal length for the greatest penetration and minimum size of the focal spot.

In all the experiments the diameter of the beam incident on a lens was held constant, as was the beam power, by siting the lens at the same distance from the resonator. The laser setup used helium and carbon dioxide delivered to the beam-material interaction zone to provide coverage of the molten metal on corrosion-resistant steel plates and carbon steel plates respectively.

The tests were made on stepped specimens with 1-mm steps s to determine an ideal focused spot position and, hence, an optimal focal length for each lens to an accuracy of 0.5 mm (Fig. 2.19). After melting, each shelf was cut into two transverse sections to estimate the arithmetic mean of the melt depth for each step. The distance from a lens to the step with the largest penetration depth was taken as the lens focal length.

Figure 2.20 illustrates experimental data for the melt depth h as a function of the focal length f of convexo-plane lenses. The variations in the melt depth with the focal length is the same for the two steel grades at different rates of melting. In all cases, a maximum penetration is found to be at a focal length of 165 mm.

For comparison, let us calculate the optimal focal length with a minimum focal spot size. Assuming the empirical coefficient e in equation (2.35) is unity

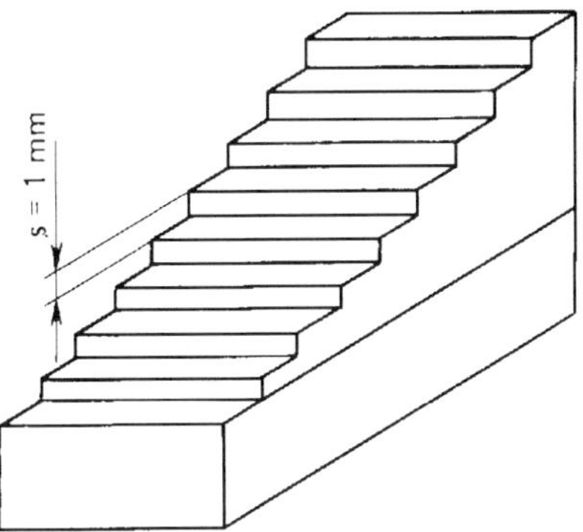

Fig. 2.19. Step-like specimen

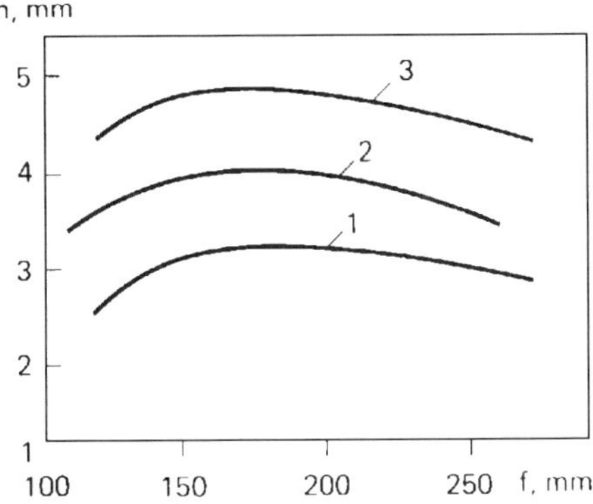

Fig. 2.20. Melt depth h in low-carbon steel specimens (curves *1* and *3*) and corrosion-resistant steel specimens (curve *2*) as a function of the focal length f of planoconvex lenses and melting speed of 16, 25, and 32 mm/s (curves *3*, *2*, and *1*, respectively)

and substituting into (2.35) the output ring with width $b \cong 7$ mm, $p' = 2.81$, $D \cong 35$ mm, $e = 1$, and $\lambda = 10.6$ μm, we obtain $f_{opt} = 160$ mm, which is close to the experimental value.

A comparison between theoretical and experimental results attests the validity of this technique for estimating the optimal focal lengths of lenses used in laser processing.

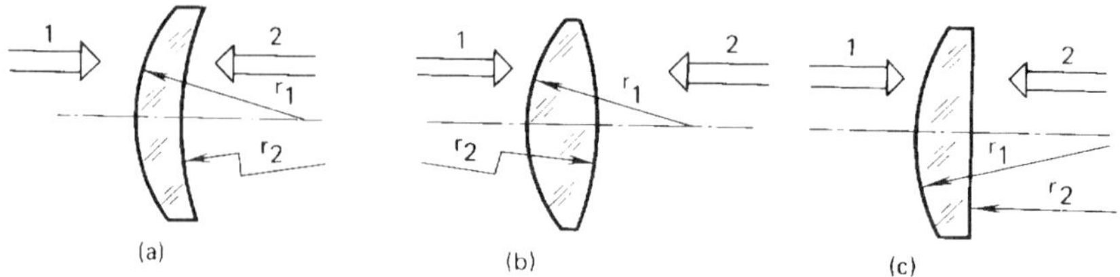

Fig. 2.21. Lenses for determining the effect of aberrations on penetration depth: (*a*) 170-mm focal length lens with $r_1 = 52$ mm, $r_2 = 160$ mm, and $p' = 7.42$ for beam *1* and $r_1 = -160$ mm, $r_2 = -52$ mm, and $p' = 22$ for beam *2*; (*b*) 170-mm focal length lens with $r_1 = 94$ mm, $r_2 = -437$ mm and $p' = 2.45$ for beam *1* and $r_1 = 437$ mm, $r_2 = -94$ mm, and $p' = 7.24$ for beam *2*; (*c*) 165-mm focal length lens with $r_1 = 75$ mm, $r_2 = \infty$, and $p' = 2.81$ for beam *1* and $r_1 = \infty$, $r_2 = -75$ mm, and $p' = 10.23$ for beam *2*

What is of much interest to process engineers is a feasible and unique ways of selecting the lens configuration that provides a minimum focal spot. In order to verify the relations worked out for design purposes, experiments were undertaken on three KCl lenses of approximately the same focal length to examine how aberrations affect the focal spot size and, hence, the melt depth (Fig. 2.21). Each lens can have two values of the parameter p' depending on which of the surfaces of the lens faces the oncoming light.

The laser welding was made on stepped specimens from low-carbon steel in a carbon dioxide atmosphere by the technique described above. Figure 2.22 gives experimental curves of the relative penetration depth h_{rel} plotted against the ratio c and parameter α_2. The graph also shows how p' varies with c and α_2. The relative depth h_{rel} was taken to be unity when a lens with a minimum value of p' gave the deepest penetration.

The graph of Fig. 2.22 demonstrates that the weld penetration is the largest at a minimum value of p' that ensures the least aberrations. This is the case for a biconvex lens with radii of curvatures of $r_1 = 94$ mm and $r_2 = -437$ mm, the more curved side of which faces the beam (points *1*). If the light enters the same lens from its less curved side, the penetration depth is smaller (points *2*). The convexo-plane lens (points *3*) provides almost the same weld penetration as the lens with the lowest aberrations. The meniscus lens (points *4*) performed less satisfactorily than the previous two lenses and the plane-convex lens (points *5*) yielded the poorest results.

The experimental results confirm the basic design relationships, so the designer can use them to select an adequate focusing lens with the desired parameters to get the highest efficiency for a laser heating process.

A practical problem is how to site accurately the selected lens with respect to the work surface. To this end, experiments were performed to estimate how the penetration depth varies with the deviation of the focal point from its

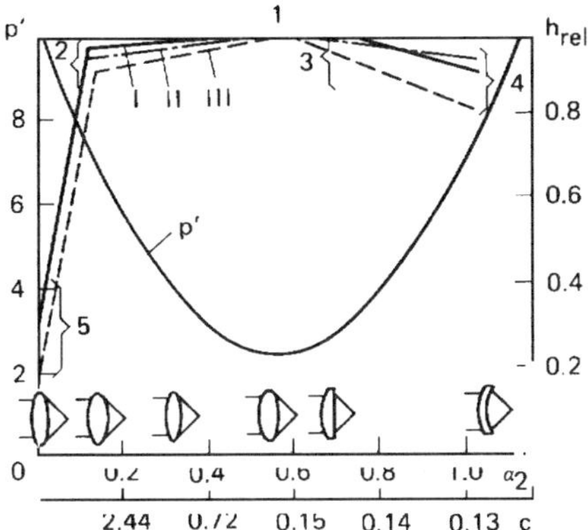

Fig. 2.22. Parameter p' and relative penetration depth h_{rel} versus c and α_2 at weld speed of 16 mm/s (curve *I*), 20 mm/s (curve *II*), and 32 mm/s (curve *III*)

optimal position. The beam was focused by plano-convex lenses with different radii of curvature and focal lengths of 165, 220, and 290 mm. The value of p' was 2.81 for all lenses. Tests were made on low-carbon steel specimens in the atmosphere of carbon dioxide at a treatment speed of 20 mm/s. Figure 2.23 displays the curves of relative melt depth h_{rel} versus deviation ξ of the focus from its optimal position. A shorter-focus lens causes the worst decrease in the melt depth. With a shift of the focal point by 5 mm, the depth of melting drops by 6, 35, and 50% for lenses with focal lengths of 290, 220, and

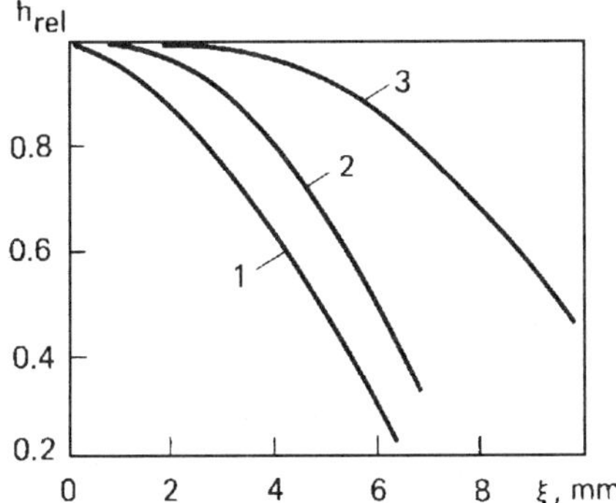

Fig. 2.23. Relative penetration depth as a function of the focus displacement from an optimal focal length of the lens of 165 mm (curve *1*), 220 mm (curve *2*), and 290 mm (curve *3*)

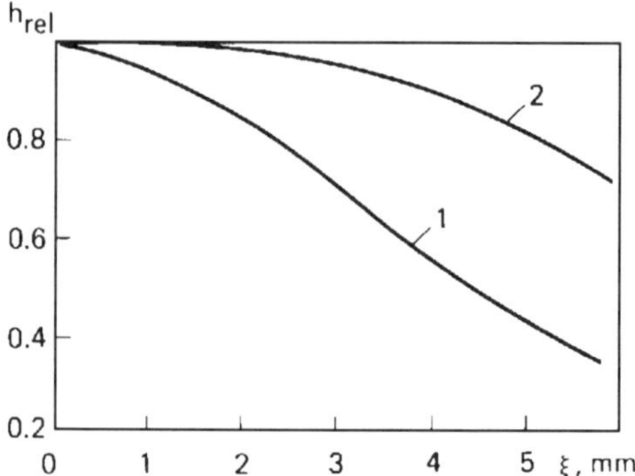

Fig. 2.24. Relative penetration depth as a function of the focus displacement from the optimal position for a plano-convex lens with f_{opt} = 140 mm, p' = 2.81 (curve *1*) and p' = 10.23 (curve *2*)

165 mm, respectively. If a change in the melt depth is to be kept within 10%, the focal point displacement for lenses with focal lengths of 165, 220, and 290 mm must not exceed 1.5, 3.2, and 5.7 mm respectively.

The deviations are less severe for lenses with higher aberration. Figure 2.24 illustrates how the penetration depth h_{rel} in low-carbon steel treated at a speed of 20 mm/s in a carbon dioxide atmosphere varies with focal point displacement ξ for lenses with f_{opt} of 140 mm. Curve *2* is the plot of h_{rel} for a plano-convex lens which has 3.64 times the aberration of a convexo-plane lens (curve *1*). The melt depth for the case of the plano-convex lens changes less with the focal point deviation.

Assuming that a permissible change in the melt depth is 10%, the focal point deviations for the convexo-plane and the plano-convex lens are 1.7 and 4 mm respectively, as follows from Fig. 2.24.

The above discussion of changes in the penetration depth should take into account the laser power density in the near-focal region. It is apparent from Fig. 2.16 that the beam has a minimum cross section (waist) in the plane of an ideal focused spot. The length l of the near-focal region is the distance between the cross sections in which the power density drops to half the power density q_0 at the waist (Fig. 2.25). The length l should be defined considering beam divergence and aberration. An approximate relation for the length l can be derived from (2.34):

$$l \cong p'D^2/16\,f + 2(e \times 1.2\lambda/b)\,f^2/D. \qquad (2.38)$$

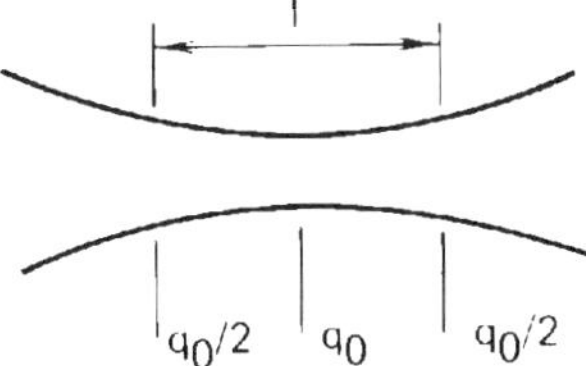

Fig. 2.25. Laser beam geometry in the near-focal region

This expression allows us to interpret uniquely the experimentally established curves of penetration depth as a function of focal length deviation. It is obvious from Fig. 2.23 that a beam focused by a short focal length lens for a given focal point deviation melts metal to a smaller depth because, other things being equal, it has a shorter length l. On the other hand, it follows from Fig. 2.24 that the melt depth changes less for a deviation of the focus of a lens that has a greater aberration (higher value of p') because it has a greater length l.

We can therefore conclude that a lens with a greater length l or a lens with a large aberration offers a more stable penetration depth. An increase in the penetration stability need not entail a decrease in either laser power density or power efficiency. In each specific case, a comparison of the focusing system used should be undertaken with due regard for the beam power, speed of heat treatment, cover gas, etc.

2.5. Focusing with Eccentric Lenses

If a lens is set not accurately with respect to the beam, an eccentricity h_e appears. This can increase aberration and reduce the process efficiency. But a shift in the lens axis relative to the beam axis leads to a corresponding shift in the beam focal point, which can be an advantage in the design of laser beam guiding systems and laser scanning systems. This phenomenon can also be applied in surface heat treatment along a circular trajectory produced by rotating an eccentric lens about the beam axis. In particular, this principle can be used to work out an effective process of producing circular weld seams run on flat and spherical surfaces.

In order that laser processing with a rotating eccentric lens should be efficient, the designer must optimize the focusing system parameters to secure a small focused spot with the highest energy density.

The shape and size of a beam spot focused by an eccentric lens are calculated taking into account spherical aberrations and diffraction divergence. The design approach is similar to that in Sec. 2.4 for coaxial lenses and involves the use of the theory of third-order aberrations. But in distinction from a coaxial lens for which the ideal focused spot depends solely on the longitudinal spherical aberration as illustrated in Fig. 2.16, an eccentric lens suffers longitudinal aberration which varies around the contour, thus rendering the ideal focused spot indeterminate. Because for laser processing it is important to obtain the least focal spot area and the highest energy density, the plane of the best focal setting is best taken to be the plane normal to the lens axis, in which the beam spot area is at a minimum.

The transverse spherical aberration of an eccentric lens is a variable quantity because the peripheral ray is incident at a varying height h above the optical axis (Fig. 2.26). According to equation (2.23), the transverse spherical aberration b_t is proportional to the cube of the height h of the ray above the lens axis and varies inversely as the square of the lens focal length f:

$$b_t = ph^3/f^2, \tag{2.39}$$

where $p = p'/2$.

From the similarity of the triangles in Fig. 2.26, we can determine the longitudinal spherical aberration

$$b_l = b_t f/(h + b_t). \tag{2.40}$$

The slope of the peripheral ray *1* coming from the lens

$$-\tan\gamma = \alpha = -b_t/b_l = -(h + b_t)/f. \tag{2.41}$$

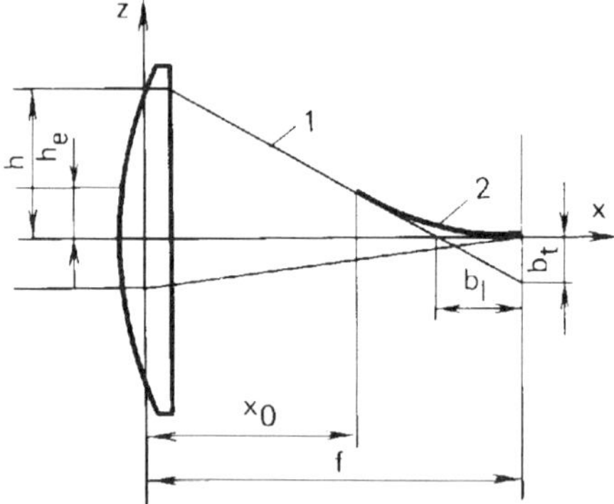

Fig. 2.26. Laser beam focusing with an eccentric lens

The equation for ray *1* assumes the form

$$z = \alpha x + h = -\ [(h + b_t)/f]\ x + h. \tag{2.42}$$

Substituting (2.39) into (2.42) yields

$$z = -\ (h/f + ph^3/f^3)x + h. \tag{2.43}$$

To search out an equation for the caustic curve *2*, we should differentiate (2.43) with respect to h and obtain a set of parametric equations

$$\begin{aligned} x &= f^3/(f^2 + 3ph^2), \\ z &= 2ph^3/(f^2 + 3ph^2). \end{aligned} \tag{2.44}$$

Eliminating h from the set of equations yields the caustic equation in explicit form

$$z = \left(\frac{f - x}{3px}\right)^{3/2} 2px. \tag{2.45}$$

Thus the paths of the rays past the lens are described by equation (2.43) of straight line *1* and equation (2.45) of curve *2*. The position of the point of transition from the straight line to the caustic curve is found by the simultaneous solution of (2.43) and the second equation in the set (2.44):

$$x_0 = \left(h - \frac{2ph^3}{f^2 + 3ph^2}\right)\frac{f^3}{hf^2 + ph^3}. \tag{2.46}$$

From equations (2.43), (2.45), and (2.46) we can estimate the beam cross section in any plane sited normal to the optical lens. Several planes within the focus zone are commonly drawn at a step length short enough to provide the desired accuracy. The plane in which the focused spot has a minimum diameter is the desired plane of best focal setting.

For the total area of the focused spot to be defined, the divergence component calculated with (2.10) or (2.11) should be added to the estimated area of the aberration-limited spot.

Given below are the results of calculations performed for low-aberration convexo-plane KCl lenses at D = 20 to 40 mm, $\theta = 1.5 \times 10^{-3}$ to 2.0×10^{-3} rad, f = 100 to 300 mm, and h_e = 2 to 10 mm [22].

The experimental curves of the focal spot area S for a 30-mm diameter beam having a divergence of 2×10^{-3} rad as a function of the focal length f at different values of h_e are given in Fig. 2.27. The arrows indicate the minimum spot areas corresponding to optimal focal lengths. Figure 2.28 illustrates the plots of f_{opt} against eccentricity h_e. The relation of f_{opt} to h_e is generally linear:

$$f_{opt} = f'_{opt} + h_e k, \tag{2.47}$$

where f'_{opt} is the optimal focal length of a coaxial lens at $h_e = 0$ found from (2.35) or the nomograph in Fig. 2.17; and k is a proportionality factor accounting for the slope of the straight line.

An analysis of the results reveals that the factor k depends solely on beam divergence and is equal to 8.2 and 8.9 at an angle of divergence of 2×10^{-3} rad and 1.5×10^{-3} rad respectively. At relative eccentricities h_e/D less than or equal to 1.5, we can calculate the focal length f_{opt} without any regard to eccentricity in the same manner as for a coaxial lens. In this case the error is less than 10%. If h_e/D is larger than 0.15, equation (2.47) should be used to estimate $f_{opt.}$

In the case of beam focusing by an eccentric lens, the ideal focused spot does not coincide with the focal plane, but lies closer to the lens, a distance Δf from this plane. Figure 2.29 depicts the plots of Δf against h_e. From the

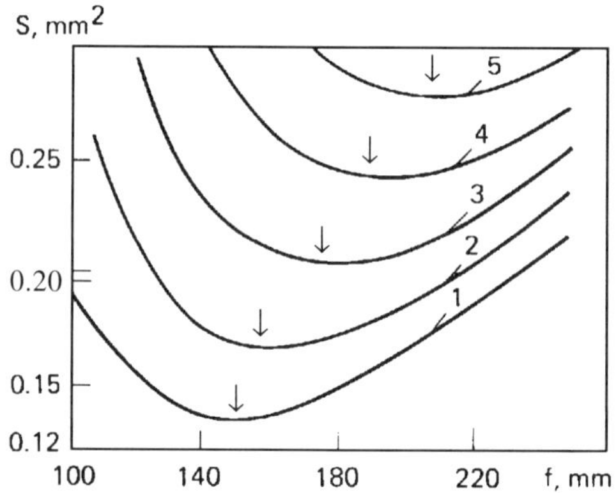

Fig. 2.27. Focal spot area S as a function of the focal length of a plano-convex lens at an eccentricity h_e of 2, 4, 6, 8 and 10 mm (curves *1*, *2*, *3*, *4*, and *5*, respectively)

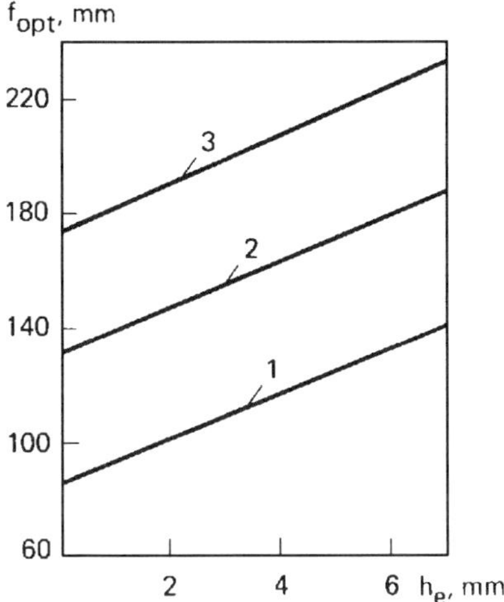

Fig. 2.28. Optimal focal length f_{opt} of a plano-convex lens vs eccentricity h_e at $\theta = 2 \times 10^{-3}$ rad and beam diameter of 20, 30, and 40 mm (curves *1*, *2*, and *3*, respectively)

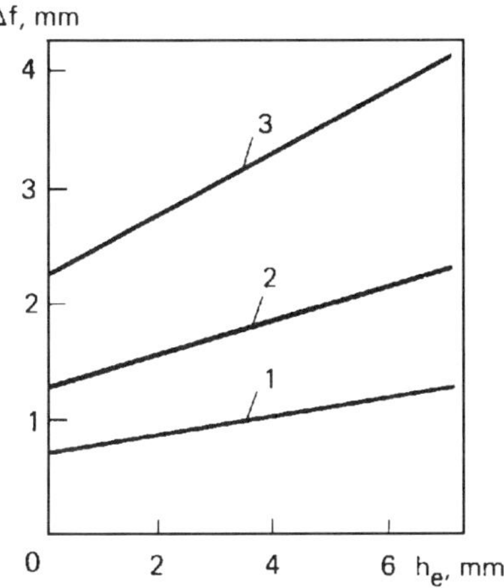

Fig. 2.29. Distance Δf between the focal plane and the plane of the least focal spot size as a function of h_e at a lens focal length of 300, 180, and 100 mm (curves *1*, *2*, and *3*, respectively)

analysis of the plots it follows that the shift Δf at large values of h_e may substantially differ from the shift specific to coaxial lenses ($h_e = 0$). The value of Δf for eccentric lenses is a function of the focal length f as well as h_e. The shorter the focal length for a constant h_e, the larger Δf. An eccentric lens focuses the beam to a minimum spot that has an irregular elongated shape (Fig. 2.30).

These theoretical statements for optimizing beam parameters with an eccentric lens were checked in the experiments on stepped stainless steel specimens shown in Fig. 2.19. The specimens were heat-treated at a rate of 29 mm/s with a 30-mm diameter beam focused by an eccentric, convexo-plane, 280-mm focal length lens. The experimental results are displayed in Fig. 2.31. The focal point shift Δf is laid off along the x-axis. The zero reference point corresponds to the position of the lowest step of the specimen. The penetration depth is generally little dependent on Δf, and so the curves are almost similar in shape and lie close to one another.

The experiments confirmed that at small values of h_e/D of about 1.5, an eccentric lens can be set in the same manner as a coaxial lens with respect to the specimen surface.

The melt zone produced by a beam focused with an eccentric convexo-plane lens has axial symmetry (Fig. 2.32). It is clear from Fig. 2.32*a*, *b* and *c*, that the cross sections of melt zones do not change significantly for small h_e. But as the eccentricity increases, the melt depth decreases rapidly (Fig. 2.32*d*).

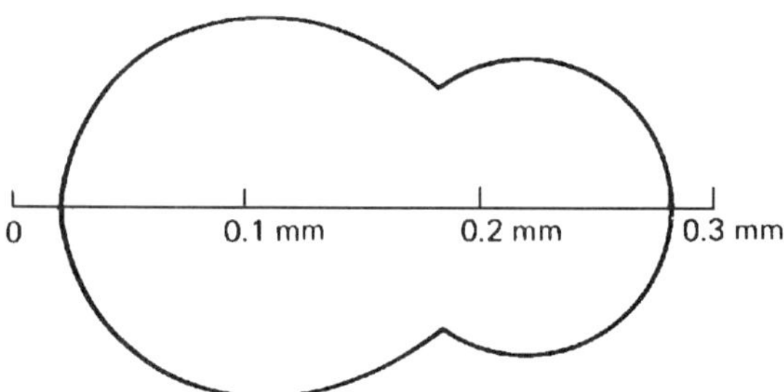

Fig. 2.30. Cross section of the beam focused by an eccentric lens in the plane of an ideal focal spot. The spot is drawn at $h_e = 6$ mm, $D = 30$ mm, and $f = 190$ mm without any regard to beam divergence

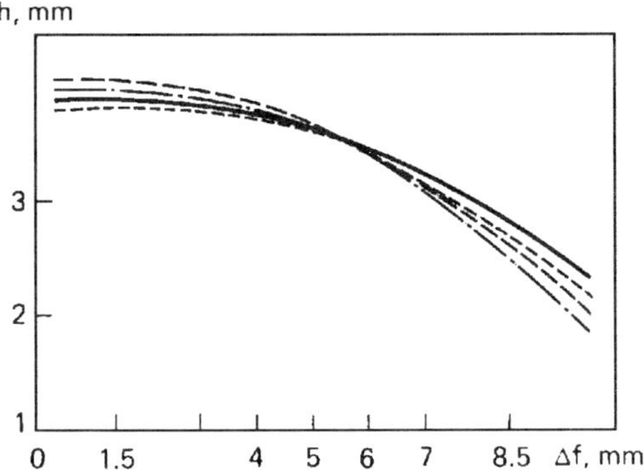

Fig. 2.31. Penetration depth h versus focal point shift. The solid line, dash line, dot-and-dash line, and dot line correspond to h_e of 0, 3, 6, and 9 mm, respectively

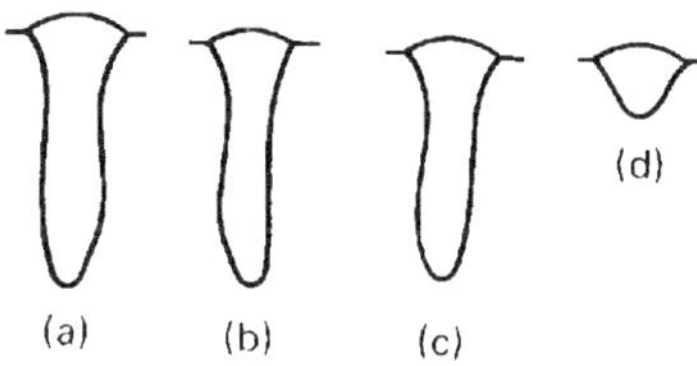

Fig. 2.32. Profiles of fusion zones produced with a 160-mm focal length lens, (*a*) through (*c*), and 220-mm focal length lens, (*d*), at eccentricity h_e of 0 mm (*a*), 3 mm (*b*), 6 mm (*c*), and 20 mm (*d*)

2.6. Analysis of Specular Reflector Parameters

It is preferable to focus high-power laser beams with the inverted Cassegrain reflectors described in Sec. 2.2. They eliminate third-order spherical aberrations using spherical, paraboloid and hyperboloid mirrors. According to the findings in [33], third-order spherical aberrations are absent if the curvature radii r_1 and r_2 for the secondary and the primary mirrors, respectively, and the distance L between these mirrors are

$$r_1 = 1.236\, f, \tag{2.48}$$

$$r_2 = 3.236\, f = 2.618\ r_1, \tag{2.49}$$

$$L = r_2 - r_1. \tag{2.50}$$

Expression (2.50) indicates that the mirrors are concentric. A reflector obeying the above relations is better than similar reflectors using spherical mirrors because it suffers the least aberrations. However, an on-axis reflector can screen the beam by the secondary mirror. It can be shown that if the above relations hold, the value of K which accounts for the central shielding of the beam is

$$K = d_{in}/D \cong 0.447, \tag{2.51}$$

where d_{in} is the inner diameter of the ring-shaped beam incident on the secondary mirror and D is the outer diameter of the output ring.

The screening effect observed in an on-axis reflector limits the application of this reflective system for focusing the ring mode output from an unstable resonator. A reflector will have no energy loss by screening if the resonator magnification obeys the condition

$$m = D/d_{in} \leq 1/K = 2.236. \tag{2.52}$$

High-power CO_2 lasers use resonators whose magnification approaches unity, and so (2.52) is valid at a short distance from the resonator. With an increase in the distance, the screening effect makes itself felt and relation (2.52) does not hold any longer. For this reason, the reflective system should be placed close to the resonator.

An off-axis Cassegrain reflective system displayed in Fig. 2.14 offers much promise for laser processing. It is free from shielding if the following condition is met:

$$h_0 > 2.618 r_l. \tag{2.53}$$

Here, h_0 is the height of the beam axis above the reflector axis, as illustrated in Fig. 2.14, and r_l is the radius of the laser beam.

In order that the entire oncoming light fall on the small mirror, the radius of curvature of the mirror surface should be

$$r_1 > 3.618 r_l. \tag{2.54}$$

Expressions (2.49), (2.53), and (2.54) apply for rough estimation of the parameters of an off-axis reflector. To obtain more accurate estimates of the parameters that provide the highest energy density at the focused spot, we must determine the focal spot size at the specified parameters of the incident beam with a consideration of the effect of the reflector aberrations and beam divergence on the spot size. An accurate spot size is found by a trigonometric calculation of the paths traveled by the rays through the reflector.

A spot focused with an off-axis reflector is generally not circular. Consequently, it is necessary to optimize the spot with regard to its area. However, calculations reveal that the spot shape well approximates an ellipse whose area is easy to determine from its axes which are the distances b_1 and b_2 between points (x_0, z_0) and (x_2, z_2) in Fig. 2.14 in the plane of the paper and normal to the paper, respectively. A minimum aberration-limited spot is a function of the distance between the point (x_0, z_0) of the intersection of the outer rays and the point (x_2, z_2) of intersection of the normal line with the caustic curve. As seen from Fig. 2.14, the plane of a minimum spot size is at an angle γ to the beam axis, and so the work surface must lie in this plane to achieve the highest energy density at the spot.

Software has been written to analyze the path lengths traveled by laser rays in an off-axis reflector. Computations for various parameters of the laser beam and reflector have made it possible to establish some rules for securing the highest power density in the focused spot.

The basic design parameters of reflectors are the distances b_1 and b_2 and focal spot area S. Figure 2.33 shows the plots of b_1, b_2, and S versus radius r_1 of curvature of the secondary mirror for a laser beam diameter $D = 2r_1$ of 60 mm and beam divergence angle θ of 2×10^{-3} rad. The curves of b_1, b_2, and S retain their shapes with a change in D. The arrows indicate the minimum values of the parameters at different values of r_1. The optimal radius $r_{1\ opt}$ is the radius that affords the smallest focal spot area and, hence, the highest energy coupled to the spot. The value of $r_{1\ opt}$ lies midway between the r_1 values at which b_1 and b_2 are at a minimum.

As seen from the graph, the S curve approaches a horizontal shape at r_1 close to $r_{1\ opt}$. This means that the designer of an off-axis reflector has a certain latitude in the choice of r_1 which can deviate from $r_{1\ opt}$ by ±10% without entailing a significant variation in the focal spot area. If there is a need from design considerations to prescribe an r_1 that deviates from $r_{1\ opt}$ in excess of ±10%, setting $r_1 > r_{1\ opt}$ is better since the spot area grows more slowly in this range than when $r_1 < r_{1\ opt}$.

Figure 2.34*a* illustrates generalized calculation data for the $r_{1\ opt}$ of an off-axis reflector versus beam diameter D at different values of beam divergence. The $r_{1\ opt}$-D relation is linear within this divergence range. Given D and θ, the values of $r_{1\ opt}$ can be found from Fig. 2.34*a*. If D and θ go beyond the scale, $r_{1\ opt}$ should be sought by interpolating and extrapolating the presented data.

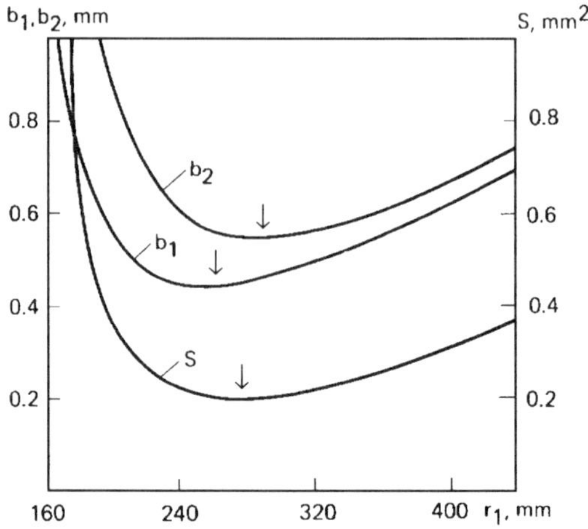

Fig. 2.33. Parameters b_1 and b_2 and focal spot area S vs r_1 of the convex mirror of an off-axis reflector

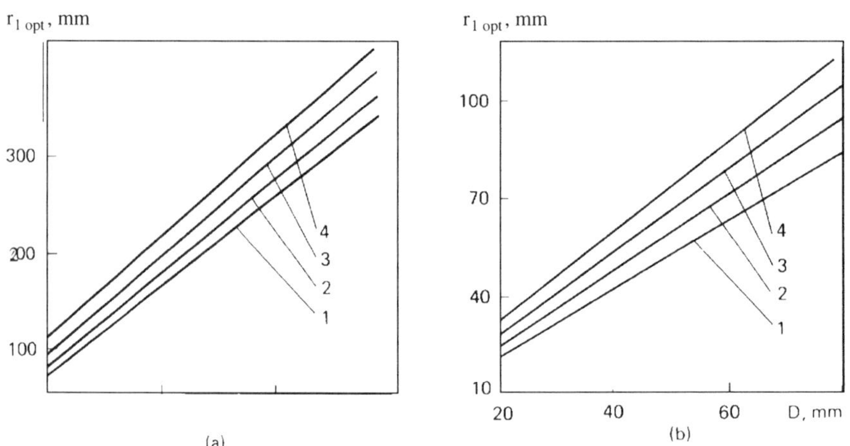

Fig. 2.34. Nomographs for estimating $r_{1\,opt}$:
(a) for off-axis reflector at θ of 3×10^{-3} (line *1*), 2×10^{-3} (line *2*), 1.5×10^{-3} (line *3*), and 10^{-3} (line *4*); (b) for center-fed reflector at θ of 5×10^{-3} (line *1*), 3×10^{-3} (line *2*), 2×10^{-3} (line *3*), and 10^{-3} (line *4*)

Let us consider some design guidelines needed to optimize on-axis reflector parameters. As mentioned earlier in the text, the axisymmetric Cassegrain reflector illustrated in Fig. 2.13 shields the central portion of the circular beam, and so it is practicable for the ring mode output.

Design optimization for an axisymmetric reflector relies on the relationships presented above for an off-axis reflector. The schematic diagram of Fig. 2.13 implies that the radius r_a of the aberration-limited focused spot is equal to the distance from the x-axis to the point of intersection of the peripheral ray with the caustic curve. The total focal spot size can be defined by adding the divergence-induced component.

Figure 2.34*b* displays the plots of calculated values of $r_{1\ \mathrm{opt}}$ for an on-axis reflector versus beam diameter at different values of beam divergence.

Comparison between design calculations for on-axis and off-axis reflectors is useful for design purposes. Given the same parameters of a laser beam, an on-axis reflector with a smaller value of r_1 and thus smaller dimensions can provide the same degree of focusing, i.e. the same focal spot size, as an off-axis reflector because the former reflector suffers much lower aberrations. In other words, the on-axis reflector of the same dimensions as the off-axis reflector provides a smaller focused spot and is more preferable in this respect. Comparison between the charts of Figs 2.34 and 2.17 reveals that on-axis and off-axis reflectors provide a higher power density than lenses.

Experimental studies of laser heat treatment in a wide range of laser powers have been made with the use of various designs of on-axis and off-axis Cassegrain reflectors for focusing a beam to a desired spot. The analysis of the experimental data attests that on-axis reflectors displaying a large angle of beam convergence are optimal focusing systems in the range of powers of up to 10 kW. Other things being equal, an advantage of these reflectors over the reflectors with smaller angles of convergence is that they provide a deeper penetration at the same power of the beam leaving the resonator. But the latter reflectors are given preference over the on-axis reflectors in the range of powers above 10 kW, although the 10-kW boundary is certainly conditional. In particular, if the work surface is to be treated with a beam whose scan speed is so high that the penetration depth is small even at a power in excess of 10 kW, the angle of convergence is of no significance. The deciding factor is yet a high power density of the beam focused by a reflector with a large angle of convergence.

Thus, axisymmetric reflectors with large angles of convergence can handle laser powers only in the limited range because the convergence angle affects the depth of melting. With an increase in the power above the upper limit, it is found advisable to change over to on-axis reflectors with small angles of convergence. But in the latter case, too, the upper limit to the power yet exists because the energy loss in the reflector rises with power. For example,

Table 2.1. Data for Selection of Some Parameters of the Reflectors Illustrated in Figs 2.13 and 2.14

Beam cross section	Type of reflector	P, kW	r_1	r_2	h_0	D_1	D_2
Ring,							
$D_{in}/D = 0.45$	On-axis	<10	From chart		—	D	4.5D
	On-axis	>10	7.1D	2.618r_1			
Arbitrary	Off-axis	<10	From chart		1.4D	1.1D	5D
	Off-axis	>10	7.1D		1.4D	1.1D	5D

in the experiments carried out with an axisymmetric reflector, such as shown in Fig. 2.13, whose angle β of convergence was merely 10°, the depth of melting was held near 25 mm despite an increase in the power of the beam that retained its sharp focus. These experiments imply that at high power it is necessary to use off-axis reflectors.

The experiments on laser melting with the beam focused by off-axis reflectors showed that the depth of melting grew monotonically with laser power. An oblique-angled reflector and a convexo-plane lens of the same focal length were used to focus a 3-kW beam and treat the surfaces of corrosion-resistant steel specimens at a rate of 15 mm/s. The reflector provided a deeper penetration than the lens. This is because a reflector affords a higher energy density in the hot spot than a lens and entails a lower power loss. The experimental results enable us to conclude that specular objectives employed in CO_2 laser setups for heat treatment of metals are a better choice because they ensure a higher efficiency of the process and have a longer life than lenses.

Table 2.1 lists the data that can be used for the calculation of the parameters of on-axis and off-axis Cassegrain reflectors with consideration for the power and cross section of the beam. These are approximate values obtained from the analysis of design and experimental data. Referring to Table 2.1, we can first select the type of reflector and then the radius r_1 of curvature of the convex mirror depending on the beam power P. For $P \leq 10$ kW the radius r_1 can be chosen from the chart of Fig. 2.34. For $P > 10$ kW this radius can be taken equal to 7.1 D. In a similar way, other parameters can be defined using the data presented in the table.

3 Laser-Induced Plasma Effects

The laser equipment intended for welding, cutting, cladding, surface hardening, and other heat treatment processes can secure a high production capacity with the use of lasers capable of delivering a very high power to workpieces. At present, a laser proves to be the source of the highest energy density and in this respect it excels an electron-beam source. But the latter can provide a deeper penetration at the same power because the electrons of the electron beam interact with metals in a different way than the photons of a laser beam. However, heat treatment with an electron beam necessitates the creation of a vacuum in the interaction zone. In contrast, a laser beam can interact with materials in any gaseous medium at the atmospheric pressure. The composition of a gaseous mixture can properly be chosen to protect the material surface from high temperatures and oxidation.

The laser beam of a high power density delivered to a workpiece for its treatment in cover gases or in air can lead to the formation of a low-temperature plasma. Investigations suggest that a low efficiency of energy coupling to the target is due largely to plasma waves which absorb and scatter the laser radiation, thereby reducing the power and the power density of the heat source on the material surface. This chapter describes briefly some mechanisms of formation of a laser plasma and presents the results of theoretical and experimental studies on plasma processes associated with laser processing in cover gases carried out in co-authorship with V.V. Marushchenko.

3.1. Plasma Formation

There are different approaches to the description of the mechanisms of plasma wave excitation. The results of fundamental studies on the mechanisms of an optical discharge in gases (laser-induced spark) are given in [34] which describes the conditions at which the optical breakdown occurs and propagates in the environment. The authors of [35] have suggested a theoretical model of plasma formation due to material vaporization from the heat affected zone and subsequent absorption of the laser radiation in vapors, attended by a fast growth of the temperature and the degree of ionization.

One more approach to the plasma mechanisms explains the plasma wave excitation due to avalanche ionization in gas layers adjacent to the target surface. But the ionization threshold decreases as a result of electron emission from the irradiated target, gas heating by the shock wave of vapors, and additional heating of the gas as it comes in contact with the target hot surface.

The factors that have an appreciable effect on the mechanisms of plasma formation are structure defects, microasperities of work surfaces, surface layer oxidation, and chemical reactions between the vapors and cover gas [35].

It should be noted that a plasma plume appears in all laser-induced processes involving metal melting with vaporization. At the initial stage of plasma plume formation it is necessary to consider ionization with regard to the density of metal vapor particles which exhibit a low ionization potential. The mechanism of development of an electron avalanche in a readily ionized mixture has been treated in [34]. Ionization is brought about by fast electrons which acquire energy as they absorb photons. Such an electron can trigger ionization as it collides with an atom and frees two electrons of low energy. The process of repeated stages of photon absorption and collisions of electrons with atoms causes an avalanche ionization, i.e. plasma formation.

At the initial stage of plasma breakdown, free electrons collide with neutral atoms of the vapor and gas, thus leading to energy dissipation. At the final stage, the plasma plume offers strong ionization and laser energy is lost through collisions of electrons with positive ions. Since the absorption of laser radiation at wavelengths covering the visible and the infrared region is negligible, laser energy dissipates significantly only at a high degree of ionization, namely, at very high temperatures of 15 000 to 20 000 K. In this temperature range, the amount of lost energy is enough to sustain the plume. At a degree of ionization of a few percent, it is the ions that mostly scatter electrons rather than atoms.

There is a somewhat different opinion in regard to the causes of plasma plume formation near the metal target surface. The authors of [37] substantiate theoretically the point that the physical nature of optical breakdown is suggestive of the mechanism of triggering the chemical reactions of burning by the hot surface.

In other words, the discharge can occur under the action of thermal ionization of metal vapors even if metal vaporization is insignificant. The threshold power density conducive to optical breakdown varies mainly with the properties of the target material such as the absorptivity, thermal conductivity, energy of vaporization, and ionization potential of vapor atoms. According to the theory advanced in [37], the properties of the cover gas and its pressure have a little effect on the threshold value of the power density.

Paper [38] has described the mechanism of laser-induced gas breakdown near a metal target due to a considerable increase in the electric field strength of the incident beam near microasperities on the target surface. Small portions of the protrusions heat up to the boiling point in an extremely short time that is a few orders of magnitude shorter than is the case for small portions of the hot spot. These protrusions form the regions of totally ionized plasma, which initiate laser-supported detonation (LSD) waves propagating in a few microseconds over the entire interaction zone.

A number of works analyzed the mechanisms of development of a laser spark and excitation of a gas ignition wave, or presented detailed investigations on laser work formation by the methods of time-base, high-speed, time-lapse photography, holographic photography, and other photography methods. As follows from the investigations, the plasma wave moves from the target surface toward the laser until the fluence remains sufficiently high. The wave develops at the instant when the power density reaches its threshold value and involves a light flash attended by a sharp growth of noise.

As the plasma cloud moves toward the laser, the plasma parameters begin to change. The rate of plasma propagation is one of the important characteristics that determines the parameters and behavior of optical breakdown. The authors of [40] have described the velocity and velocity-dependent parameters of a plasma wave and considered the process of optical breakdown stabilization by the flow of a gas in the direction of the focused beam. The theoretical and experimental studies have revealed that as the plasma plume extends upward, both the velocity of the plasma wave front and plasma temperature drop off.

The gas flow in the beam direction stabilizes optical breakdown most efficiently. At a certain power of the focused beam, the plasma wave front will shift with respect to the focal point depending on the gas flow velocity and lie in a beam cross section where the power density secures the wave front velocity that is equal to the velocity of the oncoming gas flow.

The laser power required to maintain an optical discharge varies with gas pressure. The absorption coefficient grows with gas pressure, thus causing an increase in heat release defined as the product of the power density by the absorption coefficient. To compensate for the loss by heat conduction and, hence, to maintain the optical discharge, a lower power density is sufficient at higher gas pressures.

A laser plasma appreciably changes the optical properties of a medium where laser radiation propagates. Investigators commonly attribute this effect to changes in the beam divergence. Under certain conditions, this effect can lead to a lower divergence, i.e. light self-focusing. It is common practice to treat the effect of self-focusing in terms of a dielectric waveguide assumed to appear as a result of nonlinear changes in the permittivity [41].

The power density distribution in any beam cross section exerts a strong effect on self-focusing. A decrease in power density near the beam axis may lead to focusing of the beam in a medium which generally defocuses it.

It is found from the experiments that a laser beam can converge to a smaller spot in the region of cold vapors as it passes through a highly absorbing plasma cloud. Self-focusing results from changes in the permittivity of the plasma. The distribution of the permittivity in the cross section of the plasma cloud is such that its highest value is at the center of the beam. Since the light beam always deflects toward the region of a higher permittivity, self-focusing of the beam results.

Theoretical and experimental studies help us gain insight into the nature of plasma excitation and disclose a number of the mechanisms of plasma formation in gases. However, much effort has still to be made to clarify the practical aspects concerning the effects of a plasma cloud in various gases on the depth of melting and laser beam efficiency. Of much interest to specialists is the study aimed at clarifying the mechanisms of interaction of the material vapors, gaseous environment, and laser plasma in the conditions of vigorous vaporization typical of laser welding, cutting, cladding, and heat treatment with melting.

3.2. Laser-Induced Plasma Processes in Cover Gases

Laser heat treatment at temperatures close to the melting points of materials, such as alloy steels and highly reactive metals and alloys, is applied in monoatomic inert gases, namely, argon and helium. The processes of heat treatment of low-alloy steels and constructional materials use carbon dioxide, nitrogen, and air as shielding or cover gases. Researchers of [42] thoroughly investigated the effect of a cover gas composition on the depth of laser melting, which was found to be quite appreciable.

Work [22] described the experiments on laser treatment of corrosion-resistant steel in various cover gases delivered to the interaction zone through a nozzle coaxial with the beam. The specimens were heat treated at a speed of 33 m/s using a 5-kW CO_2 laser beam focused with a 230-mm focal length lens. The laser beam excited a characteristic plasma plume in the interaction region. Table 3.1 presents the experimental results.

To clarify the effect of the composition of a cover gas on the melting efficiency in the laser beam-metal interaction zone, first consider the properties of cover gases (Table 3.2).

Molecular gases dissociate at high temperatures into components

$$N_2 \rightleftarrows 2N \quad -945\ \text{kJ/mole},\ -9.8\ \text{eV},$$

$$CO_2 \rightleftarrows CO + O \quad -532\ \text{kJ / mole},\ -7.3\ \text{eV},$$

$$CO \rightleftarrows C + O \quad -1076\ \text{kJ/mole},\ -11.1\ \text{eV}.$$

Table 3.1. Parameters of Laser Melting in Cover Gases

Cover gas	h, mm	h/b	h/h_{He}	ρ_0, kg/m^3	$Q \times 10^4$, m^3/s
He	6.1	2.8	1.00	0.178	5.00
CO_2	5.1	1.8	0.84	1.980	2.83
Air	4.9	2.1	0.77	1.290	–
N_2	4.2	1.9	0.72	1.250	2.83
Ar	1.0	0.6	0.14	1.780	2.83

Notes: b, average width of the melt zone; ρ_0, gas density at 0°C and 0.1 MPa; h/h_{He}, ratio between the depths of melting in a given gas and in helium, with h_{He} taken equal to unity; Q, gas flow rate.

Table 3.2. Thermophysical Properties of Cover Gases

Gas	Ionization potential, eV	Dissociation potential, eV	Total potential, eV	Thermal conductivity, W/m K, at T = 370 K	Heat capacity, kJ/kg K, at T = 370 K
He	24.58	–	24.58	0.1730	5.20
Ar	15.76	–	15.76	0.0210	0.52
N_2	15.58	9.80	25.38	0.0237	1.04
Air	11.70	10.00	21.70	0.0250	1.01
CO_2	13.80	18.40	32.20	0.0210	0.92

Since high-temperature processes occur in a plasma plume, the energy balance should be defined with due regard for the energy of dissociation.

Let us look at the conditions of plasma formation in the processes of laser heat treatment in the atmosphere of inert gases. In most of the processes, the beam vigorously vaporizes the metal surface zone, and so metal vapors rise above the target. The metal vapors certainly reduce the effective ionization potential V_{eff} of a vapor-gas mixture. The quantitative estimate of the effect of metal vapors can be made from V_{eff} defined as

$$V_{\text{eff}} = -\frac{T}{5800} \ln \sum_{i=1}^{k} \sqrt{c_i} \, \exp\left(-\frac{5800\, V_{\text{ion}}}{T}\right), \tag{3.1}$$

where T is the temperature, K; k is the number of components in a vapor-gas mixture; c_i is the concentration of the ith component, volume percent; and V_{ion} is the ionization potential.

Figure 3.1 illustrates how V_{eff} estimated from formula (3.1) for argon and helium varies with the iron vapor content, in volume percent, at different temperatures corresponding to 0.5–2.0 eV (1 eV is equivalent to 11 600 K). It is evident that the potentials V_{eff} for Ar and He drop heavily at an iron vapor content of less than 5%. When c_{Fe} = 5%, the difference between the values of V_{eff} for Ar and He becomes small, about 2 eV. This indicates that the conditions for emission of free electrons from the target treated in argon or helium are approximately identical.

In the general case, the conditions for steady-state optical breakdown in a pure gas or material vapors can be defined through the solution of a simplified problem since the solution of the exactly stated problem presents considerable difficulties, mainly because the theoretical model used suffers from some uncertainties associated, in particular, with an insufficient knowledge of the collisions of atoms and the excitation of atoms to different energy levels. For this reason, the estimates of the above conditions can be made with the aid of simplified expression (3.1), namely, in terms of the degree of gas ionization at a specified temperature with consideration for the energy balance.

The mechanism of optical breakdown in gases behaves in the following manner. Free electrons and positive ions move from metal vapors heated to

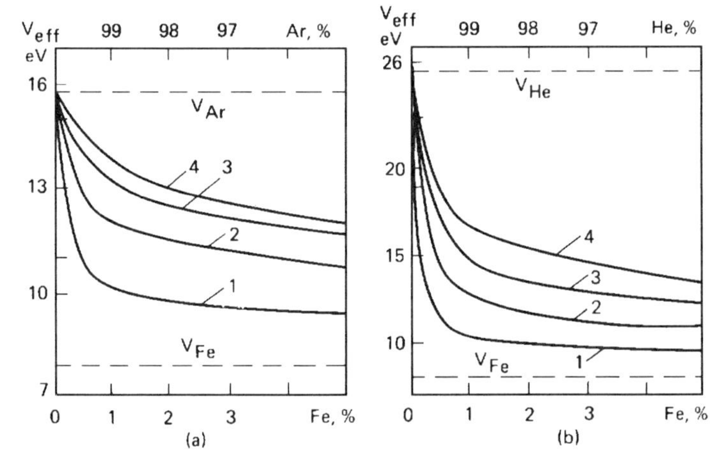

Fig. 3.1. Effective ionization potential of argon (*a*) and helium (*b*) versus iron vapor content at (*1*) 5800 K, (*2*) 11 600 K, (*3*) 17 400 K, and (*4*) 23 000 K

3 000–6 000 K and gain access to the laser interaction zone. The free electrons pick up energy in the field of positive ions and ionize neutral atoms of the vapors. The degree of vapor ionization gets higher through heat conduction and causes both the gas temperature and the absorption coefficient for laser radiation to increase. An approximate value of the threshold power density q_{th} for a CO_2 laser at which a discharge in metal vapors can occur is not less than 6×10^5 W/cm^2.

This value of the power density is easy to obtain by focusing the CO_2 laser beam whose power $P \geq 1$ kW. The plasma plume originates from metal vapors. To get an idea of the buildup of a steady-state discharge in a gas, we should consider the energy balance with regard to the ionization frequency, electronic energy loss, and loss of electrons. The ionization frequency must be sufficient to make up for the charged particles lost in the discharge region at a rate that is higher than the rate of electrons lost.

According to the findings presented in [34], a minimum laser power P_{min} required to maintain a steady-state discharge can be found from the plasma energy balance. For the discharge to be held steady in Ar and He, the approximate values of P_{min} must be no less than 670 W and 7 600 W, respectively.

To verify these theoretical findings, the experiments were performed, in which the laser beam was focused in the gas jet directed along the beam axis and absorbed in the cone of the power meter after its defocusing. A short-time arc discharge excited between the tungsten electrodes connected to a welding power source generated a plasma in the focal region. The arc discharge was then suppressed to allow the laser beam to sustain the plasma process in the gas. The laser power was smoothly reduced to a threshold value at which the optical discharge vanished. The beam was focused by a KCl lens with a focal length of 400 mm. According to the experiments, the values of P_{min} required to sustain the plasma in Ar and He are equal to about 800 W and 8 000 W, respectively, which agree with the theoretical values.

The character of changes in the penetration depth, as illustrated in Table 3.1, can thus be accounted for by a change in the threshold power needed to maintain the optical discharge in an inert gas.

The same conclusion can be drawn from the experimental results [42] pertaining to the effect of an argon-helium mixture on the penetration depth for laser welding in corrosion-resistant steel (Fig. 3.2). The penetration depth sharply decreases as the helium content of the mixture drops from 40 to 20%. But the effective ionization potential V_{eff} changes within narrow limits. The values of V_{eff} obtained from formula (3.1) are equal to about 16 eV and 16.4 eV at a helium content of the gas mixture of 20% and 40% respectively. Accordingly, the maximum absorption coefficient a for CO_2 laser radiation is near 0.66 cm^{-1} at $T \cong 18\,000$ K. At this temperature, the thermal conductivity

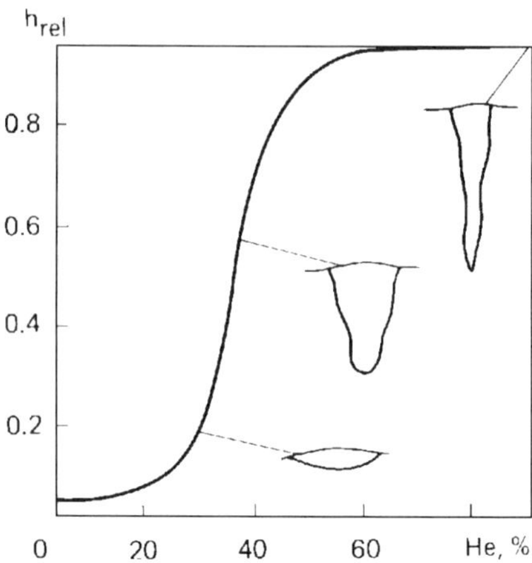

Fig. 3.2. Relative penetration depth $h_{rel} = h/h_{He}$ versus helium content of the argon-helium mixture in laser welding of corrosion-resistant steel at $v_w = 33$ mm/s and $P = 5$ kW

k of the mixture at 20% He and 40% He is equal to nearly 1.97 W/m K and 3.23 W/m K, respectively. These values are found by extrapolating the values of k to the high-temperature region. Using the data presented in work [34] and the above values of a, k, and T to estimate P_{min} required to maintain the discharge in the argon-helium mixture at 20% and 40% helium content, we get P_{min} of about 3.2 kW and 5.4 kW, respectively.

In the case under consideration, the penetration depth illustrated in Fig. 3.2 varies with a minimum threshold power P_{min}.

If the laser power P rises above the threshold typical for the gas used, the plasma generated in the beam caustic region may move toward the laser (Fig. 3.3). The cold gas on the side of the plasma cloud that faces the laser heats up rapidly and begins to absorb the beam energy so the plasma moves upward in the cone of the focused beam until the power absorbed compensates for both the power lost through heat removal from the high-temperature zone and the power lost in emission. The plasma front stops moving at a certain distance l_{pl} from the beam caustic where the laser power density is still sufficient to compensate for the power loss in the environment. An approximate value of l_{pl} can be found from the expression [34]:

$$l_{pl} \simeq \pi r_0^2 \lambda^{-1} \sqrt{P - (P_{th} + P_r)/P_r}, \tag{3.2}$$

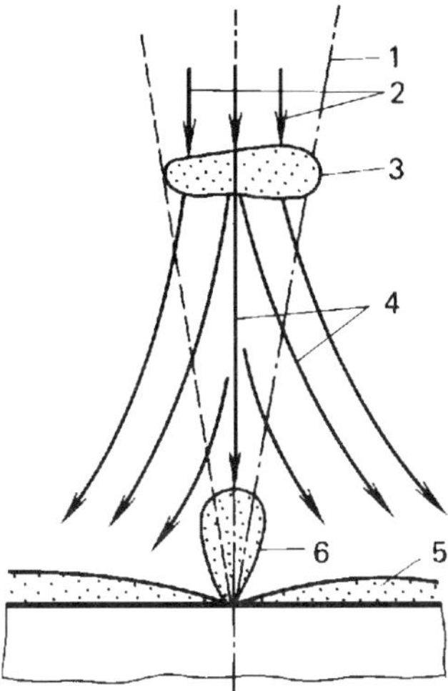

Fig. 3.3. Propagation of plasma cloud: (*1*) laser beam; (*2*) cold gas; (*3*) ionized gas; (*4*) hot gas; (*5*) boundary layer; (*6*) metal vapor

where r_0 is a minimum focused spot radius; λ is the laser wavelength; P_{th} is the power lost through thermal conduction; and P_r is the power lost in radiation.

Experiments were undertaken using a setup illustrated in Fig. 3.4 to verify the theoretical values of l_{pl} obtained from expression (3.2). An arc discharge in various cross sections of the light cone was excited between two tungsten electrodes connected to a power source of 70 V. The power source was then cut off and the laser beam was used to maintain the discharge in argon. The focal point was held at a distance of about 30 mm above the target surface to eliminate the effect of metal vapors. Figure 3.5 displays the theoretical (*1*) and experimental (*2*) values which differ somewhat from one another because the calculations did not take into account the power density distribution in the beam.

An optical gas discharge gives rise to an increased temperature gradient that is responsible for convective gas flows. Experiments reveal that the plasma propagates at different velocities in the region of converging rays [40]. The higher the power density in a cross section of the beam, the greater the plasma velocity. The cold gas flows toward the interaction zone at the speed of heat diffusion due to thermal conduction. An optical gas discharge can be thought

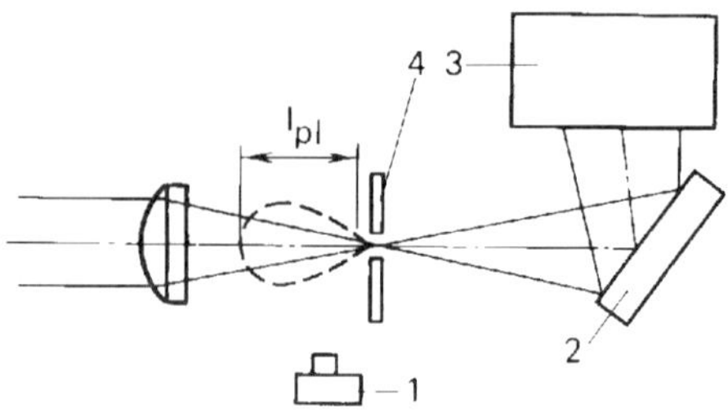

Fig. 3.4. Experimental setup for measurement of the parameters of laser-induced plasma: (*1*) camera; (*2*) mirror; (*3*) power meter; (*4*) target

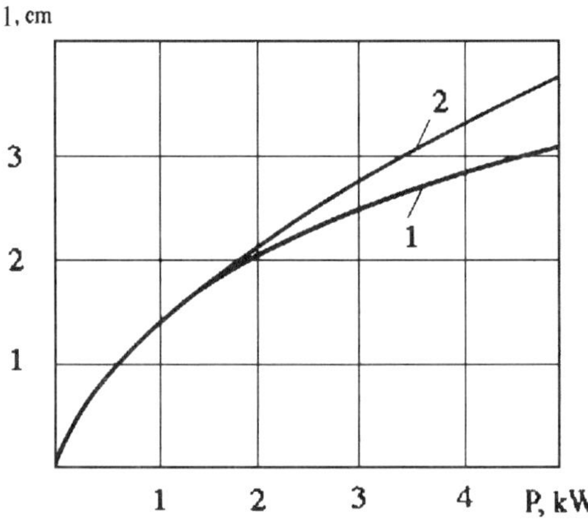

Fig. 3.5. Distance l_{pl} of the front of plasma discharge in argon from vertical beam caustic versus CW CO_2 laser power

to produce a mechanism of energy conversion. The gas heated through thermal conduction absorbs the energy of an electromagnetic field, flows upward, and heats up the cold gas. In the final analysis, the evolved energy is taken up by the gas discharge and removed from it in the form of radiant energy. In general, this mechanism of energy conversion is valid for laser welding, cutting, cladding, and other processes carried out in cover gases.

When a laser beam moves through a plasma, the focusing rays converge and increase the power density, while the plasma reduces it. The weakened beam that has passed through the plasma can generate one more optical discharge in the beam cross section where the power density reaches its threshold value. For example, if provision is made to dissipate the initial combustion wave by causing it to move from argon to air, a sequence of combustion waves will emerge and move one after the other toward the laser. These waves are difficult to observe because the brightness of a high-temperature plasma and metal vapors is very intensive. A characteristic bang is indicative of the formation of optical breakdown.

It is important to estimate correctly the rate of flow of a cover gas in the direction of the laser beam to the interaction zone. The flow rate must be such that the plasma of a steady-state discharge should move to the caustic region of the focused beam. The temperature of the gas flowing from the absorption region should be as high as possible.

Paper [43] presents a set of equations for the computer-aided estimation of the temperature field in an additional gas flow moving along the beam toward the target surface in the presence of a steady discharge. An example of the calculations for laser welding in argon at a laser power of 4 kW appears in Fig. 3.6. The temperature pattern is seen to change appreciably with the gas flow rate. As the rate of argon flow increases, the laser energy absorption region shifts toward the cavity along the laser beam axis. The plasma temperature then rises. If the temperature of the gas that enters the cavity is higher than the initial ionization temperature of the vapor-gas mixture filling the cavity, one more combustion wave appears and causes the gas temperature to jump to a high value. With a further increase in the gas flow, the optical discharge builds up in the cavity and the absorption zone above the target does not appear.

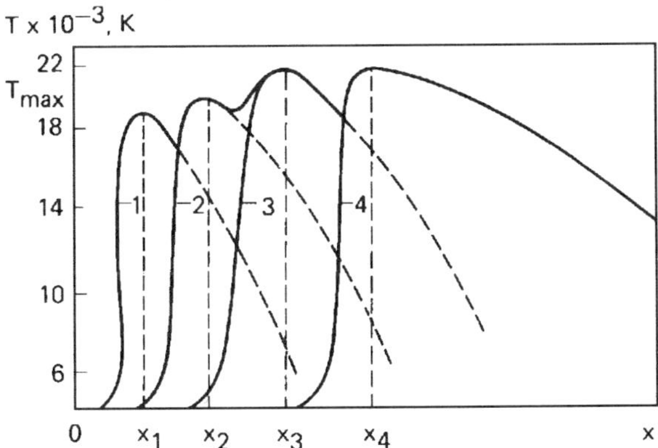

Fig. 3.6. Profiles of temperatures in argon stream at flow rates of 0, 1.0×, 1.5×, and 2.0 × 10^{-4} m^3/s (curves *1*, *2*, *3*, and *4*, respectively)

The theoretical model suggested in paper [43] is a useful tool for estimating the optimal rate of an additional gas flow, at which the laser beam absorption zone can shift to the focused beam caustic zone located on or slightly below the work surface. These conditions ensure the least axial temperature gradient in the gas, a high laser-energy coupling efficiency, and a low radiation loss in the environment. The model has been found useful in the development of a laser heating process with an additional gas flow.

3.3. Laser-Plasma Shielding Effect

A steady-state discharge excited near the target surface changes the conditions of beam focusing and laser energy. As it passes through an inhomogeneous medium, the light beam deviates from its path toward the region of a higher permittivity. In a first approximation, the relative permittivity of a plasma can be found from the equation presented in [44]:

$$\varepsilon = 1 - \omega_{pl}^2 / \left(\omega^2 + \nu^2\right), \tag{3.3}$$

where ω_{pl} is the plasma frequency; ω is the laser radiation frequency; and ν is the effective rate of collisions of electrons with heavy particles.

The plasma frequency can be written in the form [44]

$$\omega_{pl} = 5.65 \times 10^4 \, n_e^{1/2}, \tag{3.4}$$

where n_e is the electron number density.

Since the collision rate ν is much lower than ω that reaches 1.78×10^{14} rad/s for a CO_2 laser, we can neglect ν in equation (3.3) and write the expression for ε with regard to equation (3.4) in the following form:

$$\varepsilon = 1 - 10^{-19} n_e. \tag{3.5}$$

Limiting the electron number density to $n_{e\,max} \le 3 \times 10^{23}\,m^{-3}$, from equation (3.5) we obtain the condition for the plasma permittivity [44]:

$$1 \ge \varepsilon \ge 0.97. \tag{3.6}$$

Express the plasma refractive index n_{pl} in terms of the permittivity ε and permeability μ of the plasma:

$$n_{pl} = \sqrt{\varepsilon\mu} \tag{3.7}$$

Assuming μ to be constant and equal to unity, we have

$$1 \geq n_{pl} \geq 0.985. \tag{3.8}$$

Expression (3.8) permits us to estimate the highest deflection of a laser beam, or its refocusing due to refraction in the plasma. For this, we should specify the variation mechanism of electron number density in the plasma, namely, of the refractive index.

Consider two cases of the electron number density distribution in the plasma.

1. The plasma cloud has a clear-cut core, and the electron number density is constant within and sharply decreases without the core. This is the case for a small-size steady-state optical discharge located near the laser beam caustic. The angle γ_{pl} of refraction here only varies with the angle γ of laser beam incidence on the plasma surface (Fig. 3.7):

$$\sin\gamma_{pl} = \sin\gamma\, n/n_{pl}\,, \tag{3.9}$$

where n is the refractive index of a cold gas. Letting n = 1, we obtain

$$\sin\gamma_{pl} = \sin\gamma/n_{pl}\,. \tag{3.10}$$

From the above expression it follows that refraction in the laser plasma increases with the angle of incidence of the beam to the plasma surface. If this is the case, a homogeneous plasma causes the focal plane to shift along the beam axis.

2. The plasma boundaries are smeared, and the electron number density n_e has the Gaussian (normal) distribution:

$$n_e = n_{e\,max} \exp\left(-r^2/r_{pl}^2\right), \tag{3.11}$$

where r denotes the coordinates of the point of interest and r_{pl} is the plasma cloud radius.

In this case, the beam passes through a plasma layer whose refractive index varies nonuniformly in the radial direction in compliance with (3.5). Equation (3.11) implies that the electron number density is the highest along the beam

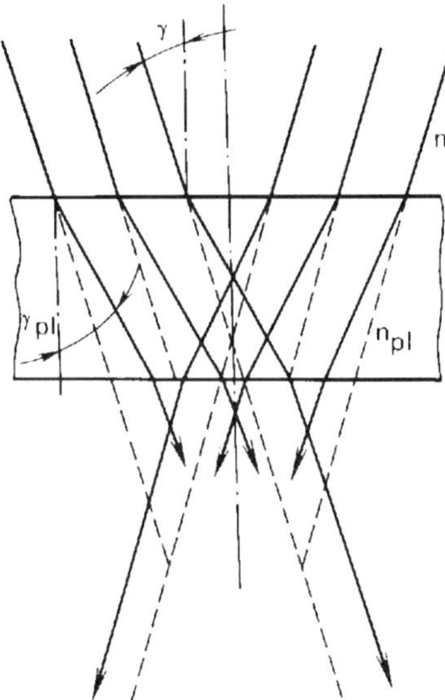

Fig. 3.7. Refraction of light in laser-induced plasma at a constant electron density

axis, and so the permittivity is the lowest on the axis but increases to unity in the radial direction at the plasma boundary. The beam thus deviates from the axis and experiences defocusing. Of course, the pattern of refraction in actual conditions is more complex. Yet, the simplified model described above allows us to define adequately the effect of beam refraction in the plasma on the power density distribution in the heat affected zone.

Laser heating with melting and vaporization gives rise to a vapor flow from the interaction zone, which exerts a noticeable influence on the optical discharge in a gas. Since the general solution to the problem of changes in the optical properties of a medium due to the interaction of the vapor flow with a steady optical discharge presents some difficulties, a simplified analysis can be undertaken to determine how the vapor flow affects the laser beam refraction.

Let us consider the mechanism of changes in the absorption coefficient A for laser radiation. Metallic vapors heavily affect the ionization potential V_{eff} of a gaseous medium, with the result that the absorption coefficient A changes accordingly (Fig. 3.8). As is obvious from the diagram, the values of A increase sharply with decreasing V_{eff} and the corresponding temperatures drop off in a similar way. This diagram allows us to analyze qualitatively the effect of a vapor flow on the refraction of CO_2 laser radiation. The absorption and refraction of the laser beam in a steady optical discharge give way to the

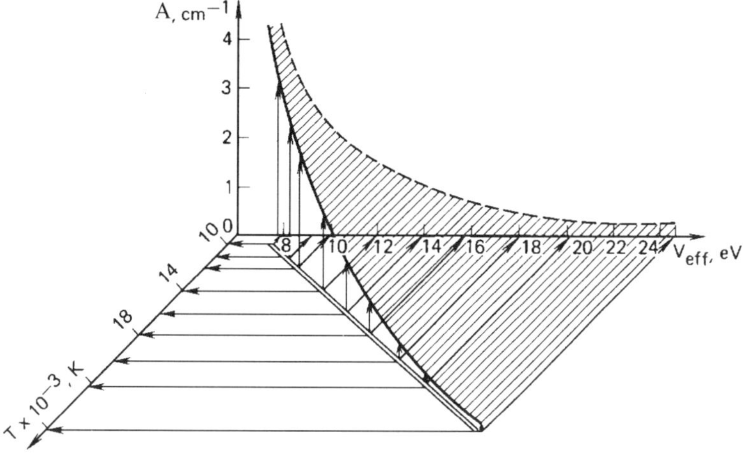

Fig. 3.8. Absorption coefficient versus ionization potential V_{eff} of a gaseous medium

absorption and refraction in an erosive plasma plume formed by the vapor flow. A vapor present in the plasma reduces V_{eff} and increases the absorption coefficient. The plasmoid that intensively absorbs the laser radiation can almost completely decouple the beam from the HAZ and sharply reduce the melt depth. The plasma transparent to the laser radiation takes a very small time to convert to the plasma that completely shields the target and discontinues vaporization.

The rate of scatter of the vapor cluster grows as the beam goes on heating the plasma plume, which then dissipates and the power of the beam incident on the target surface rises to its initial value. A steady-state discharge may again appear in the zone of a high power density. The beam obviously melts the material of the HAZ in a periodic manner. The time periods of melting and target surface shielding vary with the thermophysical properties of the target material and cover gas and also with the parameters of laser radiation.

From the analysis presented above we can conclude that a steady-state discharge heavily refracts laser rays. As the vapor cluster cools the plasma, the beam focusing becomes better, but the subsequent plasma plume maintained by the products of erosion shields the target and stops the process of material removal by vaporization.

Numerous investigations have confirmed the periodic character of laser heating with vaporization. To verify the results of theoretical studies, workers of [45] carried out experimental measurements of the conductance of a laser plasma ignited by a CW CO_2 laser beam that heated metallic specimens. The laser power was varied from 1 to 3 kW and the specimens were moved at

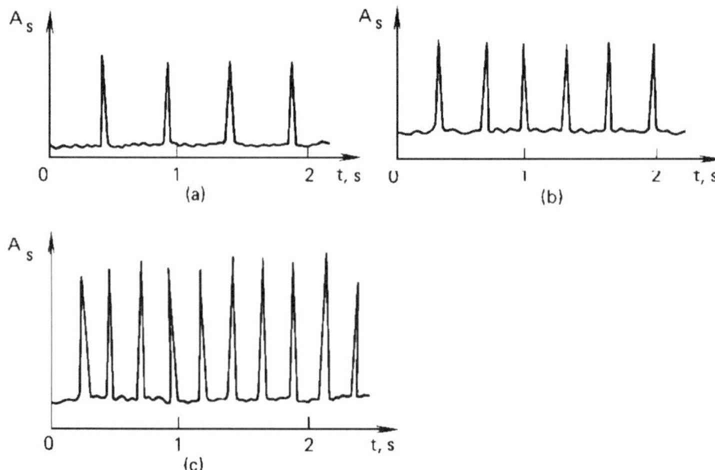

Fig. 3.9. Oscillograph records of the acoustic waves of amplitudes A_s built up in argon-shielded welding at laser powers of (*a*) 3 kW, (*b*) 4 kW, and (*c*) 4.6 kW

a constant speed of 10 mm/s. The experimental results indicate that the region of a high conductance, or high electron number density, is always present above the target surface. A decrease in the conductance and appreciable decay of the amplitude of the current occur over a small time period of about 5×10^{-3} s. As the beam power grows, the frequency f of a high pulsating current increases and reaches 100 to 300 s^{-1} at a power of 3 kW.

A combustion wave formed in a gas cover layer moves upward until it enters the air layer and decays, following which a new wave arises near the surface, and so on. It is evident from experiments that as a combustion wave starts decaying when it passes from argon to air, a shock wave appears probably because the thermal energy of the former wave converts to acoustic energy. The oscillograms of sound waves with amplitudes A_s that emerge in the process of laser welding in argon are shown in Fig. 3.9. The recurrence frequency of sound waves is found to vary from 2 to 4 s^{-1} in the range of laser powers from 3 to 5 kW.

4 Thermal Processes in Laser-Metal Interaction Zones

What lies at the basis of any process of laser treatment is the ability of a laser beam to create a heat flux of high density in a small spot on the target surface. The laser radiation absorbed on the target of any nontransparent solid produces a heat source whose heat power density is sufficient to melt and vaporize the material.

There is large number of different models for investigating the processes in the laser beam-material interaction zone. We consider the basic models that have received the widest recognition [8, 46, 47]. The laser beam incident on a target is partially reflected from the target surface and partially absorbed in a surface layer 0.1 to 1.0 μm thick by conduction electrons which gain more energy and intensively collide with one another. At the initial time instant $t_i = 10^{-11}$ s these electrons give up a small amount of absorbed energy to the crystal lattice of the metal, so the electron temperature T_e of the metal considerably differs from the lattice temperature T_{lat} .

The energy transfer from free electrons to the crystal lattice grows in intensity with time. Starting at the relaxation time $t_{rel} \cong 10^{-9}$ s, the difference between T_e and T_{lat} becomes insignificant and the metal can be thought to acquire a common temperature T on condition that the laser power density in the heat affected zone is not higher than 10^9 W/cm^2.

The main share of laser-induced heat propagates into the metal bulk by way of electron conduction [48], and so the laser-induced thermal processes are similar in nature to conventional processes of metal heating. This suggests that heat transfer in laser-treated metals can be dealt with on the basis of the classical theory of heat conduction.

To take advantage of the body of mathematics relating to the theory of heat conduction, we must first adequately describe a heat source created on or slightly below the target surface. This can only be done with due regard for the specific features of interaction of the laser beam with a solid.

4.1. Energy Transfer to a Target

The intensity with which a material reflects the laser radiation is defined by the material reflectivity that varies with the wavelength of radiation. Table 4.1 lists the values of reflectivity R obtained at room temperature and at an angle of beam incidence of 90° for some metals in the range of wavelengths typical of the most popular lasers [47]. It should be noted that the metals most heavily reflect a CO_2 laser beam.

Table 4.1. Reflectivity of Some Metals

Laser	λ, μm	Reflectivity			
		Au	Cr	Ag	Ni
Ar	0.488	0.415	0.437	0.952	0.597
Rubi	0.694	0.930	0.831	0.961	0.676
Nd-YAG	1.064	0.981	0.901	0.964	0.741
CO_2	10.600	0.975	0.984	0.989	0.942

The absorptivity $A = 1 - R$ of metals appreciably grows with temperature. The value of A for an unoxidized metal surface at an emission wavelength of 10.6 μm can be found from the expression

$$A = 112.2\sqrt{\sigma^{-1}}, \tag{4.1}$$

where σ is the dc conductivity of metals, S/m.

The electric conductivity of metals decreases with increasing temperature and the absorptivity grows accordingly. An important factor is the state of a laser-treated surface. If the laser beam heats a metal surface in air or in any other oxidizing atmosphere, an oxide film forms on the surface and adds to its absorptivity that can increase a few times at a wavelength of infrared radiation of 10.6 μm. Table 4.2 presents the values of the absorptivity of some metals with unoxidized polished surfaces and surfaces oxidized at 873 K for 2 hours.

The coupling efficiency of laser energy can rise significantly with changes in the surface finish of a target or its chemical composition. According to Y. Arata of Japan, an increase in the surface roughness Rz of corrosion-resistant steel and ingot iron from 34 to 120 μm raises the absorptivity by a factor

Table 4.2. Absorptivity of Some Metals at $\lambda = 10.6$ μm

Metal	Absorptivity	
	Unoxidized surface	Oxidized surface
Au	0.010	–
Al	0.034	0.25-0.50
Fe	0.050	0.33-0.74
Zr	0.083	0.45-0.56
Ti	0.094	0.18-0.25

of 1.2 to 1.5 and 2.5 to 2.8, respectively. A metal powder coating or a paint coating gives 2.0- to 2.5-fold increase in the absorptivity.

The approach involving an increase in the surface roughness to raise the absorption efficiency not always justifies the effort from the technological viewpoint. In contrast, the use of various absorbent coatings is undoubtedly justifiable particularly at temperatures of laser heating below the melting point T_m. The use of absorbent coatings on smooth ground surfaces has become indispensable in most laser treatments.

An applied coating must strongly absorb radiation. The absorptivity of the coating grows with its electrical conductivity. Its material must be easy to produce, inexpensive, harmless, stable in long storage, adhere well to metals, have high melting and vaporization temperatures, and readily conduct heat. The coatings that would equally well satisfy the above requirements are not yet available. Laser technology uses (1) chemical coatings obtained by manganese or zinc phosphotizing, sulfurizing, and oxidizing; (2) paints containing Al and Zn oxides; colloidal solutions of carbon in acetone and alcohol; (3) sprayed oxides of metals; and (4) soot of resin, oil, and other substances.

Whatever the type of coating material, it cannot be completely free of shortcomings. For example, the soot deposited on the surface does not provide a uniform coat and colloidal solutions give coats that poorly adhere to metals. Also, oxides sprayed on metal surfaces call for a complex and costly spraying process.

The absorptivity of coatings may change considerably with the conditions of laser heat treatment. To varify how these conditions affect the absorptivity of coatings, researchers of [49] used a CW CO_2 laser to harden the surfaces of carbon steel specimens covered with three types of coating: (1) $MH_3(PO_4)_2$, Fe_2S_3, and Cr coatings 2 to 10 μm thick, obtained by phosphatizing, sulfurizing,

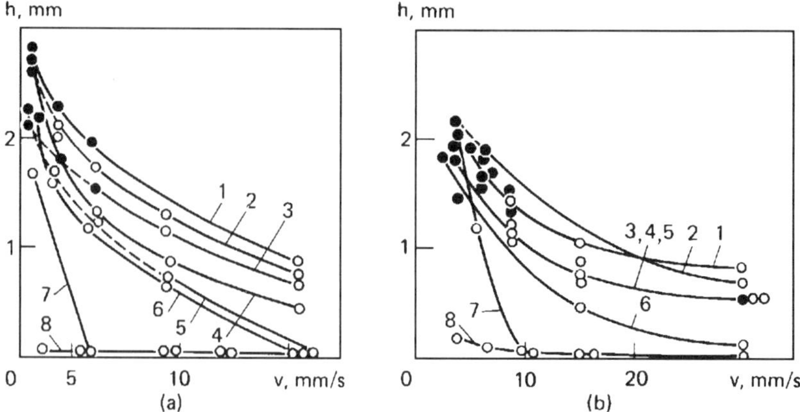

Fig. 4.1. Laser-hardened depth as a function of heat treatment speed *v* in absorbent-coated polished steel specimens at a laser power density of (*a*) 20 MW/m^2 and (*b*) 50 MW/m^2: (*1*) gas black; (*2*) Al_2O_3 with organic binder; (*3*) aqueous solution of ZnO; (*4*) ZnO with organic binder; (*5*) aqueous solution of carbon black; (*6*) solution of graphite in acetone; (*7*) sulfide; (*8*) electrodeposited chromium. (Dark and light circles identify heating with and without melting, respectively.)

and electroplating respectively; (2) carbon solution coatings 10 to 60 μm thick, applied by brushing, spraying, and dipping; and (3) paint coatings 10 to 60 μm thick, containing highly absorbing oxides of metals. The treated specimens were cut into macrosections to estimate the hardened depth along the track of hardening.

Figure 4.1 illustrates the results of experiments on polished specimens covered with various coatings. The hardened depth appreciably varies with the hardening rate *v* since it determines the time of beam interaction with the coating surface and its temperature. With an increase in the speed *v* above 6 mm/s, the beam fails to melt the surface. The coating containing the Al and Zn oxides and also soot aerosol coatings offer the best result (curves *1* through *4*).

For comparison, Fig. 4.2 illustrates the result of laser hardening of the ground specimens from the same steel. The pattern of changes in the heat affected zone *h* is seen to be similar to that illustrated in Fig. 4.1. The hardened depth in phosphate-coated specimens is comparable with that observed in specimens with Al_2O_3 and ZnO coatings. The hardened depth in a ground specimen is greater than that in a polished specimen, probably because the coating adheres better to the polished surface.

The above example of application of coatings does not certainly cover all the cases of laser heat treatment with absorbent coatings. Much work has still to be done to develop more efficient, cheaper, and more practicable coatings.

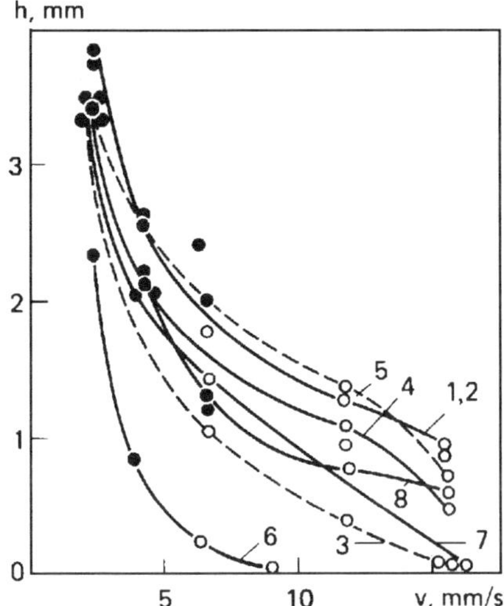

Fig. 4.2. Hardened depth versus rate of hardening at 20 MW/m^2 in steel specimens covered with various coatings:
(*1*) gas black; (*2*) Al_2O_3 with organic binder; (*3*) water-soluble color based on oxide mixtures; (*4*) ZnO with organic binder; (*5*) aqueous solution of carbon black; (*6*) solution of graphite in acetone; (*7* and *8*) manganese phosphate layers 5 and 10 μm thick respectively. (Dark and light circles stand for heating with and without melting, respectively.)

The processes of laser welding and cutting commonly dispense with coatings. The shape of a surface, however, has a noticeable effect on its absorptivity in the process of deep penetration welding associated with keyholding—a phenomenon that enables the beam to form a deep keyhole or cavity [50]. The intensity of heating in a deep melting process increases because the walls of a narrow and deep hole absorb the laser radiation. An increase in the depth-to-width ratio (aspect ratio) from 1.0 to 3.0 raises the absorptivity from 0.38 to 0.80 [51].

Other factors affecting the absorptivity of metals are the laser power density distribution in a focused spot, beam collimation (directionality), angle of beam convergence, and laser power density in the heat affected zone, the latter being the governing factor.

The basic laser heating processes, in the order of increasing amount of the energy per unit volume, are surface hardening, welding, and gas-assist cutting. The clear-cut energy boundary between these processes does not exist. They are amenable to the qualitative analysis on the basis of common thermophysical concepts.

The character and efficiency of any kind of laser processing primarily depend on the laser power density q delivered to a workpiece. A laser power density of 10^4 to 10^5 W/cm^2 causes the heat affected zone to acquire a steady temperature state in which the material does not yet vaporize or disintegrate at a high rate. For the case in hand we can assume that an adequately focused beam creates a point heat source or a plane heat source on the target surface, the heat power density of which is

$$q_h = Aq. \tag{4.2}$$

This type of heat source is commonly applied in surface hardening with and without melting, cladding, alloying, and welding of thin parts. In these processes, the absorptivity of metals without absorbent coatings is about 0.35. In order to estimate the temperature fields typical of the heat source under consideration, we can use requisite expressions obtained from the solution of a linear heat conduction equation [52].

The threshold value of laser power density q_{th} above which most of the metals begin to melt and vaporize intensively varies from 10^5 to 10^7 W/cm^2 depending on the amount of heat of vaporization, thermal conductivity, and interaction (dwell) time. The approximate values of q_{th} for some metals heat-treated in air are the following [47]:

Metal	Ag	Al	Au	Cr	Cu	Fe	Mg	Ferrite
$q_{th} \times 10^{-4}$, W/cm^2	640	240	350	22	260	30	97	4

If the power density is well in excess of the threshold value, a steady state of metal disintegration will set in [48, 53], so that a major share of the energy coupled into the target will be spent on intensive vaporization and scatter of vapors and droplets. The amount of heat expended here in melting the material is relatively small and the interaction zone is generally free of the liquid phase. These conditions of heating are too rigorous for welding but quite suited for cutting.

The efficiency of laser cutting considerably increases in the atmosphere of assist gases such as O_2, Ar, CO_2, and N, which help produce a narrow kerf with parallel edges. The absorptivity A is commonly equal to 0.5 or 0.6 and sometimes approaches unity [50] owing to the effect of beam self-focusing in the hole [53]. This effect causes a periodic increase in the power density along the beam axis due to a multiple reflection of the laser radiation from the walls of a deep and narrow cavity [54]. The limiting values of the metal absorptivity are sometimes independent of the physical properties of a metal.

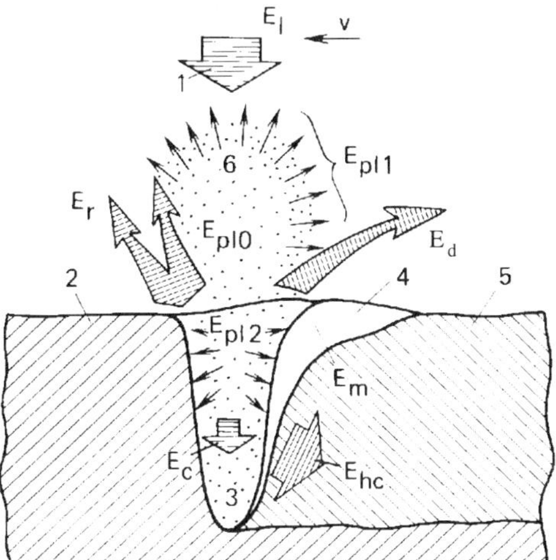

Fig. 4.3. Laser energy distribution in the steady conditions of metal melting:
(*1*) focused beam; (*2*) base metal; (*3*) cavity; (*4*) liquid metal; (*5*) weld metal; (*6*) plasma plume

the metal surface state, and the gas composition. At power densities of 10^5 to 10^6 W/cm^2 that are slightly below the threshold value, the liquid phase in the HAZ sharply increases in volume and forms a weld pool since the released energy rapidly melts the walls of the cavity to its entire depth.

As the laser beam moves on the target surface and forms a cavity, the molten metal flows along the walls to the rear portion of the weld pool under the effect of vapor pressure and difference in the surface tension forces between the central and the rear portion of the molten pool. The molten metal then cools and crystallizes to form a weld joint.

Figure 4.3 illustrates the laser energy distribution in the steady conditions of melting [55]. The energy balance equations below establish the relationship between the energy E_l coupled from the laser to the target and other kinds of energy produced by the beam in the interaction zone:

$$\begin{aligned} &E_l - (E_{pl0} + E_r + E_d) = E_c, \\ &E_{pl0} = E_{pl1} + E_{pl2}, \\ &E_c + E_{pl2} = E_m + E_{hc}, \end{aligned} \tag{4.3}$$

where E_l is the laser energy of the focused beam; E_{pl0} is the energy of the plasma-vapor plume; E_{pl1} is the radiant energy emitted from the plume into the environment; E_{pl2} is the fraction of plume energy absorbed by cavity walls through convection and radiant exchange; E_r is the fraction of laser energy reflected from the base metal and cavity bottom; E_d is the energy of disintegration particles blown off by the vapor-gas jet; E_c is the fraction of laser energy absorbed by cavity walls in the process of electron-photon collisions; E_m is the energy of heat of the molten metal pool; and E_{hc} is the energy of heat delivered to the base metal and parent metal by way of heat conduction.

The efficiency of laser energy transfer to a target depends on the effective absorptivity A_{eff} which actually determines the net laser-beam absorption efficiency η_n or coupling coefficient for laser radiation. An approximate expression for A_{eff}, derived from equations (4.3), takes the form

$$A_{eff} = \eta_n = \left(E_c + E_{p12}\right)/E_l \,. \tag{4.4}$$

There is a great body of theoretical and experimental information on the absorption efficiency of laser treatment processes, the analysis of which permits us to establish the relations between the power density and the limiting temperature in a heat affected zone.

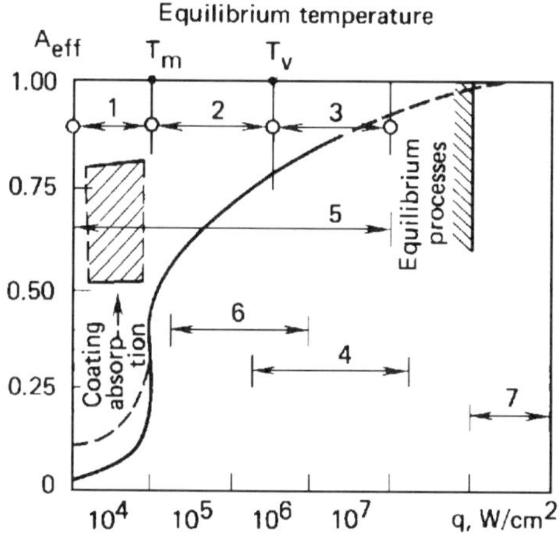

Fig. 4.4. Thermal diagram illustrative of the effective absorptivity as a function of laser power density in heat treatment of metals:
(*1*) surface hardening; (*2*) welding; (*3*) gas-assist cutting; (*4*) hole piercing; (*5*) precision treatment; (*6* and *7*) other processes; T_m, melting point; T_v, vaporization temperature

Present-day industrial lasers work materials at power densities of 10^4 to 10^8 W/cm^2. Over this range, it is possible to determine how the theoretical or experimental values of A_{eff} vary with power densities and limiting temperatures [48, 51].

Figure 4.4 displays a simplified thermal diagram constructed on the basis of the analytical data, from which we can readily estimate the tentative values of the thermal efficiency of any laser heating process at a given power density. Knowing the effective heat power density q_h, we can then apply an appropriate model to perform a further analysis of the thermal processes.

The simplest approach to estimating the temperature fields is to solve the differential heat-conduction equation reduced to a linear form [52]. The basic design equations for temperature distribution patterns will be dealt with later in the text. More complex models involving nonlinear relations for the processes of deep penetration welding will be treated in Sec. 4.4.

4.2. Laser Beam Efficiency

Proceeding from theory of thermal processes in welding [52], we can express the laser-beam absorption efficiency in terms of the net efficiency η_n given by equation (4.4). The effective heat power P_{eff} refers to the amount of heat that the beam of power P can deliver to a target in a unit time:

$$P_{eff} = \eta_n P. \tag{4.5}$$

The net efficiency η_n defined earlier as the ratio of absorbed power to incident power characterizes the effectiveness of laser energy conversion and heat exchange. This coefficient has the same meaning as the absorptivity A in heat treatment without melting and as the absorptivity A_{eff} in heating with melting.

For the processes of laser heating with melting, the effective utilization of heat energy coupled to a metal target is defined by the thermal efficiency η_{th} which is the ratio of the amount of heat required to melt a metal to the total amount of heat delivered to the metal target [52]:

$$\eta_{th} = vS_m C_m / P_{eff}, \tag{4.6}$$

where v is the speed of heat treatment; S_m is the cross-sectional area of a molten metal pool; and C_m is the heat content per unit volume of a molten metal, including the latent heat of vaporization.

Using the notions of net absorption efficiency η_n and thermal efficiency η_h, we can unambiguously specify the optimal conditions of laser heat treatment. As noted is Sec. 4.1, absorbent coatings can raise the laser beam efficiency η_n to a rather high value. More studies have still to be made to obtain adequate values of the efficiency of laser welding. The estimates of η_{th} given in the literature on the subject are rather discrepant.

The theory of movable heat sources is a useful tool in the investigations of the thermal efficiency of laser welding [56]. Paper [57] describes a generalized thermal model of deep penetration welding with a linear heat source, which relates the depth-to-diameter ratio h/D to the laser power P and weld rate v_w. The normalized curves presented in work [57] display the results of the solution of heat conduction equations, which help estimate how the absorption efficiency and melting rate vary with the weld speed and laser power. The efficiency of coupling of the laser beam energy to a target for welding conditions selected experimentally is found to be the following: up to 100%, deep penetration at 20 kW of laser power; above 50%, deep penetration at 3.8 to 8.0 kW of laser power; and less than 20% in welding thin plates with a power of 250 W.

The theoretical values of the net efficiency η_n calculated in terms of the model of [57] are approximate and require experimental verification. The available experiments data on the thermal efficiency of laser welding are hardly reliable since they are poorly consistent. For example, the experimental values of η_{th} presented in paper [51] for laser welding in corrosion-resistant steel vary from 0.28 to 0.52 depending on the weld speed, whereas this efficiency given in works [58, 59] reaches 0.8 or 0.9 under similar conditions of welding. Since the discrepancy between the results is significant, it is difficult to generalize the obtained data. The same is also true for the values of the thermal efficiency presented in papers [60, 61].

We have carried out experimental studies of the laser welding efficiency using a continuous 5-kW CO_2 laser setup [22]. Preliminarily, we have worked out the calorimetric measurement technique for a fast and reliable estimation of the net heat power. The experiments involved the butt welding of 150- by 100-mm specimens 2.0 to 4.5 mm thick, produced from various materials such as titanium alloy, corrosion-resistant steel, and low-carbon steel.

According to the experimental results, the absorption efficiency heavily depends on the conditions of welding. With welding performed at a laser power close to a maximum value, an increase in the weld speed considerably changes the net efficiency. For example, in the process of welding 2-mm titanium alloy plates in helium, with argon applied on the opposite side of the specimen to protect the weld root, the net efficiency η_n at a laser power raised to 5.0 kW increases 2.5 to 2.8 times at the same weld speed and about 10-fold at a higher weld speed (Fig. 4.5, curve *1*). When the welding process occurs at a minimum

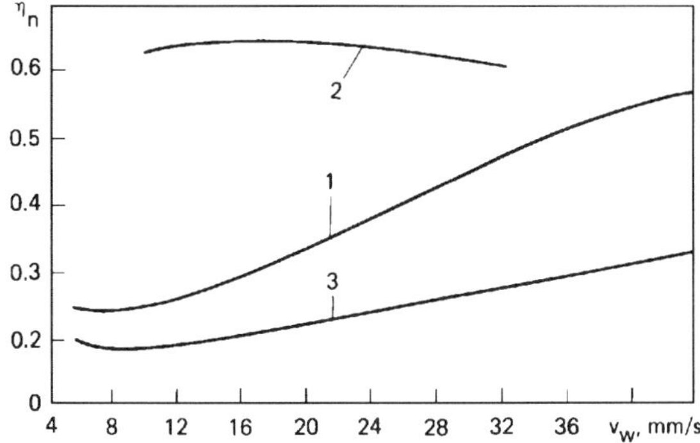

Fig. 4.5. Net laser-beam efficiency vs rate of welding of titanium plates 2 mm thick at different powers of the laser beam focused by a 215-mm KCl lens:
(*1*) at 5.0 kW; (*2*) at constant penetration depth; (*3*) net efficiency calculated with the model given in [58]

possible laser power which still provides for the full penetration at a specified weld speed, the beam efficiency η_n generally remains the same (curve *2*) and is likely to approach the upper limit since its steady value almost coincides with the extremum which the curve *1* approaches asymptotically.

In the process of welding 3.5-mm thick low-carbon steel plates the pattern of changes in η_n is similar. The net efficiency strongly depends on the weld speed since the cavity profile varies with weld speed and affects the absorptivity (Fig. 4.6). An increased laser power delivered to a target (points I) causes the metal to melt very fast, thus forming a through hole, in which case a fraction of the laser radiation passes through the hole without transferring the energy to the hole wall. It is this lost energy that accounts for a low thermal efficiency of the welding process, although the weld joint formed proves acceptable.

As the weld speed gradually increases, the through hole becomes narrower at the bottom (points II) until it closes at the bottom portion (points III) to form a keyhole conducive to the highest efficiency of the welding process. With a further increase in the weld speed (points IV), the weld penetration decreases somewhat because the beam delivers a less amount of energy to the cavity and η_n decreases somewhat.

Where the welding conditions ensure a minimum loss of laser energy and provide for a regular full penetration, an increased weld speed insignificantly reduces the weld width. The hole profiles shown in Fig. 4.7 illustrate the point.

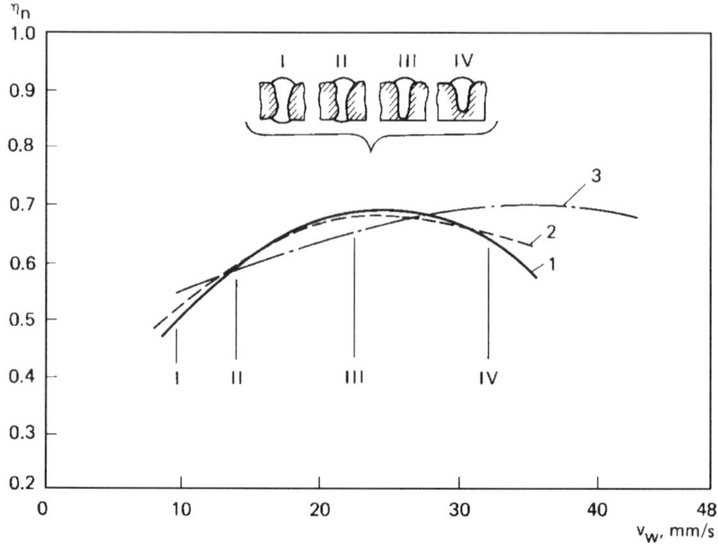

Fig. 4.6. Net laser-beam efficiency vs rate of CO_2-shielded welding of 3.5-mm low-carbon steel plates at a power of 5 kW of the laser beam focused by a 215-mm KCl lens: (*1*) plates after machining; (*2*) as-received plates; (*3*) plates after chemical etching

The aspect ratio of the weld changes little, and so the beam efficiency generally remains the same.

Proceeding from the above described regularities, we can decide on the choice of a laser power required for welding. In the general case, an intensive welding process offering the highest production rate at a minimum laser power is the best choice. The position of an extremum on the curve of η_n determines the choice of a weld speed. Where the weld speed is to be held slow to ensure, for example, the desired strength of the metal, the welding process can be run at a lower laser power in the conditions of a steady absorption efficiency with a minimum loss of laser energy.

In order to estimate the thermal efficiency η_{th} from equation (4.6), specimens were cut into macrosections to check the fusion zones.

As noted earlier, the coefficient η_{th} determines the share of heat energy expended in metal melting and thus reveals the total heat loss by both heat transfer in the solid phase and liquid metal overheating. Hence, the pattern of η_{th} variation with the weld speed is predictable to a certain extent. If the heat source moves at a slow speed, a large amount of heat energy has time to propagate into the bulk of metal and in the direction of motion of the heat source, which heavily heats up the solid phase enclosing the fusion zone. The

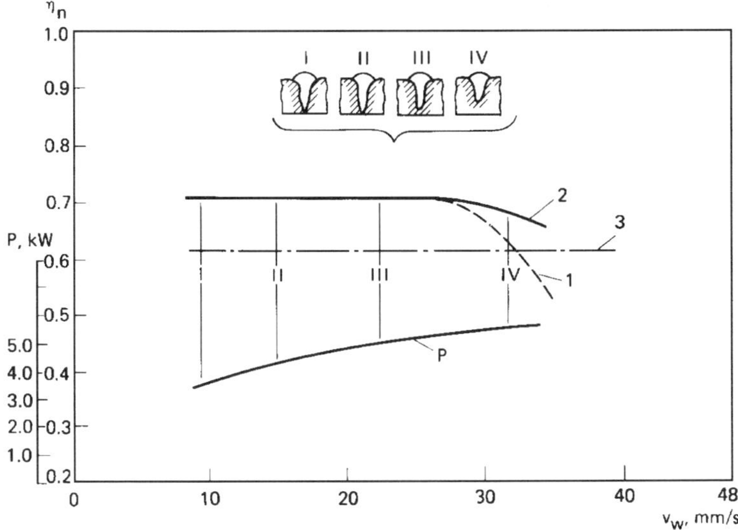

Fig. 4.7. Net laser-beam efficiency vs rate of welding of 3.5-mm low-carbon steel plates at a minimum power of the laser beam focused by a 215-mm KCl lens that ensures constant penetration:
(*1*) after machining, in helium; (*2*) as-received, in carbon dioxide; (*3*) after chemical etching, in argon-helium mixture

melting efficiency is low at such a speed. With an increase in the weld speed, the main heat fluxes begin to propagate in the direction normal to the weld seam, so the amount of heat lost to conduction gets smaller and the thermal efficiency starts growing.

Our experimental data have confirmed this pattern of changes in η_{th} (Figs. 4.8, 4.9, 4.10). As is evident from curve *1* in Fig. 4.8, the thermal efficiency of welding of titanium alloy plates ranges from 0.23 to 0.26 at a weld speed of 6 to 8 mm/s, but rises to 0.4 at v_w of up to 36 mm/s. The curve *2* in Fig. 4.8 and the curves in Figs. 4.9 and 4.10 display a similar change in η_{th} which reaches 0.44 to 0.46 at a weld speed of about 36 mm/s. These values of η_{th} are close to a limiting value of 0.484 calculated for a thin plate melted with a linear high-power fast-moving heat source [56]. So, it is safe to say that, irrespective of the level of laser power, the region of optimal welding conditions lies in the range of weld speeds from 20 to 40 mm/s. The thermal efficiency of laser welding lies on the average between 0.35 and 0.45, whereas the value of η_{th} in conventional fusion welding does not exceed 0.22 [62].

The plots of Figs. 4.6, 4.7, 4.9 and 4.10 take into account the effect of the cover gas used and the physicochemical states of the surfaces of low-carbon

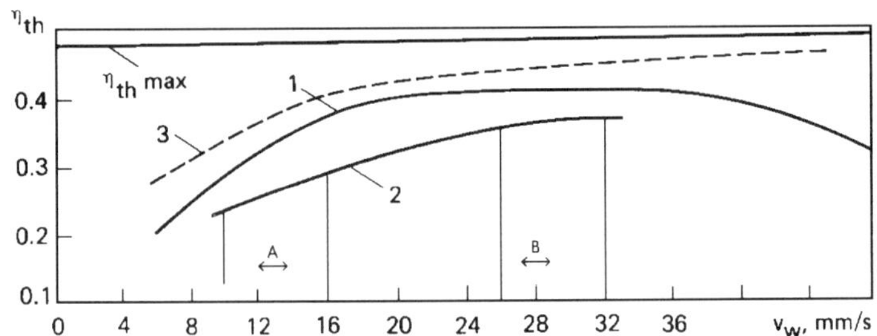

Fig. 4.8. Thermal efficiency of laser welding of 2-mm thick titanium alloy as a function of weld rate and laser power. For comparison, see the legend of Fig. 4.5

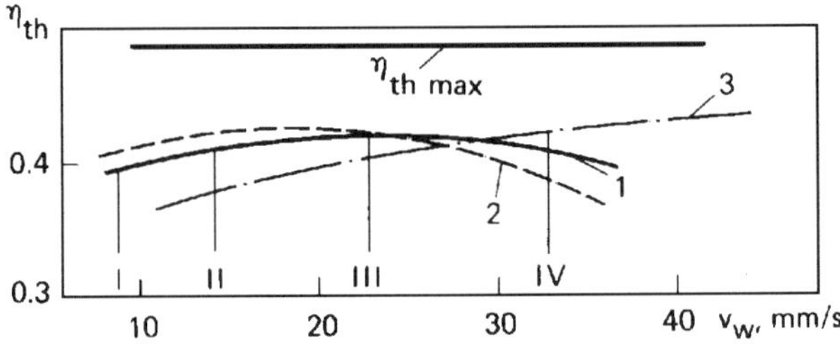

Fig. 4.9. Thermal efficiency vs weld rate for 3.5-mm thick low-carbon steel at $P = 5$ kW. For comparison, see Fig. 4.6

steel specimens on the laser beam efficiency and thermal efficiency. The efficiency of absorption of infrared radiation turns out to be little dependent on the chemical state of surfaces. The plots of Figs. 4.6 and 4.7 indicate that the values of the net efficiency for as-received specimens and machined specimens are generally the same, although the quality of weld beads in machined specimens is better.

Profile irregularities on the welded plates little affect the beam efficiency. As seen from Fig. 4.6, the highest values of η_n for specimens with three different states of the surfaces are largely the same and equal to about 0.7.

The effect of the kind of cover gas used on the laser beam efficiency is more perceptible. Changing over from CO_2 to the argon-helium mixture leads to a decrease in the beam efficiency under the same welding conditions, as is obvious from curves *3* in Figs. 4.7 and 4.8. This is because the plasma processes exert a different effect on the absorption efficiency in various gases.

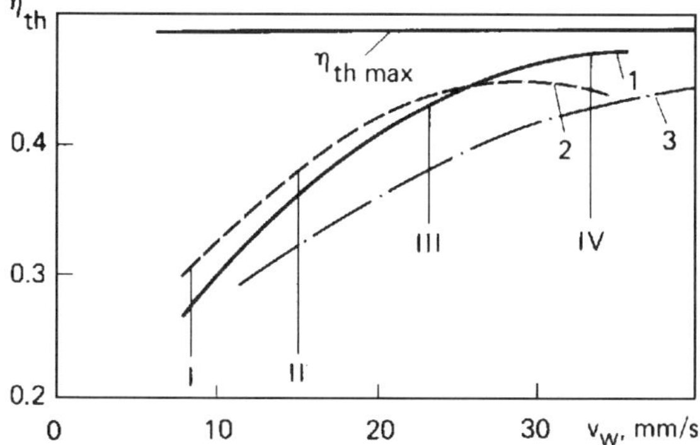

Fig. 4.10. Thermal efficiency of welding of 3.5-mm low-carbon steel plates at a minimum laser power ensuring constant weld penetration as a function of the weld rate and physicochemical state of the plate surface. For comparison, see Fig. 4.7

In implementing a process of laser welding in workshop conditions, it is of importance to verify how the deviations in the laser equipment parameters affect the formation and quality of the weld joint. An inaccurate alignment of the edges of the parts to be welded changes the position of a focal plane relative to the target surface, thus altering the amount of laser energy coupled to the target and the character of plasma processes.

Shifting the focal plane from its optimal position by ± 2 mm reduces η_n from 0.63 to 0.42 in welding as-received low-carbon steel plates 3.5 mm thick (Fig. 4.11). This efficiency drops to the least value when the shift Δf of the focal point below the surface reaches 0.3 mm.

Reciprocating the focusing lens with respect to a target can appreciably increase the weld efficiency [63]. This causes the focal point below the target surface to change its position in depth at a frequency ν and amplitude y (Fig. 4.12). This shift of the focal point position, Δf, must be in agreement with the propagation rate of the fusion front and vaporization rate.

In welding steels, titanium alloys, and aluminium alloys at the same laser power, a periodic shift of the focal position along the beam axis causes a 60 to 80% increase in the thermal efficiency, but only a small percentage increase in the net absorption efficiency [63]. The penetration depth grows by 40% as against that achieved in welding with the beam whose focal point remains fixed in position.

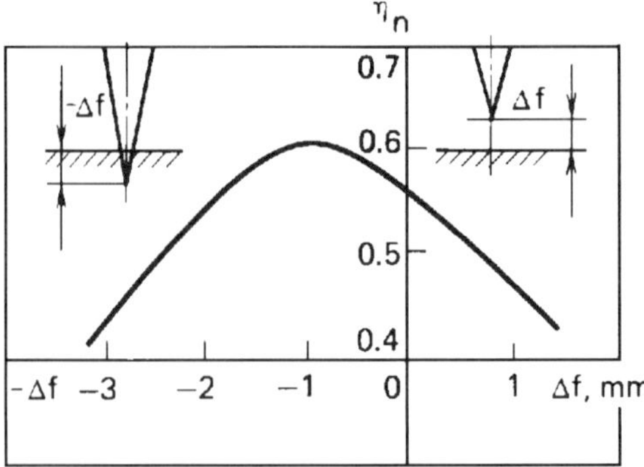

Fig. 4.11. Thermal efficiency of laser welding as a function of the position of the focal point relative to the target surface. Laser power, 5.0 kW; weld speed, 16 mm/s; focal length of lens, $f = 230$ mm

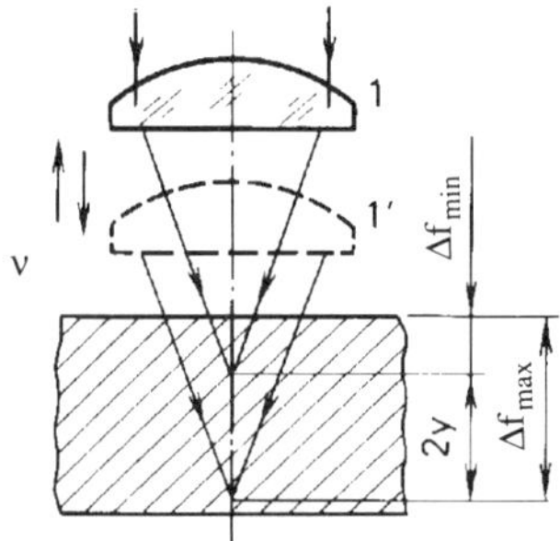

Fig. 4.12. Schematic of the system of reciprocating the lens to change the focal point position below the target surface; ν, rate of change of focal point position; $2y$, peak-to-peak shift; Δf, depth of focal point below surface

4.3. Temperature Field Models

The simplest way of defining the temperature fields is to solve analytical expressions [52] derived from linear differential heat-conduction equations under linear boundary conditions. This means that the thermal coefficients such as the thermal conductivity K, volume specific heat $c\rho$, and heat transfer coefficient α are thought to be independent of temperature.

Consider some solutions of the equations [52], which are suitable for simple calculations. In the analysis of heat treatment with a pulsed laser or a CW laser that interacts with a target over short-time periods, use can be made of instantaneous heat sources.

When a point heat source gives up heat Q at time t = 0 to the surface of a semi-infinite body, the surface temperature at point 0 is

$$T(r,t) = \frac{2Q}{c\rho(4\pi kt)^{3/2}} \exp\left(-r^2/4kt\right), \tag{4.7}$$

where $c\rho$ is the specific heat; ρ is the mass density; $k = K/c\rho$ is the thermal diffusivity; $r^2 = x^2 + y^2 + z^2$ is the square of the distance from the heat source to a solid body point (x, y, z).

The above formula applies for the approximate calculation of temperatures during a short-time interaction of the laser beam with the surface of a massive part.

In the case of a linear heat source that uniformly heats a thin plate across its thickness δ, the temperature at point 0 is given by

$$T(r,t) = \frac{Q}{4\pi K\delta t} \exp\left(-r^2/4kt - \alpha' t\right), \tag{4.8}$$

where $r^2 = x^2 + y^2$ is the square of the distance from the heat source to a point (x, y); $\alpha' = 2\alpha/c\rho\ \delta$ is the factor accounting for the surface heat transfer to the environment.

From equation (4.8) we can obtain approximate values of the temperature at a spot heated for a short time with a pulsed beam or CW beam used, for example, in spot welding of plates. If a point source or a linear source heats the target for a longer time t, the temperature can be found using the principle of temperature superposition that comes down to the integration of equation (4.7) or (4.8) with respect to t.

The same principle of superposition works where it is necessary to derive heat transfer equations for a moving heat source. In this case, we need to integrate an appropriate equation for the moving source that acts during an arbitrary time. But it is also possible to obtain analytical expressions for a moving source that heats the target for a long time in the limiting quasi-steady conditions of heat propagation.

The same principle of superposition works where it is necessary to derive heat transfer equations for a moving heat source. In this case, we need to integrate an appropriate equation for the moving source that acts during an arbitrary time. But it is also possible to obtain analytical expressions for a moving source that heats the target for a long time in the limiting quasi-steady conditions of heat propagation.

For the quasi-steady state of the process of heat propagation from a point source of constant heat power P that moves at a constant speed v along the surface of a semi-infinite body, the equation of temperature distribution assumes the form

$$T(r,x)=\frac{P}{2\pi Kr}\exp\left[(-vx-vr)/2k\right], \tag{4.9}$$

where r is a constant radius vector in the moving system of coordinates, namely, the distance from the origin of moving coordinates to the point of interest; and x is the abscissa of that point.

From equation (4.9) we can obtain approximate values of the temperature field in a massive body heated with a high-power source.

The equation for the quasi-steady state of heat distribution in a plate heated with a linear source of constant heat power P that moves at a constant speed v takes the form

$$T(r,x)=\frac{P}{2\pi K\delta}\exp(-vx/2k)\,J_0 r\left(\frac{v^2}{4k^2}+\frac{\alpha'}{k}\right)^{1/2}, \tag{4.10}$$

where r is the linear radius vector in the moving coordinate system and J_0 is a Bessel function of an imaginary argument of the second kind of order zero.

Equation (4.10) is suitable for description of the temperature fields of heat affected zones in the processes of butt welding and cutting of thin plates.

Most of the processes of laser heating are run at high speed, so it is safe to simplify somewhat the equations by employing the model of a powerful fast-moving heat source [52].

The equation for the limiting case of heat propagation from a source rapidly moving in a semi-infinite solid is written as

$$T(y_0,z_0,t)=\frac{P}{2\pi Kvt}\exp\left[-\left(y_0^2+z_0^2\right)/4kt\right], \tag{4.11}$$

where t is the time counted off from the moment when the heat source traverses the plane y_0Oz_0 normal to the axis of source motion through the point of interest;

and y_0 and z_0 are the stationary coordinates of the point on the target that are coincident with the moving coordinates y and z.

For the limiting case of the process of heat propagation involving a linear fast-moving heat source, the equation below holds:

$$T(y_0,t)=\frac{P}{v\delta(4\pi K c\rho t)^{1/2}}\exp\left(-y_0^2/4kt-\alpha' t\right). \tag{4.12}$$

The above equations describe the sources of high heat-power densities. The theory of heat propagation from high-power heat sources allows for defining the temperature fields of the zones lying at a distance from the source that is 3 to 5 times the laser spot diameter d_s. The processes of heat transfer in the zones located closer to heat sources can be described for certain only with due regard for the character of power density distribution over the cross section of the beam.

Consider a simple case for a circular beam with its energy uniformly distributed over the hot spot of radius r_s. For surface treatment with short pulses of duration $t_p \le r_s^2/k$, the heat transfer problem can be reduced to a unidimensional problem, with the main heat flux assumed to propagate along the Oz axis normal to the target surface (Fig. 4.13).

The unidimensional temperature field produced along the Oz axis in a semi-infinite body by a circular beam for a dwell time $t > t_p$ is given by the equation

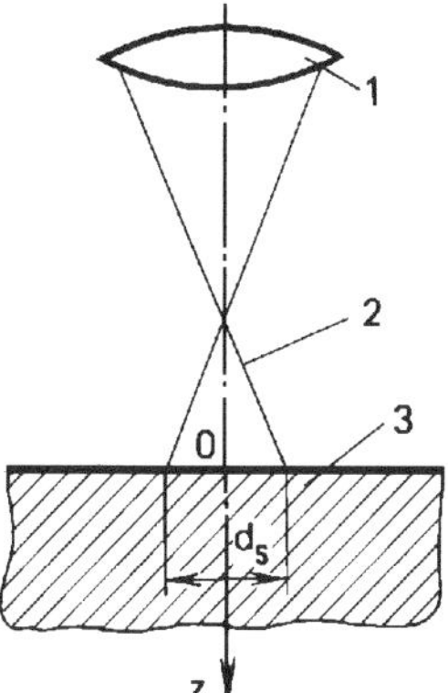

Fig. 4.13. Schematic representation of a circular beam interacting with a target material: (*1*) focusing lens; (*2*) laser beam; (*3*) workpiece

$$T(z,t) = \frac{2q_\mathrm{h}\sqrt{k}}{K}\left[\sqrt{t}\ \mathrm{ierfc}\left(\frac{z}{2\sqrt{kt}}\right)\right.$$
$$\left. -\sqrt{t-t_\mathrm{p}}\ \mathrm{ierfc}\left(\sqrt{z^2+r_\mathrm{s}^2}\ \ 2\sqrt{kt-t_\mathrm{p}}\right)\right], \tag{4.13}$$

where z is the distance along the Oz axis; $q_\mathrm{h} = \eta_\mathrm{n} q$ is the heat power density; q is the laser power density; r_s is the hot spot radius; and ierfc is the integral of the complementary error function.

For $0 < t < t_\mathrm{p}$, equation (4.13) reduces to a simpler form

$$T(z,t) = \frac{2q_\mathrm{h}}{K}\sqrt{kt}\ \mathrm{ierfc}\left(z\ \ 2\sqrt{kt}\right). \tag{4.14}$$

If it is a pulsed laser beam that interacts with a workpiece, the power density delivered to the hot spot of radius r_s is $q = P_\mathrm{p}/\pi r_\mathrm{s}^2$, where $P_\mathrm{p} = E_\mathrm{p}/t_\mathrm{p}$ is the pulse power and E_p is the energy of the pulse of length t_p.

Substituting $z = 0$ into equation (4.13) yields the expression for the surface temperature

$$T(0,t) = \frac{2q_\mathrm{h}\sqrt{k}}{K}\left[\left(\sqrt{t}\ \ \pi\right) - \sqrt{t-t_\mathrm{p}}\right.$$
$$\left. \times\ \mathrm{ierfc}\left(\frac{r_\mathrm{s}}{2\sqrt{k\left(t-t_\mathrm{p}\right)}}\right)\right]. \tag{4.15}$$

For $0 < t < t_\mathrm{p}$, equation (4.15) for a surface temperature at point $z = 0$ gets simpler and reduces to the form

$$T(0,t) = \left(2q_\mathrm{h}/K\right)\sqrt{kt/\pi}\,. \tag{4.16}$$

In order to estimate the extent to which the fusion front propagates below the surface and heats the zone to a temperature T, it is well to use the expression set up in an implicit form under the assumption that $r_\mathrm{s} >> \sqrt{kt}$:

$$T(z,t) \approx \left(q_\mathrm{h}/K\right)\left(2\sqrt{kt/\pi} - z\right), \tag{4.17}$$

where z is the depth at which the temperature is $T(z, t)$.

From expression (4.17) we can derive a simple expression for estimating the depth of heating to a specified temperature:

$$z \approx 2\sqrt{kt_p/\pi} - TK/q_h \,. \tag{4.18}$$

Work [64] presents similar analytical solutions of the temperature distribution problems which, however, deal with a different character of power density distribution over the beam cross section and power density changes in the interaction zone.

The analytical expressions are simple and easy to use for the description of thermal processes involved in laser heat treatment. But the results obtained from the solution of these expressions sometimes poorly agree with experimental data because thermal properties actually vary with temperature. The use of the above expressions in the linear form is practicable for the tentative estimation of temperatures below the melting point.

To obtain more accurate results, account should be taken of changes in the thermophysical properties of a material with fast heating and cooling. For this, we need to solve differential heat conduction equations in the nonlinear form. What complicates still more the computation procedure here is that the absorptivity varies with temperature and adds to the nonlinearity of the differential equations. Also, since the boundary conditions are commonly nonlinear, the heat transfer equations, as applied to the analysis of laser heating, form a complex set of nonlinear differential equations.

The theory of heat conduction does not contain general methods for the accurate solution of nonlinear heat transfer problems. Researchers commonly resort to different approximate methods to search for the solution on the basis of analytical statements, numerical analysis, simulation modeling, statistical methods, and other approaches.

In the analytical methods, the final solution to an analytical expression is often made by a numerical method. But where changes in the thermophysical properties can be defined by lower-order expressions, with boundary conditions specified in a simpler form, the analytical solution to a specific problem and subsequent processing of the results on a computer may prove to be a more expeditious approach. Paper [65] presents the analytical solutions of nonlinear problems that reduce to simple equations under assigned assumptions. The procedure involves the solution of the known analytical equations with added terms intended to account for the basic nonlinear conditions.

The direct methods of mathematical analysis, such as the Kantorovich method, are available for solving nonlinear heat-transfer problems with due regard for nonlinear conditions.

In wide use is the method of integral transformations for solving heat transfer problems under complex boundary conditions. The solutions of temperature field problems in laser welding are available, which take into account the thermophysical properties of welded materials. If a solid can be thought of as extending to infinity at least along one coordinate, the Fourier analysis is quite an appropriate solution method.

A number of the analytical solutions to heat transfer problems, which are applicable in the analysis of laser heating, are given in papers [66, 67]. The authors of [56] have treated some analytical solutions of nonlinear heat transfer problems, which help describe quite adequately some processes of laser heating.

The analytical methods are suitable for solving relatively simple problems and establishing qualitative relationships from their solutions. The numerical methods which have become a very powerful tool with the advent of large-powered computers can handle complex nonlinear heat-conduction problems. These are the finite difference, finite element, relaxation, and other methods, the former being most popular. What limits the application range of the finite difference method is the lack of simple procedures that would enable ease of handling the problems.

In a large number of recent papers, researchers have put forward more elaborate versions of the variational-difference method, which use discrete models to represent the solutions of differential equations in the variational form. The solution here is most often sought by the finite element method which generally involves a calculation procedure relying on a discrete model that substitutes for the continual representation of a problem [68, 69]. However, the finite element method is inferior to the finite difference method in efficiency and ease of solving the heat-conduction problems.

The relaxation method is a version of the finite difference method, which offers an efficient way to solve linear heat-conduction problems [70]. However, as mentioned above, the linear statement is a simplified formulation of a complex nonlinear process of laser heat treating.

Of interest is a modified numerical method which replaces the original differential equation by a set of differential equations with a reduced number of arguments and represents certain derivatives in the finite difference form. The method serves as an effective tool in solving the problems with nonlinear boundary conditions and the Stefan problems concerned with heat release in phase transitions.

The above numerical methods relate to determinate approach widely employed for the analysis of thermal processes. Statistical or probabilistic methods, such as the well known Monte Carlo method and its versions [72], work most effectively where the properties of, say, composite materials vary in a random way at arbitrary points in the calculation region.

The simulation methods establish similarity between an actual process under study and its physical model. Thermal processes are analyzed in terms of electric- or hydrodynamic-type models through an adequate description of the process and its model by differential equations of the same form. Electric simulation offers ease of solving nonlinear heat transfer problems on an analog computer.

In some cases, the methods mentioned above cannot be used directly to estimate the temperature fields of heat affected zones. Before applying a particular method, consideration should be given to a number of specific factors. Since the laser beam rapidly heats a target to the vaporization temperature, the heat transfer equations of the model must include nonlinear terms. To estimate adequately the heat equivalent of laser radiation and, hence, the temperature field, the model must certainly account for the amount of radiant energy absorbed and re-emitted by the plasma plume. For the processes of heating with vaporization, the procedure of determining the temperature fields should involve the analysis of the kinetics of keyholing and the dynamics of mass and heat transfer.

Only account of the above factors can lead to a comprehensive statement of the problem, the common solution of which will describe some specific cases of laser heating. This type of problem statement includes a complete set of nonlinearities met with in the theory of transient heat conduction [56], such as the nonlinearities of the first, second, and third order for the differential equations, boundary conditions, and heat sources, respectively.

In some works, researchers suggest different approaches to allow for the effect of keyholing. In work [57], for example, the authors specify the penetration depth analytically, assuming that a linear heat source heats up a thick plate over a portion of its thickness.

Work [73] presents a model that most fully covers the basic physical effects involved in the formation of a cavity. The authors describe the calculation procedure of thermal processes and give some calculation results pertaining to the interaction of a stationary laser with a metal surface. The model considers the laser interaction with the plasma and heat exchange with the cover gas flow. However, this model has a rather limited potential since it deals with a stationary source, neglects the latent heat and the nonlinearity of the heat transfer coefficient, and does not consider the thermal balance in the ignited plasma.

Section 4.4 below presents a more general model of deep penetration, which describes mathematically the laser energy absorption in a gaseous phase and in a cavity, the thermal effect of a gas-vapor cloud on the cavity walls, and the energy loss in vapors and by radiation as the fusion front and vapors periodically move along the front portion of the cavity and the molten metal flows back into the rear portion of the hole.

4.4. Thermal Processes in Deep Laser Welding

The laser power density q required for welding must ensure the conditions at which the rate of energy input exceeds the rate of energy propagation through heat conduction.

Under these conditions, the energy coupled to a target rapidly heats the material to its vaporization temperature and thus produces a molten pool necessary to form the weld joint. The portion of the vaporized material blown off the target surface is rather small as compared to the volume of the liquid phase.

The power density q must not be in excess of the threshold power density q_{th} (see Sec. 4.1) at which the rate of vaporization is comparable with the rate of heat removal. The power density in excess of q_{th} initiates volumetric vaporization which essentially precludes the formation of a molten pool. This process of material disintegration is typical of hole piercing, cutting, and scribing.

To specify a lower limit to q, consider the deformation of a free melt surface. In the conditions of surface vaporization typical of the process of welding, the liquid-vapor interface deforms under the reactive vapor pressure [74]

$$p_r = 0.5\, CT\rho(T), \tag{4.19}$$

where C is the Boltzmann constant and ρ is the saturated vapor density at temperature T.

The equation for q over a Gaussian beam incident on the target and causing the melt surface to deform is written as [74]

$$q(r) = q_0 \exp\left(-r^2/r_0^2\right), \tag{4.20}$$

where r is the running value of the beam radius r_0 and q_0 is the laser power density at the center of the Gaussian spot.

Figure 4.14 illustrates how the relative deformation z/z_0 of the melt surface varies in the z direction with r/R at different values of R/r_p

$$R/r_p = \left(\frac{Aq_0 r_0 c_v}{4\sqrt{\pi}\, KT_m^2}\right)^{1/2}, \tag{4.21}$$

where R is the radius of a molten metal spot; r_p is the running value of R, at which p_r changes in magnitude; A is the absorptivity; c_v is the specific heat

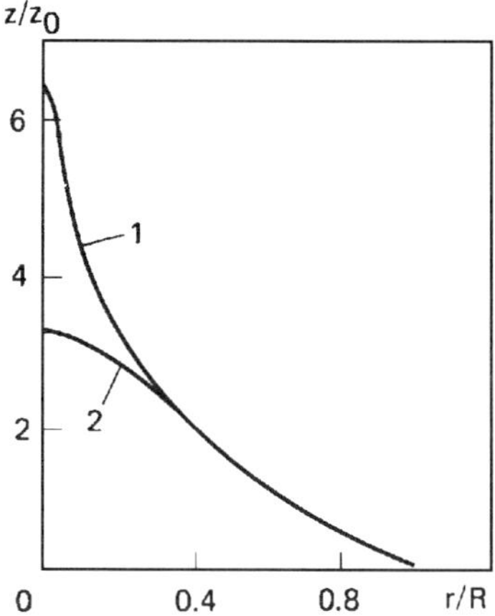

Fig. 4.14. Variations in z/z_0 with r/R at $\beta = 19.6$ (curve *1*) and $\beta = 3.8$ (curve *2*)

of vaporization; K is the thermal conductivity; and T_m is the melting temperature.

From the curves of Fig. 4.14 it follows that the depth of depression near the center of the hot spot sharply increases at large values of R/r_p and results in a deep cavity. The critical power density conducive to the cavity formation [74]

$$q_c = \frac{2c_v K}{ACr_0\sqrt{\pi}}\left(\ln\frac{m^{3/2}\nu_0 r_0}{\sigma c_v^{1/2}}\right)^{-1}, \tag{4.22}$$

where m is the atomic mass; ν_0 is the Einstein frequency; and σ is the surface tension coefficient.

The hole produced in a target allows the beam to penetrate deep into a part, thereby initiating the process of keyholing. The critical value of q_c estimated from equation (4.22) defines a lower limit to the laser power density q. So, the value of q for deep-penetration laser welding obeys the conditions $q_c \le q < q_{th}$. At values of q below q_c, the target material melts to a small depth which is adequate for surface hardening and cladding. Hole piercing, cutting, and other processes of heating with vaporization require the power densities in excess of q_{th}.

Experimental studies on the mechanism of deep penetration. In order to gain insight into the laser-induced process of keyholing, it is well to resort to the known concepts of cavity formation in electron-beam welding which is in essence similar to laser welding. The aspect of mass transfer in the keyhole is given treatment in a large number of works. Paper [75] describes a model of droplet transfer of the molten metal from the front wall of the cavity to its back wall across the bottom. The workers of [76] hold to the idea that the molten metal predominantly flows back along the side walls of the cavity. In papers [77, 78], the researchers have thoroughly considered the phenomenon of melting spikes at the weld root and ascribed its origin to the periodic transfer of molten metal along the front wall, and also to changes in the plasma transparency, scatter of clusters, and other cyclic processes occurring in the plasma plume and gaseous medium of the cavity. Work [76] has presented the results of theoretical and experimental studies on the cavity geometry. Generalized concepts of the physical processes in electron-beam welding are given in work [79].

Unlike electron-beam welding, laser welding is made in a gaseous atmosphere and attended by strong reflection of the light flux from a work surface. This explains why the dynamics of laser melting displays quite specific features [9, 80, 81]. Fast filming of laser melted quartz plates and quartz-metal interface zones has revealed a region of maximum heat release on the front wall and periodic changes in both the plume emittance and the angle of vapor scatter. But since the volume of the liquid phase is very small because of the fast sublimation of quartz, this type of modeling is more appropriate for the analysis of laser cutting.

Researchers of [80] have verified the pattern of liquid phase transfer from the movement of the molten metal of thin pins fastened at various heights

Table 4.3. Experimental Results of Laser Welding

Parameter	Glass ceramic		Glass ceramic-metal		Metal	
Weld speed, mm/s	11	27.8	11	27.8	11	27.8
Penetration depth, mm	10.9	7.0	5.5	3.8	4.4	3.2
Fusion front propagation frequency, Hz	18	46	27.5	76	–	–
Fusion front propagation rate, mm/s	200	320	150	290	–	–
Depth of melting spike, mm	0.8	2.6	0.5	1.8	0.5	1.2
Spike step, mm	–	0.8	–	0.4	–	0.3
Depth of beam waist below surface, mm	3.0	3.0	1.5	1.5	1.5	1.5

on a metal specimen. The liquid phase is found to move along the side walls of the cavity in its upper and medium portion and also across the cavity bottom. The authors of [81, 82] have described the discrete character of molten material transfer from the front wall of the cavity and revealed local zones of laser-material interaction on the front wall.

In an effort to clarify the mechanism of keyholing, we have carried out the welding experiments on glass ceramic and glass ceramic-metal specimens using a 3.5-kW CO_2 laser beam focused by a 160-mm KCl lens. A high-speed cine camera recorded fusion zone profiles through red and orange filters on black-and-white cine film at a scan speed of 180 to 1 000 frames per second. The experimental results are listed in Table 4.3.

Glass ceramic is highly stable to light because its linear expansion coefficient is close to zero. The fusion zone profile in glass ceramic is similar to that in metal since the former has a low content of the vitreous phase and is stable to sublimation. The photographs taken at arbitrary moments of laser heating permit us to judge the geometry of the cavity and molten pool in the cross section along the seam axis (Fig. 4.15).

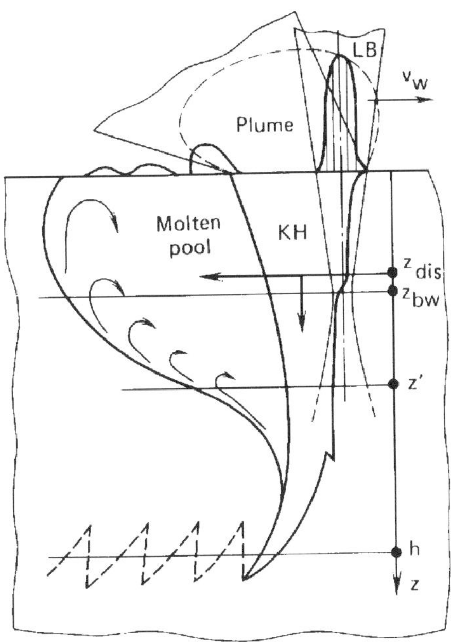

Fig. 4.15. Geometry of the keyhole (KH) and molten pool in cross section along the seam axis. LB, laser beam; *h*, penetration depth; z_{dis}, depth of direct irradiation spot; z_{bw}, depth of beam waist; *z*, boundary between regions of stable and unstable melting

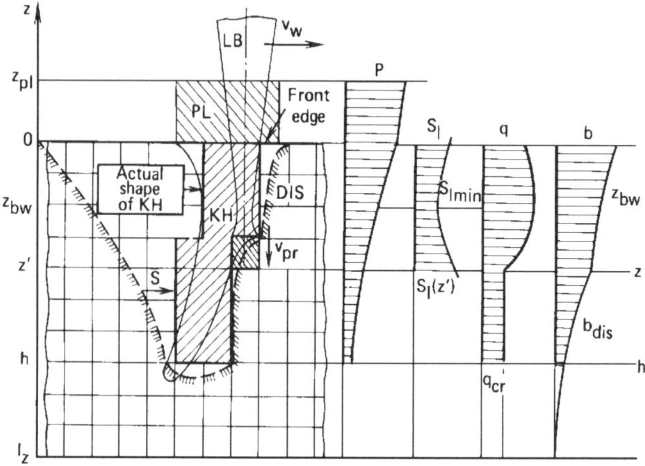

Fig. 4.16. Model of deep-penetration laser welding. KH, keyhole; PL, plume; DIS, direct irradiation spot; S_l, cross-section area of laser beam; q_{cr}, critical power density; b, cavity width; h, penetration depth

The experiments have revealed a number of regularities of the welding process. On the photographs of weld zones, the molten pool displays a clear-cut front wall of the cavity which converges downward and bends backward in the opposite direction to welding direction (Fig. 4.16). The fusion front moves into a specimen at a certain propagation rate v_{pr} and frequency f_{pr}. With an increase in the weld speed v_w from 11 to 27.8 mm/s, v_{pr} and f_{pr} grow on the average from 200 to 320 mm/s and 18 to 46 Hz, respectively. The liquid material flows back to the rear wall in small portions predominantly in the horizontal direction at a speed that is much higher than the speed of the liquid in the rear portion of the molten pool. The frequency f_t of liquid phase transfer to the rear of the cavity is 3 to 10 times as high as $f_{pr.}$

The molten pool illustrated in Fig. 4.15 consists of two clearly defined regions, the upper region of stable penetration that extends to the point z' and the lower region in which melting spikes occur at a frequency comparable in the order of magnitude to the fusion front penetration frequency. The heat source releases a maximum amount of heat energy at the depth z_{bw} of the focused beam waist below the surface.

Shifting the focal point with respect to the target surface considerably changes the parameters of welding. With the focal point located at an optimal depth z_{opt}, the frequencies f_{pr} and f_t decrease and the melting spikes at the root become smoother. A further increase in the depth of the focal point below the target surface disturbs the mechanism of weld penetration, so that the molten material clogs the hole and the efficiency of melting drops off.

At the initial moment of keyholing, the fusion zone width can be taken equal to the diameter of the focused spot on the target surface. As the penetration depth increases, the cavity width gets smaller.

The mechanism of periodic transfer of the molten material from the front to the rear wall of the cavity causes a whirl-like flow of the liquid from the bottom upward along the rear wall.

From the analysis of photograms it was difficult to ascertain the character of molten material transfer in the lower region of the pool below the point z'. But it can be said for certain that the volume of the liquid phase during the melting spike is small and the liquid flows upward at the instant when the cavity grows in depth.

In the experiments conducted on glass ceramic-metal specimens, the heat track was run along the edge of the metal plate close to the glass ceramic-metal interface so that the weld pool profile on the side of the glass ceramic plate could remain transparent. However, these experiments cannot simulate exactly the process of metal welding because it is impossible to prevent glass ceramic from melting and to create the same conditions of heat release from both materials. Nevertheless, the experiments provide a means for verifying the kinetics of metal melting and revealing some regularities.

The results listed in Table 4.3 attest that the processes of melting in both types of specimens are qualitatively similar. But since there is a difference between the thermal properties of the two materials, the penetration depth in the composite specimen is smaller than it is in a glass ceramic specimen, the frequency of melt ripples is higher, the ripple height relative to the penetration depth changes little, and the rate of penetration is somewhat slower.

The physical model of deep penetration welding. The text below describes a theoretical model of the thermal processes in deep-penetration laser welding.

In accordance with this model illustrated in Fig. 4.16, the focused laser beam moves along the surface of a thick plate at a weld speed v_w and produces a cavity propagating along the seam in step with the beam. The cavity is filled with both metal vapors and the plasma arising from the optical discharge of the vapor-gas medium. The cavity and the region above the cavity differ in the density and temperature of metal vapors and, hence, in absorptivity. According to the adopted concepts [8, 70], the mechanism of absorption in the vapor-plasma medium obeys the Bouger-Lambert law. For the plume above a target surface, the incident laser beam power $P(z_0)$ at the entry to the cavity is given by

$$P(z_0) = P\exp\left(-I_{\mathrm{pl},}\right), \tag{4.23}$$

where $I_{pl} = \int_{z_{pl}}^{0} a_{pl} dz$ is the absorption intensity of the plume; z_{pl} is the plume height; and a_{pl} is the absorption coefficient of the plume.

The laser power delivered to an arbitrary depth z of the cavity is

$$P(z) = \eta P(z_0) \exp(-I_c), \tag{4.24}$$

where $I_{cw} = \int_{z}^{0} a_{cav}\, dz$ is the absorption intensity of the cavity; η is the factor accounting for a decrease in the laser power due to a horizontal displacement of the beam from the direct irradiation spot (DIS); and a_{cav} is the absorption coefficient of the cavity.

Let us consider the beam penetration into the cavity. The focusing system provides direct paths of rays, but as the beam moves deeper into the target, the rays undergo multiple reflection from the cavity walls and refraction due to the inhomogeneity of the vapor-gas medium. These effects need be taken into account, although it is rather difficult to define them quantitatively. The model under discussion deals with the limiting case of the vapor-gas crater illustrated in Fig. 4.17, which consists of an upper and a lower portion separated by a critical cross section at depth z' where the effects of refraction and defocusing are insignificant and do not divert the rays from their original paths. The diagram of cross sectional areas $S(z)$ of the beam at any depth z of the cavity is drawn for a beam focused by an off-axis Cassegrain reflector with a focal length f = 90 mm at an aperture D = 40 mm and beam divergence $\theta = 2 \times 10^{-3}$ rad. The laser power density in the upper portion of the cavity in any cross section above the point z' is found from the expression

$$q(z) = P(z)/S(z). \tag{4.25}$$

The power density over the critical cross section at depth z' is

$$q(z') = q_c = P(z')/S(z'). \tag{4.26}$$

In the lower portion below the critical depth z', the effects of refraction and reflection completely disturb the original profile of the focused beam. In accordance with the theoretical statements presented in work [78], we can average the laser power density q in this portion of the cavity over its cross section and assume it to be constant and equal to a critical value of q_c, as illustrated in Fig. 4.16. The power density $q(0)$ at a target surface, i.e. at depth $z = 0$, must be higher than the critical value to ensure deep penetration.

Consider the geometry of the cavity. At the critical depth z', the cavity width b is equal to the beam width. The width b of the cavity in its lower portion is found on the assumption that the power density is constant due to the self-sustained character of laser beam interaction, in which case the width varies in an exponential manner with the power density [78]. In the upper portion, the width also varies in the same manner from $b(0)$ to $b(z')$,

The cavity width

$$b = P(z)/q_c h_{\text{dis}} \quad \text{at } z \le z' \quad \text{and}$$
$$b = b(0)\left[b(z')/b(0)\right]^{z/z'} \quad \text{at} \quad z > z', \tag{4.27}$$

where h_{dis} is the depth to the direct irradiation spot (DIS) below the target surface.

In order to define the propagation rate v_{pr} of the direct irradiation spot on the front wall of the cavity, we can use the known concepts of the fusion front propagation for the case of instantaneous removal of the molten material and partial evaporation [8, 79]. The propagation rate v_{pr} for the beam moving at a weld speed v_w along the target surface is found from the solution of a unidimensional equation of heat transfer involving a moving ablation front. Introducing the factor C given in equation (4.24) and the correction factor β accounting for the power loss due to heat conduction and radiation, the equation for v_{pr} takes the form

$$v_{\text{pr}}(z) = \frac{a_c(1-\beta)q(z)C}{\rho_m(c_m + Vc_v)}, \tag{4.28}$$

where a_c is the absorption coefficient of the front wall of the cavity; ρ is the molten metal density; c_m is the specific heat of melting (fusion); c_v is the specific heat of vaporization; and V is the volume of the vaporized metal.

As noted earlier, the liquid metal flows periodically from the front to the back wall of the cavity. So long as the liquid layer is thin, the surface forces retain the liquid metal as they counterbalance the recoil forces of vapors. When the layer reaches a certain thickness, the forces acting on the liquid cause it to move back along the side walls. The model certainly considers an instantaneous transfer of the liquid at the end of each time period. If the liquid phase moves at a high speed, the model can disregard both the heat exchange with side walls and the volume of vaporized metal.

As the fusion front propagates into the material, the beam travels in the horizontal direction relative to the target surface at a weld speed v_w. From the data of Table 4.3 it follows that the propagation rate v_{pr} of the fusion front is 15 to 20 times as fast as v_w. The coordinate x for the laser beam can thus be held invariable during the propagation time required to form a cavity. The factor C given in equation (4.24) to account for the reduction of the laser power as the beam passes the zone of direct irradiation is written as

$$C = 1 - t/t_{\text{dis}}, \tag{4.29}$$

where t is the time of fusion front propagation, $0 \le t \le t_{dis}$; $t_{\text{dis}} = 2\sqrt{S(0)/\pi}/v_w$ is the time taken for the direct irradiation spot (DIS) to move from the surface to the cavity bottom; and $S(0)$ is the cross-sectional area of the beam spot on the target surface.

At $t = t_{dis}$, the beam forms the next direct irradiation spot along the x axis, which penetrates into the material to the required depth.

The amount of laser energy coupled to the front wall of the cavity is proportional to $1 - C$. The plasma plume emits energy into the environment and acts as a surface source of the power density

$$q_{\text{pl}} = \frac{(P - P_0)d_{\text{pl}}/(2d_{\text{pl}} + 4z')}{\pi(d_{\text{pl}}/2)^2}, \tag{4.30}$$

where β is the power of incident radiation; P_0 is the power delivered to the target surface; and d_{pl} is the plume diameter determined by the beam spot on the surface.

The laser energy absorbed in the cavity by the vapors and plasma and the energy reflected from the direct irradiation spot thus create a surface heat source on the cavity walls, the heat power density of which is equal to

$$q_{\text{h}} = \beta(P_0 - a_c P_{\text{dis}})/S, \tag{4.31}$$

where β is the factor used to account for the total loss of energy removed from the cavity, P_{dis} is the power delivered to the direct irradiation spot at depth h_{dis} below the target surface; and S is the area of cavity walls.

The experiments have revealed that the temperature of the cavity walls changes little in depth, although the amount of energy absorbed in the cavity decreases exponentially with increasing depth, as follows from equation (4.24).

The model resolves this seeming contradiction by introducing relation (4.28) for v_{pr} as a function of the cavity depth. In the lower portion of the cavity, where the laser power is smaller, the propagation rate v_{pr} decreases and so the spot of direct irradiation heats up the material for a longer time.

The physical model under study accounts for the temperature dependence of such thermal parameters as the specific heat, thermal conductivity, and reflectivity, allows for the thermal effects involved in phase transformations when handling Stefan problems, and considers the plasma to be homogeneous and constant in temperature. The model thus covers the modern concepts of complex physical processes.

The text below describes some statements of a mathematical model of deep penetration welding, which has formed the basis for developing the algorithm and the program adapted to solve problems on a computer.

The mathematical model of deep penetration welding. In the general case, the equation for a three-dimensional nonstationary heat transfer problem assumes the form

$$c\rho(T)\frac{\partial T}{\partial t} + \operatorname{div}\mathbf{W} = Q;\ (x, y, z) \in G, \tag{4.32}$$

where $c\rho$ is the volume specific heat; $\mathbf{W} = -K(T)$ grad T is the heat flux; K is the thermal conductivity; Q is the amount of heat evolved; and G is the calculation region.

The problem is symmetric relative to the longitudinal plane of motion of the beam, so the half-space of a specimen is taken as the calculation region G. The model considers the following regions:

$\xi(x, y, z, t)$, region of direct irradiation spot (DIS);
$\pi(x, y, z, t)$, region adjacent to the preceding region of DIS;
$\psi(x, y, z, t)$, region of surface plume source;
$\nu(x, y, z, t)$, cavity region.

The amount of heat evolved is found from the heat balance equation:

$$Q = Q_1 + Q_2 + Q_3 + Q_4, \tag{4.33}$$

where

$$\begin{aligned} Q_1 &= P_{dis}/V; && (x, y, z) \in \xi, \\ &= 0; && (x, y, z) \notin \xi, \end{aligned}$$

$$\begin{aligned} Q_2 &= P_{pl}S/V; && (x, y, z) \in \psi, \\ &= 0; && (x, y, z) \notin \psi, \end{aligned}$$

$$Q_3 = P/V; \quad (x, y, z) \in \pi,$$
$$= 0; \quad (x, y, z) \notin \pi,$$

$$Q_4 = P_c/V; \quad (x, y, z) \in \nu,$$
$$= 0; \quad (x, y, z) \notin \nu.$$

Here, S is the elementary surface area and V is the elementary volume.

From equations (4.23), (4.24), (4.30), and (4.31) we estimate the power of heat fluxes:

$$P_{\rm DIS} = a_c P_0 \exp(-I_c)C, \tag{4.34}$$

$$P = a_c P_0 (1 - C), \tag{4.35}$$

$$P_{\rm pl} = a_c (P - P_0) / [(\pi/2) d_{\rm pl} (d_{\rm pl} + 2z')], \tag{4.36}$$

$$P_c = \beta[P_0 - a_c P_0 \exp(-I_c)]/S. \tag{4.37}$$

For a problem to be stated uniquely, the model should include boundary conditions which, in our case, correspond to a third boundary-value problem:

$$\mathbf{W} = \alpha_s(T)(T_0 - T); \ (x, y, z) \in \partial G, \tag{4.38}$$

where α_s is the surface heat transfer coefficient; T_0 is the ambient temperature; and ∂G is the boundary of a solid. According to the condition of symmetry, the heat flux $\mathbf{W} = 0$ and $y = 0$. In the initial conditions, the temperature $T = T_0$ is taken constant and the penetration time t is equal to zero.

Equation (4.32) describes the propagation of heat within either a solid or a liquid phase. The heat fluxes at the phase interface are given as

$$W_2 - W_1 = L_m \left(\frac{d\mathbf{R}}{dt}, \frac{\text{grad } F}{|\text{grad } F|} \right),$$
$$\tag{4.39}$$
$$W_{1-2} = -\left(K \text{ grad } T, \frac{\text{grad } F}{|\text{grad } F|} \right)_{1-2}.$$

Here, L_m is the latent heat of vaporization; $\mathbf{R}(t)$ is the position of the phase boundary; $F[T(R, t)]$ is the phase boundary condition at $T = T'$, where T'

is the phase transition temperature, and subscripts 1 and 2 relate to the phases with temperature T which is lower and higher than T', respectively.

Equation (4.32) with conditions (4.39) at the interface can be written as a homogeneous equation of the form

$$\left[c\rho(T) + L_m\delta(T - T')\right]\partial T / \partial t + \text{div}\,\mathbf{W} = 0, \tag{4.40}$$

where $\delta(x)$ is the Dirac delta function and div $\mathbf{W}$ is the divergence of a heat flux.

This mathematical model is at the root of the model worked out in numerical form with the use of a package of programs described in work [83]. Some features of the calculation model and the algorithms and programs designed to solve the problems of deep penetration welding are given detailed treatment in work [84].

The text below presents the results of solving some specific problems for estimating the kinetics of thermal processes in partial penetration welding of corrosion-resistant steel in argon with a CO_2 laser emitting at a wavelength of 10.6 μm. The parameters used in calculations are as follows: the laser power P = 3 to 5 kW; weld speed v_w = 10 to 40 mm/s; depth of focused beam waist, z_{bw} = 0.5 to 2.5 mm; beam aperture D = 35 mm; aberation parameter of the lens, A = 1.4084; focal length of the lens, f = 150 mm; and angle of divergence, $\theta = 2.0 \times 10^{-3}$ rad.

The values of beam cross sections calculated by the technique described in Sec. 4.2 are shown in Fig. 4.17, where curve *2* illustrates the data used

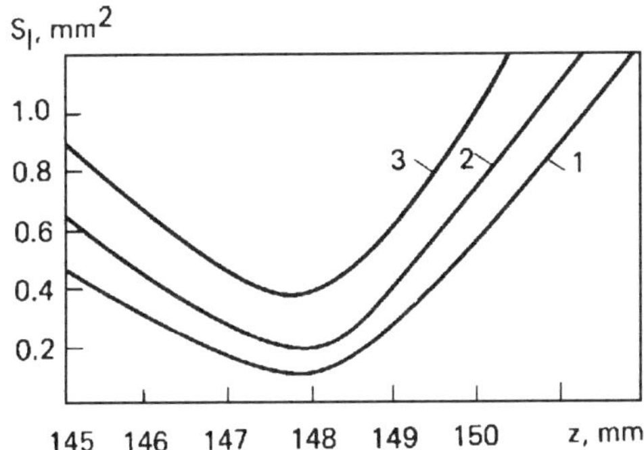

Fig. 4.17. Cross-sectional area of the laser beam in the focal plane at divergence angles θ of 1×10^{-3}, 2×10^{-3}, and 3×10^{-3} rad (curves *1*, *2*, and *3* respectively)

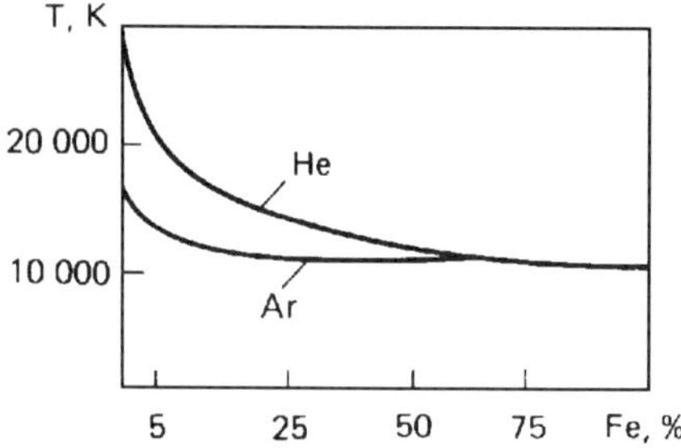

Fig. 4.18. Plasma temperature versus iron vapor content in helium and argon

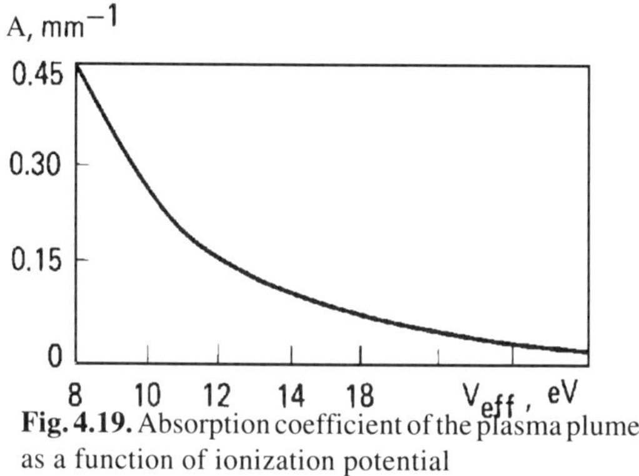

Fig. 4.19. Absorption coefficient of the plasma plume as a function of ionization potential

in the solutions of problems. Figures 4.18 and 4.19 display the experimental results used to calculate the temperatures and absorption coefficients of the plasma in and above the cavity. The values obtained for the specified conditions at q_{th} = 10^5 W/cm^2 are the following: T_c = 15 000 K, T_{pl} = 18 000 K, a_c = 0.5 mm^{-1}, and a_{pl} = 0.1 mm^{-1}.

The height of the shielding plasma cloud above the target surface was taken equal to the diameter of the beam spot on the target surface [85]. The time step covered the period from the instant of beam interaction with the target to the instant of quasistationary temperature distribution. At each time step, the temperatures in the target regions of interest were estimated to analyze the processes of heat saturation and transition to the quasistationary state. To verify the analytical model for its validity, we compared the theoretical data on the profile and dimensions of the melt pool with the experimental data obtained by the technique of jet-like melt flushing from the pool at the end

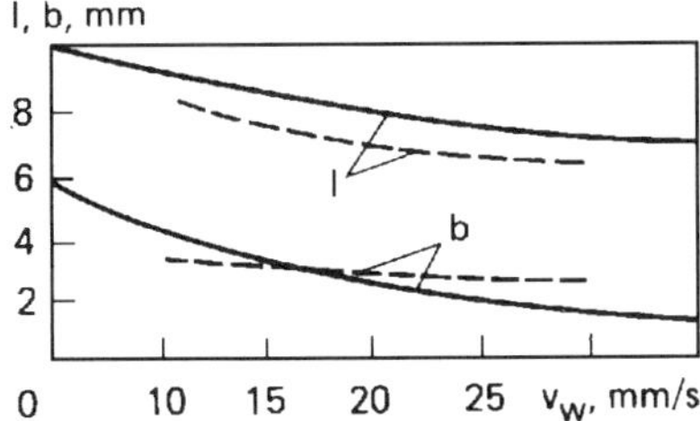

Fig. 4.20. Theoretical values (dash curves) and experimental (full-line curves) values of the length and width of the weld pool as functions of the weld speed

of the dwell time. The molten material was blown off both in the direction of welding and in the opposite direction to uncover the front wall and the back wall of the cavity respectively.

The theoretical data agree well with the experimental results obtained for the same conditions of welding (Fig. 4.20). A certain discrepancy between the results arises obviously from the fact that the model does not account for the dynamics of molten metal stirring.

The weld penetration depth h is one of the basic parameters which determine the quality of a weld and welding efficiency. From the solutions to three-dimensional heat transfer problems and the analysis of cine film records we can conclude that the weld penetration actually depends on the keyhole depth.

It is of interest to compare the penetration depths obtained from the solutions of three-dimensional and one-dimensional heat transfer problems.

Knowing the penetration rate v_{pr} of the fusion front, we can readily determine the penetration depth achieved for the time t_{pr}:

$$h = \int_0^{t_{pr}} v_{pr}(z)\, dt. \tag{4.41}$$

Figure 4.21 illustrates the curves of v_{pr} obtained from the solutions of a three-dimensional and a one-dimensional heat transfer problems at different values of the power loss factor β entering equation (4.28). The volumetric heat transfer is seen to have some effect on v_{pr} only at the final stage of weld penetration.

Comparing the solutions of three-dimensional problems with the solutions of equation (4.28), we can determine the desired values of the correction factor

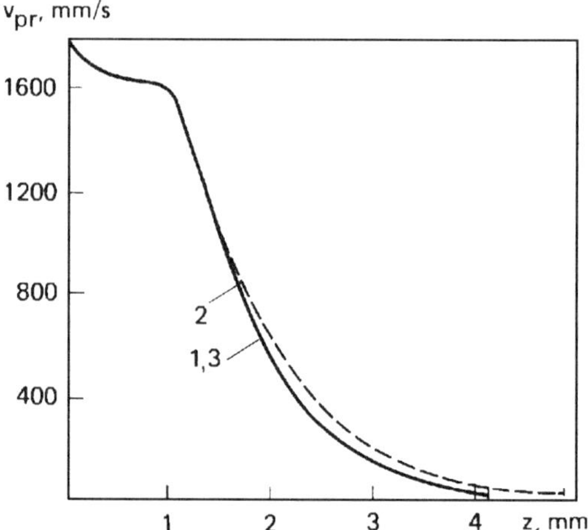

Fig. 4.21. Fusion penetration rate at $v_w = 20$ mm/s vs penetration depth:
curve *1*, three-dimensional heat-transfer model; curve *2*, unidimensional model, at $\beta = 0$; curve *3*, unidimensional model, at $\beta = 0.25$

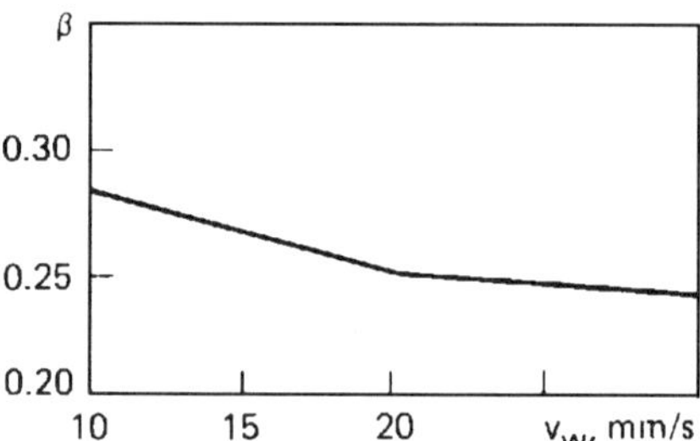

Fig. 4.22. Correction factor β vs weld rate

β at which the penetration depths are the same and the profiles of v_{pr} coincide with one another, for example, curves *1* and *3* in Fig. 4.21. The correction factor changes somewhat at low values of v_w and drops to a steady value of 0.25 at v_w in excess of 20 mm/s (Fig. 4.22).

Figures 4.23 through 4.25 compare the theoretical curves with the experimental curves of penetration depth as a function of the weld rate, laser power, and beam waist depth below the target surface. The results agree well which points to the validity of the adopted spatial model.

The theoretical results of deep penetration welding reveal some regularities of the process of heat release. Although the temperature of the front wall of the cavity in the DIS reaches the highest value, the amount of heat delivered

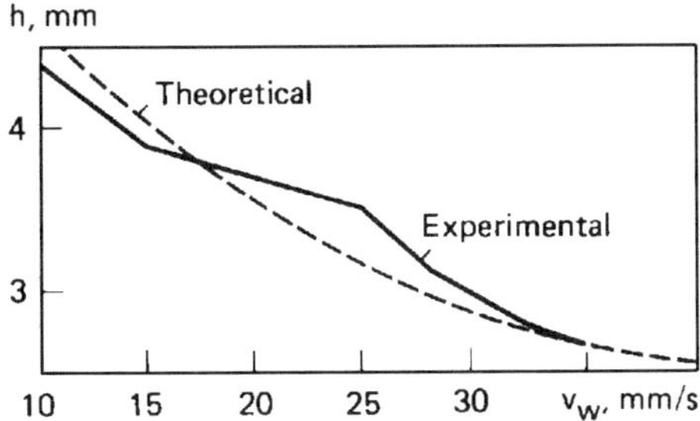

Fig. 4.23. Penetration depth vs weld rate at $P = 3.5$ kW

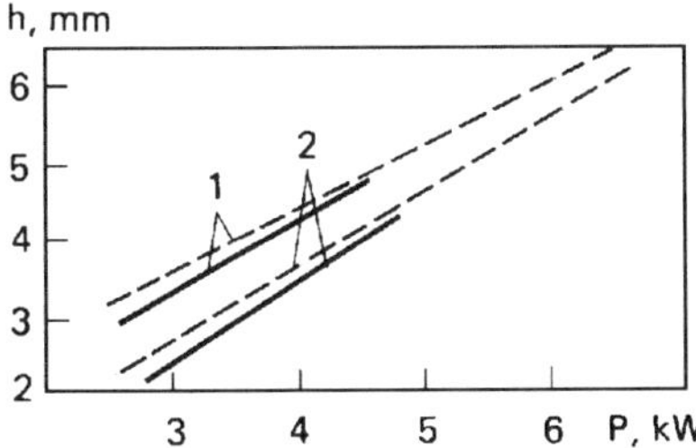

Fig. 4.24. Penetration depth vs laser power at weld rates of 16.6 mm/s (curve *1*) and 27.8 mm/s (curve *2*). Solid and dashed lines represent experimental and theoretical results respectively

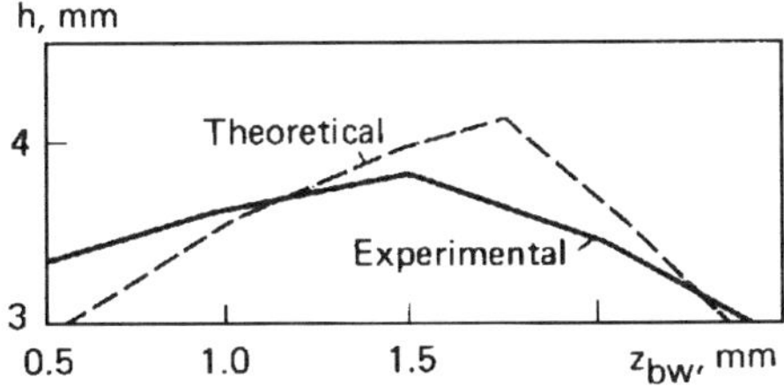

Fig. 4.25. Penetration depth vs depth of the beam waist below the target surface at $P = 3.5$ kW, $v_w = 16.6$ mm/s, and $f = 150$ mm (solid line, experiment; dashed line, calculation)

to the metal from this region is about 10 percent of the total heat energy since the DIS is small. A major share of heat energy, up to 80 percent is delivered by the vapor-plasma medium filling the cavity.

The total loss of energy from the beam is the sum of energies lost in the plume, due to radiation, and with the vapors emitted from the cavity. The amount of energy loss from the cavity reaches 40 percent according to the estimates predicted by the model. The total loss ranges from 30 to 50 percent.

The above analysis allows us to single out the most important elements of the general model and use them in design calculations of thermal processes.

4.5. Heat-Source Problem Statement

The theoretical heat transfer model discussed in Sec. 4.4 covers complex physical processes occurring in the laser beam-material interaction zone. Although such a model presents certain difficulties in calculations, the general statement of the problem involved enables the designer to establish certain regularities and single out the most important features of the thermal process.

This mathematical model is easy to adapt for practical purposes so that the user should be able to solve a variety of heat-source problems. The statement of an applied problem must cover multidimensional, nonlinear, and nonstationary features and also account for the heat of phase transition and the material absorptivity. In order to simplify an applied problem, we need to specify the dimensions and critical power density of a heat source. The procedure performed for the purpose does not involve difficulties as regards the surface heat sources used in laser surface hardening. This is not the case with laser welding. For the process of full-penetration butt welding of thin plates, the weld penetration can be considered uniform and so the heat source can be taken linear. Knowing the laser power and considering that the power density of a heat source is very high, we can estimate the effective power of a linear heat source.

It is more difficult to specify the parameters of a heat source in deep penetration welding, which has to distribute heat through the entire thickness of the plates welded. The model of a linear heat source for the keyholing process cannot afford an adequate picture of heat distribution in the heat affected zone. The thermodynamic analysis of the melting efficiency of the beam enables us to determine the weld penetration and the profile of the heat affected zone both from the specified parameters of welding and beam focusing and from the given thermal constants of metals.

Let us look at some features of the mathematical models designed to solve applied problems.

The laser beam creates a high-heat-power density source which is responsible for increased temperature gradients in space and time, so the accepted model must have a fine mesh pattern with a small time space.

The general model described in Sec. 4.4 relies on an implicit calculation diagram which ensures an absolute stability of count. With an increase in the time step above the value typical of the actual process, the solution accuracy decreases because of the error in the differential approximation and omission of some steps of heat source motion in the calculation procedure. This aspect is given a strict mathematical treatment in papers [86, 87].

The solutions of numerous heat-transfer problems point to the need of selecting a grid mesh size of no more than 0.5 to 0.8 mm and a time step of 1.0×10^{-2} to 2.0×10^{-2} s. But too fine a step necessitates less stringent requirements on both the accuracy of differential approximations and the linearization of equations. In solving applied problems, it is more advisable to use an explicit interpolation grid pattern that requires a fine time step, but relies on a simple cyclic algorithm that involves a small number of count operations at each step.

The problems of general modeling deal with a calculation region whose grid pattern has a variable step along z and y axes, which increases with distance from the source axis in a manner of geometric progression. For the models involving implicit calculations, this type of spacing is efficient, but it is not so for the explicit calculation procedure since a variable step leads to a less stable solution and a higher approximation error, which generates a need to reduce the time step still more.

Based on the above reasoning, the preference is given to a uniform grid in dealing with applied problems. The spatial steps laid out along Cartesian coordinate axes are equal to

$$s_x = s_y = s_z = s. \tag{4.42}$$

The difference approximation of the first derivative of heat conduction at two nodal points is written as

$$\frac{\partial T}{\partial \begin{cases} x \\ y \\ z \end{cases}} \cong \frac{1}{2s}\left[T_{\begin{cases} i+1,j,k \\ i,j+1,k \\ i,j,k+1 \end{cases}} - T_{\begin{cases} i-1,j,k \\ i,j-1,k \\ i,j,k-1 \end{cases}} \right]. \tag{4.43}$$

The second derivative approximated at least in three points takes the form

$$\frac{\partial^2 T}{\partial \begin{cases} x^2 \\ y \\ z \end{cases}} \cong \frac{1}{s^2} \left[T_{\begin{cases} i+1,j,k \\ i,j+1,k \\ i,j,k+1 \end{cases}} - 2T_{i,j,k} - T_{\begin{cases} i-1,j,k \\ i,j-1,k \\ i,j,k-1 \end{cases}} \right]. \tag{4.44}$$

Approximation of the time derivative is made with respect to two successive values:

$$\frac{\partial T}{\partial t} = \frac{1}{\Delta t_s}\left(T^{t+1} - T^t\right), \tag{4.45}$$

where $t + 1$ is the current time interval and t is the preceding time step interval.

Transformiung heat transfer equation (4.32) and applying conditions (4.42) through (4.45), we obtain the following equation:

$$T_{i,j,k}^{t+1} = T_{i,j,k} + \sum_{g=1}^{B} R_{\begin{cases} x \\ y \\ z \end{cases}} \left[T_{\begin{cases} i-1,j,k \\ i,j-1,k \\ i,j,k-1 \end{cases}} - 2T_{i,j,k} + T_{\begin{cases} i+1,j,k \\ i,j+1,k \\ i,j,k+1 \end{cases}} \right] + \frac{P_{\mathrm{h}} P_{i,j,k} t_s}{c\left(T_{i,j,k}\right)\rho\left(T_{i,j,k}\right)\Delta V_{i.j.k}}, \tag{4.46}$$

where

$$R_{\begin{cases} x \\ y \\ z \end{cases}} = \frac{K\left(T_{i,j,k}\right)t_s}{c\left(T_{i,j,k}\right)\rho\left(T_{i,j,k}\right)^2_{\begin{cases} h_x \\ h_y \\ h_z \end{cases}}}, \tag{4.47}$$

$$\Delta V_{i,j,k} = s_x s_y s_z. \tag{4.48}$$

Here, B represents the dimensions of a solid; g is the number of a unit vector; P_{h} is the heat power delivered to the target; $P_{i,j,k}$ is the amount of heat power coupled into an element (i, j, k) in time t_s between the instant t and instant $t + 1$; c is the specific heat; ρ is the mass density; and k is the thermal conductivity.

Equation (4.46) describes the temperature field at all points within the region G:

$$\begin{Bmatrix} i \\ j \\ k \end{Bmatrix} = 2,3,\ldots \begin{Bmatrix} M-1 \\ N-1 \\ L-1 \end{Bmatrix}, \tag{4.49}$$

where $\{M, N, L\}$ identify an index array.

The boundary surfaces of laser-treated parts are mostly free and in contact with the environment whose temperature is taken constant. Heat exchange with the environment is preset by third-kind boundary conditions which obtain the following differential form:

$$\alpha_{\text{s}}\left(T_G - T_0\right) = -K\left(T_G\right)\frac{\partial T}{\partial n}\bigg|_G, \tag{4.50}$$

where α_{s} is the surface heat transfer coefficient and n is the normal to the surface of region G.

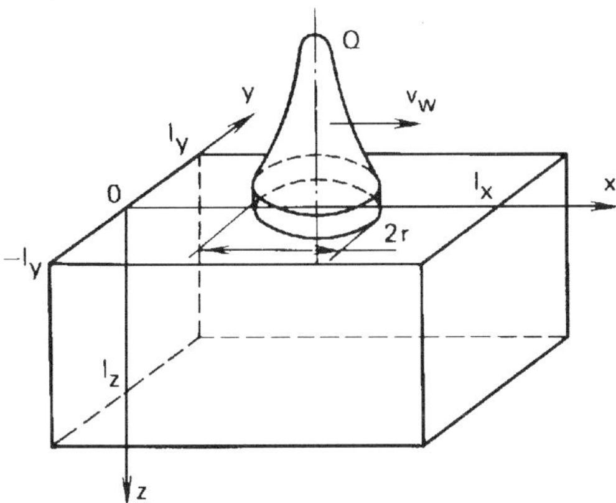

Fig. 4.26. Geometry of calculation region

Examine the equation of heat transfer from the boundary surfaces of a body at $x = 0$ and $x = l_x$ (Fig. 4.26). Transform equation (4.50) to a difference equation and simplify the latter to the form

$$T^{t+1}_{\left\{\begin{matrix}1\\M\end{matrix}\right\},j,k} \cong T^{t+1}_{\left\{\begin{matrix}2\\M-1\end{matrix}\right\},j,k} + T_0 \frac{\alpha_s \left[T^t_{\left\{\begin{matrix}2\\M-1\end{matrix}\right\},j,k} \right]^{s_x}}{K\left[T^t_{\left\{\begin{matrix}2\\M-1\end{matrix}\right\},j,k} \right]} \times \left[1 + \frac{\alpha_s \left[T^t_{\left\{\begin{matrix}2\\M-1\end{matrix}\right\},j,k} \right]^{s_x}}{K\left[T^t_{\left\{\begin{matrix}2\\M-1\end{matrix}\right\},j,k} \right]} \right]. \tag{4.51}$$

The equations of heat transfer along the y and z axes are similar in form.

One of the basic conditions to be met when stating a heat transfer problem is to account for changes in the thermal parameters of a material with temperature. If the parameters c, k , ρ, and α_s vary monotonically, functional approximation is sufficient to allow for changes. Variations in the thermal conductivity of a material can well be approximated by a quadratic polynomial of the form

$$K(T) = K_0 + K_1 T + K_2 T^2. \tag{4.52}$$

Changes in the mass density can be adequately allowed for by a linear function

$$\rho(T) = \rho_0 + \rho_1 T. \tag{4.53}$$

The surface heat transfer coefficient as a function of temperature takes the form of a fourth-degree polynomial

$$\alpha_s(T) = \alpha_0 + \alpha_1 T + \alpha_2 T^2 + \alpha_3 T^3 + \alpha_4 T^4 \tag{4.54}$$

or an exponential function

$$\alpha_s(T) = \alpha_0 + nT^{\,x}, \tag{4.55}$$

where n is the approximation coefficient.

It is most difficult to define the pattern of changes in specific heat with temperature. For metals free of polymorphic transformations in the course of heating to the vaporization point, changes in the specific heat assume a smooth monotonic pattern well defined by a quadratic polynomial

$$c(T) = c_0 + c_1 T + c_2 T^2. \tag{4.56}$$

If a material undergoes a structural transformation in heating such that the c-T curve exhibits a bend, we must assign different values of c_0, c_1, and c_2 to each of the structural states. But if the specific heat changes in an abrupt manner, for example, during phase transitions, a list of the values of c need be preset. Certainly, these values complicate computations and should apply for the cases where the thermal estimates of phase and structural transformations become of much importance.

The tabulated values of thermal constants and corresponding reference temperatures are put down in the array of initial data. In the simplest case, the linear interpolation of the tabulated values can give the value of any parameter U:

$$U\left(T_{i,j,k}\right) = U\left(T^p\right) + \left[U\left(T^{p+1}\right) - U\left(T^p\right)\right]\left(T_{i,j,k} - T^p\right) / \left(T^{p+1} - T^p\right), \tag{4.57}$$

where p is the number of a reference point.

In performing explicit calculations, it is more convenient to estimate the values of parameters from the preceding time step and then refine them in the process of direct iterations [86]. The cycle of two or three iterations is sufficient for the purpose.

It is of importance to account for the heat of phase transitions. As is known, the specific heat of pure metals changes sharply in the processes of melting and crystallization as a result of absorption and evolution of latent heat respectively. Alloys convert from one state of aggregation to another over a certain temperature interval. Here, the function $c(T)$ assumes the form of the Dirac delta function.

The discontinuous function $c(T)$ is unsuitable for the adopted calculation since it disturbs the convergence of iterations. To obviate the difficulty, recourse is made to an effective smoothed value of the specific heat, $\tilde{c}$, that permits us to replace an abrupt change in the specific heat by a corresponding function in the smoothing interval of temperature, $2\Delta T = T_{liq} - T_{sol}$:

$$\begin{aligned} & c_{sol}(T); \quad T < T_{av} - \Delta T, \\ & \tilde{c}(T) = c + L_m \delta(T - T_{av}); \ | T - T_{av} | < \Delta T, \\ & c_{liq}(T); \quad T > T_{av} + \Delta T, \end{aligned} \tag{4.58}$$

where T_{liq} is the liquidus temperature; T_{sol} is the solidus temperature; $c = 1/2(c_{sol} + c_{liq})$; T_{av} is the average temperature for the phase transition; L_m is the latent heat of fusion; and δ is the Dirac delta function.

Figure 4.27 illustrates a diagram for estimating the effective specific heat, where the trapesoid $ABCD$ with a ratio of $a = BC/AD = 1.5$ to 3 represents the delta function for simplicity. The values of c_A and c_D at points A and D can be found in pertinent reference books. Given the condition at which the hatched area below the trapesoid corresponds to the latent heat of a phase transition, we can define the values of c_B and c_C:

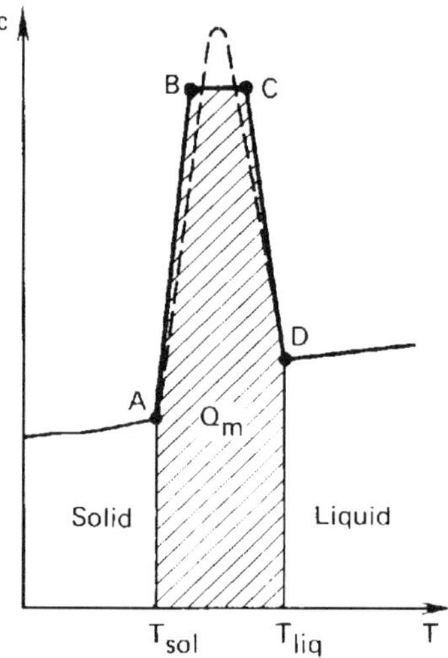

Fig. 4.27. Diagram for calculation of effective smoothed values of specific heat

$$c_B = c_C = \frac{Q_m - \Delta T(c_A + c_D)(1-a)/2}{\Delta T(1+a)}. \tag{4.59}$$

Using the values of c_A, c_B, c_C and c_D enables us to transform equation (4.32) for phases and phase interfaces to a homogeneous form. This appreciably facilitates the procedure of computation by a through-count technique without isolating a phase transition, although accounting for its latent heat.

It is essential that the explicit calculation technique should yield a stable solution. The differential approximation is considered adequate and the solution regular if with a decrease in the spatial step this solution inevitably converges to the exact solution of the initial difference equation.

For three-dimensional nonlinear parabolic equations to which differential heat transfer equations belong, the region of a stable solution by the difference method lies within the time step defined as [86]

$$t_{i,j,k} \le \frac{c(T_{i,j,k})\rho(T_{i,j,k})}{\sum_{g=1}^{B} \frac{1}{s^2_{\left\{\begin{smallmatrix} x \\ y \\ z \end{smallmatrix}\right.}} \left[T_{\left\{\begin{smallmatrix} i-1,j,k \\ i,j-1,k \\ i,j,k-1 \end{smallmatrix}\right.} + K_{\left\{\begin{smallmatrix} i+1,j,k \\ i,j+1,k \\ i,j,k+1 \end{smallmatrix}\right.} \right]}. \tag{4.60}$$

If condition (4.60) is valid, the solution is stable at all calculation points. This means that a permissible time step T_{per} must be equal to the least value of $t_{i,j,k}$.

The calculation procedure becomes much more complex if we have to correct the time step obtained from equation (4.60) for each point within the calculation region. For the sake of simplicity, it is advisable to solve applied problems at a constant time step calculated with the equation

$$t_{per} = \frac{c_{min}\rho_{min}}{2K_{max}\left(s_x^{-2} + s_y^{-2} + s_z^{-2}\right)_{max}}, \tag{4.61}$$

The values of c_{min}, ρ_{min}, and K_{max} are found from the graphs of thermal constants in the expected interval of temperatures. The value of t_{per} determined from equation (4.61) is found to be somewhat understated and can be increased to optimal values in the course of computations.

4.6. Applied Heat-Source Problems

This section describes some applied problems stated according to the rules outlined in Sec. 4.5 and adapted to be run on a computer.

Butt welding of thin plates. Laser welding setups of up to 5 kW of power are now available, which can butt weld parts 1 mm in thickness and over in a single pass, mostly forming rectilinear weld joints.

Before deciding on the choice of a laser welding process, it is commonly expedient to compare laser welding with conventional processes of arc welding. For this reason, the text below considers a heat transfer problem relating to full-penetration butt welding in a single pass with a laser beam and arc. This is a plane problem, which disregards temperature variations in depth. An example of the coordinate grid of a plate laid out into a number of finite elements to model a welding process appears in Fig. 4.28. The model does not consider the joint gap between the parts because it generally has no effect on the temperature field at a high weld speed.

In our case, we can consider the heat source to be linear for sure, which propagates heat along the x and y axes in an arbitrary fashion.

The model treats a continuous motion of the heat source at a constant speed v_w in terms of discrete actions at the nodes of the reference grid. For the case under analysis, the initial differential heat-transfer equation takes the form [56]

$$c(T)\rho(T)\frac{\partial T}{\partial t} = \frac{\partial}{\partial x}\left[K(T)\frac{\partial T}{\partial x}\right] + \frac{\partial}{\partial y}\left[K(T)\frac{\partial T}{\partial y}\right] + P - 2\alpha_s(T)(T - T_0)/\delta. \tag{4.62}$$

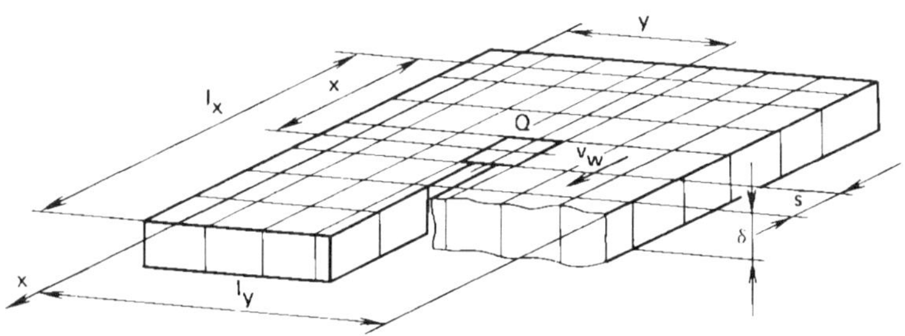

Fig. 4.28. Mapping of plate elements for modeling a process of welding

Converting equation (4.62) into a difference equation and performing requisite transformations, we obtain an interpolation formula for calculating the temperatures at the nodal points of the grid:

$$
\begin{aligned}
{}^{t+1}_{i,j} = T_{i,j} + \frac{K(T_{i,j})t_{\mathrm{s}}}{c\left(T_{i,j}\right)\rho\left(T_{i,j}\right)s^{2}} \Big\{ T_{i+1,j} + T_{i-1,j} \\
+ T_{i,j+1} + T_{i,j-1} - 4T_{i,j} + \frac{K\left(T_{i+1,j}\right) - K\left(T_{i-1,j}\right)}{4K\left(T_{i,j}\right)} \\
\times \left(T_{i+1,j} - T_{i-1,j}\right) + \frac{K\left(T_{i,j+1}\right) - K\left(T_{i,j-1}\right)}{4K\left(T_{i,j}\right)} \left(T_{i,j+1} - T_{i,j-1}\right) \\
+ \frac{P_{i,j}P_{\mathrm{h}}}{K\left(T_{i,j}\right)\delta} + \frac{2\alpha_{\mathrm{s}}\left(T_{i,j}\right)s^{2}}{K\left(T_{i,j}\right)\delta}\left(T_{i,j} - T_{0}\right).
\end{aligned}
\tag{4.63}
$$

The temperature at the boundaries along the axes $x(0, l_x)$ and y $(\pm l_y)$ is found from equation (4.51).

The model of thermal processes in arc welding assumes that the heat source has a Gaussian distribution of heat within a circular spot on the plate surface. The parameters of the Gaussian heat source produced in arc welding can be defined by the technique presented in work [52], which is widely used to calculate thermal processes in laser welding of such materials as structural steels and alloys of titanium, aluminum, and nickel. The calculated results help establish adequate welding conditions, estimate the strength of metals during welding, and analyze the structure of weld joints.

It is of interest to compare the results obtained from the solution of a nonlinear plane problem and a linear plane problem related to a fast-moving linear heat source defined by equation (4.12).

Figure 4.29 illustrates the thermal cycles calculated from a linear and a nonlinear model for laser welding of titanium alloy plates. Temperature field isotherms for a quasistationary state appear in Fig. 4.30. As seen from Fig. 4.30, the temperature profiles estimated by the two models differ to a large extent, especially in the temperature range above 900 K where the lengths of isothermal lines pertaining to the nonlinear model are by a factor of 1.5 to 3 shorter.

Given the same conditions of welding, a characteristic length of a weld pool, determined by isotherms of Fig. 4.30*a* and *b* at T_{m} of about 1 960 K, is found

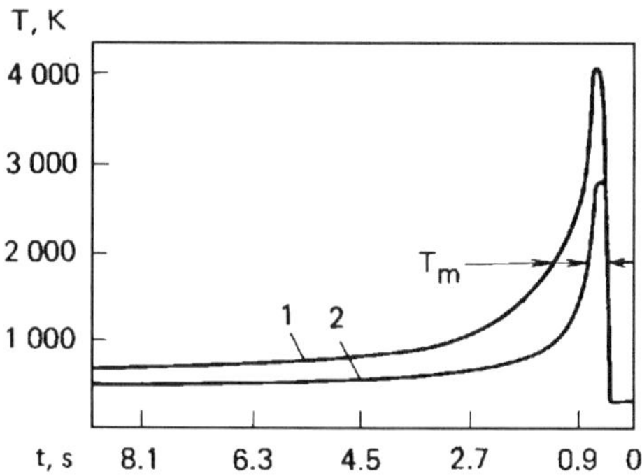

Fig. 4.29. Thermal cycles along the axis of the weld formed between 2.0-mm titanium alloy plates at a heat power density q_h of 3×10^8 W/m^2 and weld speed v_w of 26.7 mm/s. Curves *1* and *2* relate to linear and nonlinear models respectively

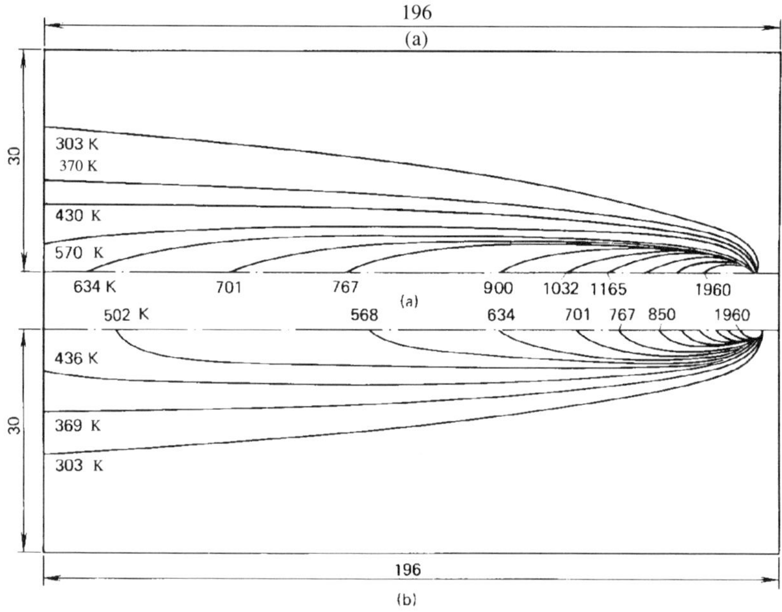

Fig. 4.30. Isotherms estimated from a linear model (*a*) and nonlinear model (*b*) for the quasistationary state set up in laser welding of 2.0-mm thick titanium alloy at $q_h = 2.75 \times 10^8$ W/m^2 and $v_w = 20$ mm/s

to be no less than 12 mm and 5 mm, respectively. The experimental values obtained by the technique of jet flushing of the molten metal under the same conditions of welding range from 3 to 7 mm.

The weld pool lengths calculated with a linear and nonlinear model and those obtained from experiments are equal to about 14.5, 4.5, and 5.7 mm, respectively. The nonlinear model allows for changes in the thermal constants of a material, and so yields more accurate results.

Figure 4.31 illustrates how the cooling rate v_{cl} of a full penetration weld in a 1.5-mm steel plate varies along the weld seam axis in a temperature range from 1 673 to 1 473 K with changes in the weld speed from 4.16 to 55.5 mm/s and laser power from 2 to 4.6 kW. The theoretical curve is the plot of data obtained from the solution of nonlinear problem. The experimental curve illustrates the results of experiments using thermocouples placed into the back portion of a melt pool to register the cooling rate.

Numerous experiments attest to a good agreement between the experimental data and theoretical results obtained with the aid of a nonlinear model which can serve as a reliable tool for the analysis of full penetration welding with consideration for the latent heat involved in phase transitions.

Partial penetration welding of thick parts. Since the heat source here cannot uniformly propagate into a thick plate, it is well to resort to a three-dimensional model to analyze the thermal processes in a heat-affected zone.

Partial-penetration butt welds come in longitudinal and circular configurations. If the laser power is only sufficient to provide a weld penetration slightly in excess of half the plate thickness, longitudinal weld seams are run on both sides of the weldment. This is also the case for a key weld with part edges acting as back-up plates.

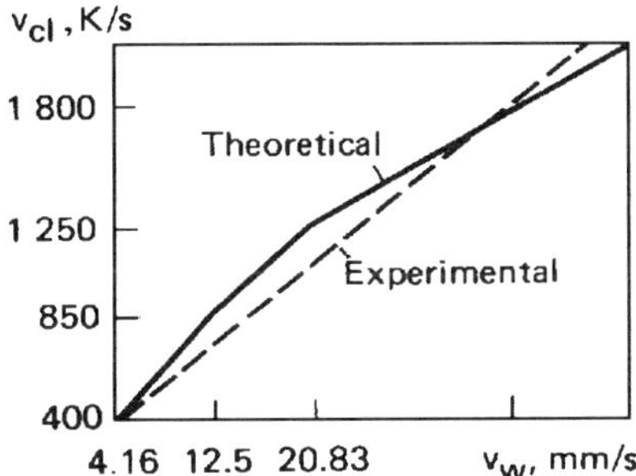

Fig. 4.31. Rate of cooling of a laser weld in steel over the range from 1673 to 1473 K as a function of weld speed v_w

Circular welds serve for joining tubes, flanges, and other cylindrical parts to thick-wall parts. A typical example of such a laser weld widely used in heat exchangers is a seam weld joining a thin-wall tube to the wall of a hole drilled in a thick plate (Fig. 4.32).

As seen from Fig. 4.32, the region of the walls to be welded is broken down into a number of finite elements to form a reference grid for estimating the thermal processes. The model here deals with the problem of a moving heat source and considers the thermal processes in the remolten region where the seam closes on itself. This model is suitable for calculating the programmed welding conditions with the aim to achieve a desired temperature field at a constant weld penetration.

The problem is more convenient to solve in cylindrical coordinates using difference equation (4.46) which now includes Lame's constants and allows for the bending of the coordinate grid.

In its final form, the difference equation for calculating the temperatures at nodal points is written as

$$T_{i,j,k}^{t+1} = T_{i,j,k} + \sum_{n=1}^{3} \left\{ \frac{K\left(T_{i,j,k}\right)t_s}{c\left(T_{i,j,k}\right)\rho\left(T_{i,j,k}\right) \begin{Bmatrix} s_r \\ s_\varphi R \\ s_z \end{Bmatrix}^2} \times \left[\xi_{\begin{cases} i-1,j,k \\ i,j-1,k \\ i,j,k-1 \end{cases}} T_{\begin{cases} i-1,j,k \\ i,.j-1,k \\ i,j,k-1 \end{cases}} - 2\xi_{i,j,k} T_{i,j,k} + \xi_{\begin{cases} i+1,j,k \\ i,j+1,k \\ i,j,k+1 \end{cases}} T_{\begin{cases} i+1,j,k \\ i,j+1,k \\ i,j,k+1 \end{cases}} \right] \right\} + \frac{P(t)C_{i,j,k}t_s}{c\left(T_{i,j,k}\right)\rho\left(T_{i,j,k}\right)\Delta V_{i,j,k}}, \tag{4.64}$$

where $\xi_{i-1,j,k} = 1 - s_r / 2R$, $\xi_{i+1,j,k} = 1 + s_r / 2R$, $\xi_{i,j,k} = 1 + (s_r / s_\varphi R)^2$, $\xi_{i,j-1,k} = \xi_{i,j+1,k} = (s_r / s_\varphi R)^2$, $\xi_{i,j,k-1} = \xi_{i,j,k+1} = 1$ are Lame's constants; $\Delta V_{i,j,k} = s_r s_\varphi R s_z$ is the elementary volume; and $C_{i,j,k}$ is the heat transfer factor accounting for

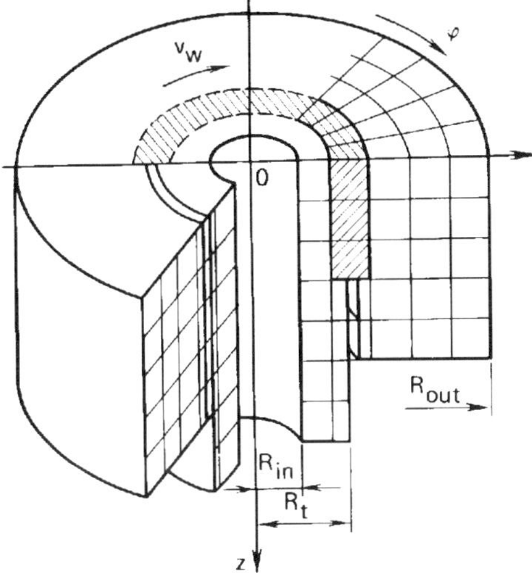

Fig. 4.32. Layout of the domain for modeling the process of welding the tube to the hole wall

the pattern of heat distribution. Equation (4.51) is suitable for estimating the temperature in boundary elements at $R = (R_{out}, R_{in})$ and $z = (0, l_z)$ (see Fig. 4.32).

The boundary was closed with respect to angle φ by superimposing the planes $\varphi = 0$ and $\varphi = 2\pi$.

Using the above described model, we calculated the optimal parameters of welding at the same penetration depth in the region of overlap of the seam ends. For this, the model included a data array used to determine the temperature field at the instant the seam closed on itself. This temperature profile provided the initial data for calculating the temperature of the overlap region formed by the beam as it moved along the earlier welded portion. The laser power and the length of the overlap region were varied to ensure a constant value of penetration.

Figure 4.33 displays the temperature profiles (*a*) at the instant the seam closes on itself and the thermal cycles (*b*) of welding a tube having an outer diameter of 11 mm and inner diameter of 4 mm to the walls of a hole drilled in a plate 20 mm thick.

The calculations were made with consideration for the latent heat of phase transitions and temperature dependence of thermal parameters. In the experiments, a mirror-galvanometer oscillograph recorded the temperatures measured with 0.15-mm diameter chrome-alumel thermocouples.

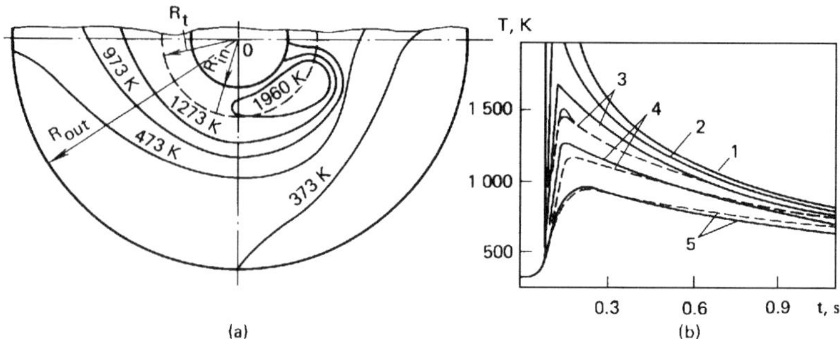

Fig. 4.33. Thermal diagrams illustrative of laser welding of the tube of radius R_t to the wall of the hole in a titanium alloy plate at P = 3.5 kW, v_w = 23.3 mm/s, and f = 149 mm: (*a*) temperature field at instant of closure of circular seam; (*b*) temperatures on the upper surface (z = 0) at points located 0, 0.6, 1.2, 1.8 and 2.8 mm away from the seam axis in the radial direction (curves *1*, *2*, *3*, *4*, and *5* respectively). Full-line and dashed curves illustrate theoretical and experimental values respectively

The curves of Fig. 4.33*b* point to a good agreement between the experimental and theoretical results at temperatures of up to 1000 K. A certain discrepancy at higher temperatures is due to inaccurate experimental estimates of heat distribution in depth and the time lag of the temperature measurement device.

Heat treatment of flat and cylindrical surfaces. In heat treatment of target surfaces, the laser beam produces a distributed surface heat source that rapidly propagates heat, particularly in metal parts noted for high heat conduction. A three-dimensional model provides more accurate results than a plane model that is suitable for the analysis of heat treatment with a fast scanning beam.

Laser hardening of plate surfaces. The model used for the analysis of this process of laser heating is similar to the model of partial penetration welding, the difference being that the former model considers a surface heat source. The model can deal with any character of laser power density distribution on the target surface, which depends on the type of laser beam used and conditions of beam focusing. The calculation technique involving a Gaussian heat source is given in work [52].

In the analysis of heat treatment of the surface covered with an absorbent, the model should account for the coat absorptivity to calculate adequately the effective heat power.

Heat transfer equation (4.46) for a volume heat source and equation (4.51) are suitable for estimating the temperatures at the nodal points of the grid and at the boundaries of elements respectively (Fig. 4.34*a*). The surface heat source model permits us to determine the temperature field and, hence, to

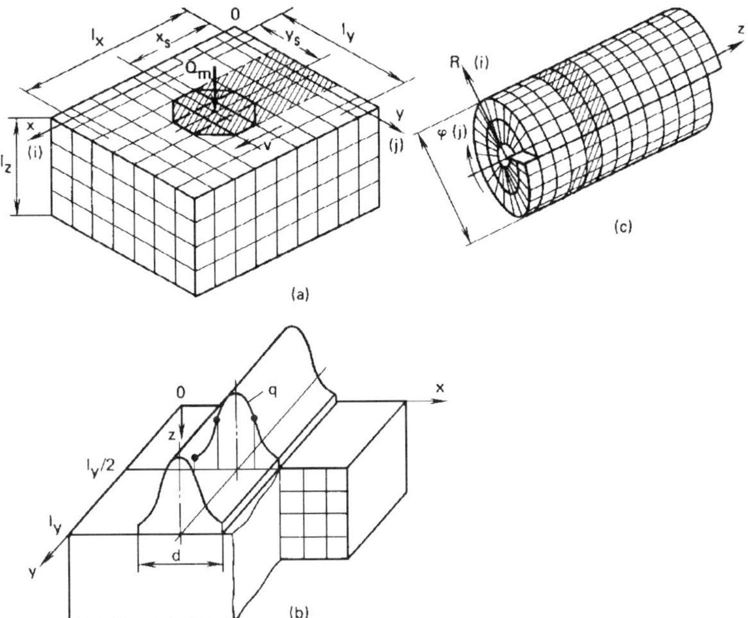

Fig. 4.34. Coordinate grids for modeling the process of laser surface hardening of a plate with a linear heat source (*a*), a shaft along the helical line (*b*), and a plate with the beam dithering perpendicular to the track

estimate the dimensions of structural transformation regions and mechanical properties of the hardened zones.

Laser hardening of cylinder surfaces. A common approach to laser surface hardening of the bodies of revolution is to rotate a part and simultaneously move the part or the beam in the longitudinal direction, thereby producing spiral heat-treat tracks with or without overlapping.

In laser surface hardening of a cylinder over 100 mm in diameter, each successive track generally covers the cooled band of the preceding track. In this case, the model of plate surface hardening is applicable for calculations. As for the cylinders of smaller dimensions, the calculation technique is more complex since it must allow for the heat left in the preceding track.

In Fig. 4.34*b* is shown the reference grid laid out on a shaft to model the process of surface hardening along the helical line through the solution of a three-dimensional model using cylindrical coordinates for convenience. Difference equation (4.64) is used to estimate the temperatures at the interior points of the domain. In the calculation region of Fig. 4.34*b*, the shift of boundary planes through an angle φ from 0 to 360° occurs at a step s along the spiral line. The temperature in the cross section $(i,\ 1,\ k)$ is given by

$$T_{i,1,k}^{t+1} = T_{i,1,k} + \sum_{n=1}^{3} \left\{ \frac{K\left(T_{i,j,k}\right)t_s}{c\left(T_{i,1,k}\right)\rho\left(T_{i,1,k}\right) \begin{Bmatrix} s_r \\ s_\varphi R \\ s_z \end{Bmatrix}^2} \right.$$

$$\times \left[\xi_{\begin{Bmatrix} i-1,1,k \\ i,N-1,k+n \\ i,1,k-1 \end{Bmatrix}} T_{\begin{Bmatrix} i-1,1,k \\ i,N-1,k+n \\ i,1,k-1 \end{Bmatrix}} - 2\xi_{i,1,k} T_{i,1,k} \right.$$

$$\left.\left. + \xi_{\begin{Bmatrix} i+1,1,k \\ i,2,k \\ i,1,k+1 \end{Bmatrix}} T_{\begin{Bmatrix} i+1,1,k \\ i,2,k \\ i,1,k+1 \end{Bmatrix}} \right] \right\}. \tag{4.65}$$

The shift of boundary planes through 360° at a step s is made by reassigning the indexes:

$$T_{i,N,k} = T_{i,1,k-n}, \tag{4.66}$$

where $n = s/s_z$ is an integer.

Heat treatment with a dithering beam. This method produces a wide zigzag pattern of the hardened track in a single pass by dithering the beam perpendicular to the track. Since the solution to a heat transfer problem stated with consideration for a complex trajectory of the beam will be time-consuming and complex, recourse is made to the model of a plane heat source to obtain approximate estimates of the thermal processes.

Because of the thermal lag due to a high frequency of beam oscillation, the quasidistributed heat source produced takes the shape of a band the width of which is equal to the double amplitude of oscillations.

Figure 4.34*c* is an illustration of the technique of surface hardening by dithering the beam. Neglecting the edge effects, we can consider a cross section at $y = l_y/z$, where the heat flux propagates along the x and z axes and does not flow along the y axis, so the boundary planes in the y direction can be taken adiabatic. Since we deal with a plane heat source problem, equation (4.63) applies for temperature calculations. As is evident from Fig. 4.34*c*, the source distributes power in Gaussian fashion at the nodal points. The calculated value of heat power is

$$P'' = P\eta/l_y, \tag{4.67}$$

where η is the ratio of the absorbed to the incident power.

Axisymmetric heating of bodies of revolution. In laser cladding, hardening, or trimming of the bodies of revolution, such as tubes and flanges, the axisymmetric beam falling on a fast revolving part can be assumed to produce a quasidistributed circular heat source that emits heat in the radial direction R and along the z axis in thickness (Fig. 4.35). The heat transfer problem

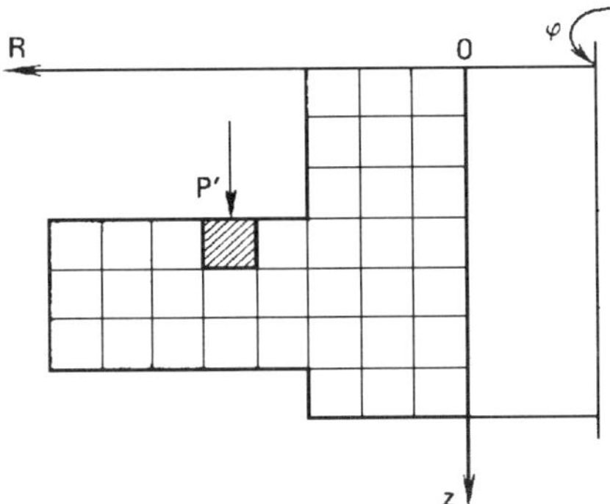

Fig. 4.35. Axisymmetric heating of a body of revolution

then reduces to a plane heat source problem, and a radial cross section at any angle φ can be taken as a calculation region.

In a similar way we can calculate a temperature field. Depending on the kind of heat treatment under study, the heat source can be thought to distribute heat either in the radial direction on the surface, for example, in a Gaussian fashion, or in any cross section along the z axis. The calculated heat power is

$$P' = \eta/\pi D, \tag{4.68}$$

where D is the diameter of the hot spot on a heat-treat track.

The above described models have provided the basis for developing a package of programs written in Fortran. For a program to be run on a computer, it is sufficient to study the list of input parameters and readout information and then introduce desired data into the program.

5 Laser-Induced Stresses and Strains in Metals

5.1. General Concepts

A solid uniformly heated over its volume expands freely without developing internal stresses. In the course of nonuniform heating of a solid, the bonds between its hot and cold portions retard thermal expansion and lead to inherent temperature stresses. These are temporal stresses which develop during a laser treatment process at the stages of metal heating, temperature leveling, and quenching.

Along with temperature stresses and strains, elastic and plastic strains can arise in a solid with changes in its volume in the course of phase and structural transformations. Plastic strains observed in a cooled solid result from inherent residual stresses.

Let us look at the mechanism of developing temporal and residual stresses by a heat source moving along the edge of a long plate [88]. As it travels along the x-axis, the heat source with the highest temperature at its central point O produces a steady temperature field which exhibits the hot and the cold portion in front of and behind the source (Fig. 5.1). The temperature gradient in the plate gives rise to temporal stresses σ_x in the direction of travel of the heat source (Fig. 5.1*a*), which are the compression stresses at the stages of heating and cooling. Assuming the yield strength σ' to be infinitly large, the maximum values of stresses will not reach the yield point, and so the deformation process will occur in the elastic region without entailing plastic strains. The stresses will then drop to zero after plate cooling.

Assume that the yield strength σ' is equal to 400 MPa and does not change under the action of heating. The stresses σ_x at the stage of heating will then reach the yield limit at point A_1 (Fig. 5.1*b*). The plate undergoes shrinkage within section A_1B_1. Disregarding the effect of hardening, we have $\sigma_x = \sigma' = 400$ MPa. Starting at point B_1, the compression stresses decrease along the curve $B_1C_1D_1$ which is equidistant from the curve BD. The stresses σ_x drop to zero at point C_1, and away from this point tensile stresses arise. After its complete cooling at point D_1, the plate will exhibit residual tensile stresses σ_{xr} below the yield strength.

Figure 5.1*c* illustrates the same case, but under the assumption that the yield point lies at 200 MPa. The compression stresses σ_x reach yield point A_2 at

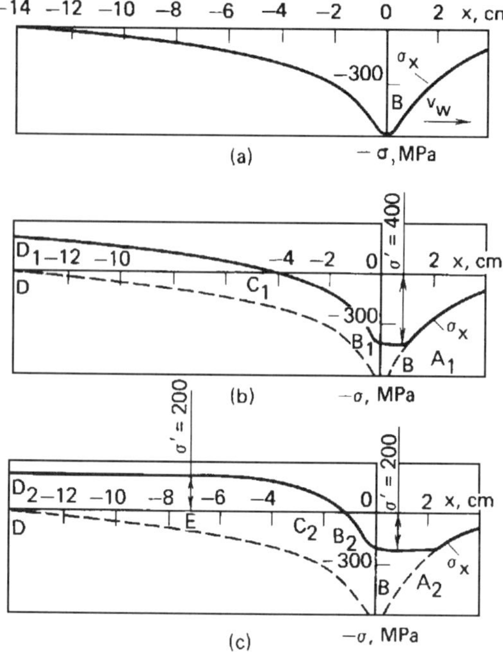

Fig. 5.1. Temporal and residual stresses set up along the edge of a plate heated by a moving source at a yield strength σ′ of the metal extending to infinity (a), 400 MPa (b), and 200 MPa (c)

the stage of heating. The section A_2B_2 is the region of shrinkage under plastic strains. At the stage of cooling, the compression stresses gradually decrease to zero at point C_2. The tensile stresses now begin to grow until they reach the yield limit at point E. The section $ED_2$of the curve is the region where plastic tensile strains are likely to appear. The plate cools down completely at point D_2 and exhibits residual stresses equal to the yield strength.

Consequently, at a relatively low temperature of heating the yield strength remains invariable and the developed stresses do not rise above this limit. The deformation process here occurs in the elastic region at a maximal temperature of less than 573 K. However, the heating temperature in the actual processes of laser heat treatment is much higher, and stresses and strains arise in the elastoplastic deformation conditions.

Inherent stresses of temporal and residual types are always held in equilibrium whatever the character of their distribution can be over the volume

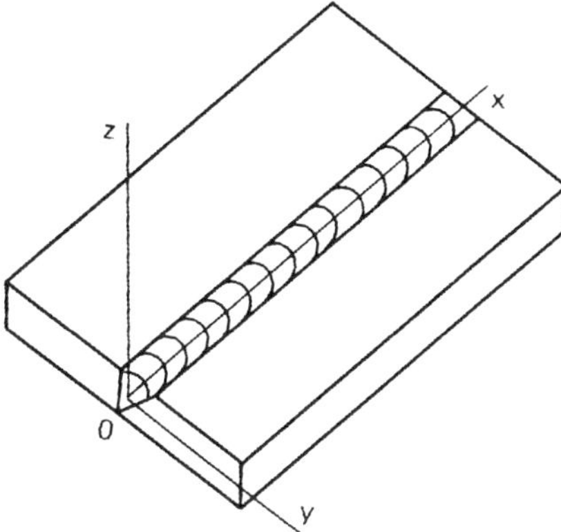

Fig. 5.2. The coordinate axes for defining the components of stresses

of a solid. The inherent stresses balanced out in a large volume, in a volume of one or a few grains, and in a volume comparable with the crystal lattice size fall into the categories of stresses of the first, second, and third order, respectively. First-order stresses are responsible for the change in the shape and size of a solid. The stresses of the second and the third order initiate the processes of material disintegration. These stresses can be uniaxial or linear, plane or biaxial, and triaxial or volumetric, depending on the process of laser heating and the character of the action of a laser beam. In laser cladding, hardening, welding, and cutting along narrow tracks, the stresses can be assumed to be uniaxial. Plane and volumetric stresses arise in the processes of surface hardening and also in welding and cutting thick plates.

The normal components of stresses are laid off along the coordinate axes, of which the longitudinal *x*-axis is coincident with the direction of laser heating (Fig. 5.2). These are longitudinal stresses σ_x, transverse stresses σ_y acting in the plane *xoy*, and stresses x_z acting in the thickness direction. Along with normal components of stresses, tangential stresses τ_{xy}, τ_{yz}, and τ_{zx} can develop in a laser-heated solid.

In analogy with stresses, we can differentiate between the normal components of strains, ε_x, ε_y, and ε_z , and the shear strains γ_{xy}, γ_{yz}, and γ_{zx} , which change the linear and angular dimensions of solids.

5.2. Theoretical Description of Stresses and Strains

Graphical methods. Uniaxial stresses σ_x and strains ε_x of the temporal and permanent types are most readily amenable to the mathematical description, for example, by the graphical method presented in [89]. The basic statements of this method rely on the following assumptions: (1) the cross sections are flat and do not bend in the process of heat treatment; (2) the workpiece only develops longitudinal stresses σ_x; (3) the target is an ideal elastoplastic body whose yield strength σ' remains constant up to 773 K and then linearly drops to zero at 873 K; and (4) other thermal and mechanical properties do not vary with temperature.

To illustrate how this method works, consider an example of the analysis of stresses and strains developed in the process of deep-penetration butt welding of two long plates of width b each, with the laser beam travelling along the plate edges at a certain welding speed v_w (Fig. 5.3). The heat source here is a linear source moving at a definite speed. For the case in hand, we can use equation (4.10) or (4.12) to define the quasistationary temperature state of the plates.

At the stage of heating, stresses and strains are defined in the cross section *a-a* (Fig. 5.3*a*) where the width of the weld zone heated to 873 K is the largest. This temperature determines the position of the section *a-a*.

Figure 5.3*b* illustrates the temperature profile in the section *a-a* and the corresponding temperature strains αT, where α is the heat transfer coefficient. The metal portions located in this cross section would elongate in proportion to αT if there were no bonds between them. Since we assume that the cross sections are flat, at the stage of heating the tensile strains ε_h in all the portions will be the same as illustrated by the straight line a_1a_1. The portions of the weld in cross sections undergo both elastic and plastic strains. The plus and minus signs in the figure stand for elastic tensile and shrinkage strains, respectively.

An elastic stress is defined as the product of the appropriate elastic strain by the modulus of elasticity, E. At temperatures of up to 773 K, the values of stresses lie in the region bounded by the yield strength σ'. In the interval from 773 to 873 K, the stresses linearly drop to zero in accordance with changes in σ'.

At the stage of heating the plastic shrinkage strains $\varepsilon_{h,\,sh}$ within the cross-hatched area are taken equal both at 873 K and above this temperature. The width of the plastic strain region is $2b_p$. These strains cause residual stresses.

The equilibrium condition for stresses σ_x over the plate width, i.e., the equality of the areas bounded by the curves of tensile and compression

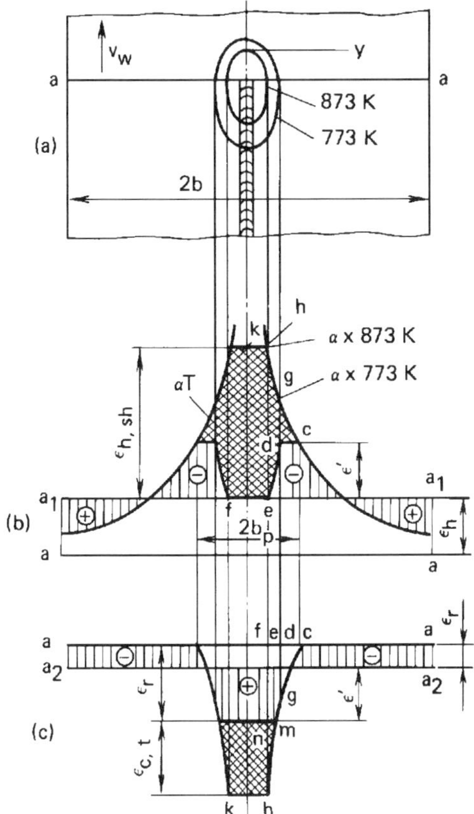

Fig. 5.3. Diagrams for estimating longitudinal stresses and strains by the graphical method: (a) temperature field; (b) strains at the stage of heating; (c) strains at the stage of complete cooling

stresses,

$$\int_0^{2b} \sigma_x dy = 0 \tag{5.1}$$

determines the actual position of the straight lines a_1a_1 which is commonly preset by the method of successive approximations.

Figure 5.3*c* is an illustration of stresses and strains developed at the stage of cooling. Should the cooling process occur without the effect of metallic bonds, the metal parts would shorten in keeping with the diagram *cdefkh*. However, the section *a-a* will now take the position denoted by a_2-a_2, since the residual strains ε_r cause the plates to shrink. The area hatched with straight lines represents elastic strains, and the cross-hatched area illustrates plastic tensile strains $\varepsilon_{c,t}$ developed after cooling. The residual plastic strains ε_r within the area *cfnm* are shrinkage strains.

Multiplying the elastic strains by the elasticity modulus E yields the values of residual stresses σ_{xr}. The position of the straight line a_2-a_2 conforms to the equilibrium condition of residual stresses σ_{xr}. As is obvious from Fig. 5.3*c*, the stresses σ_{xr} in the weld fusion zone and heat affected zone are equal to the yield strength, $\sigma_{xr} = E\varepsilon' = \sigma'$, where ε' is the limiting value of the strain. These results well agree with experimental data for low-carbon steel.

In a similar manner we can construct diagrams for the analysis of laser cladding or laser surface hardening. But here, apart from equilibrium condition (5.1), consideration should be given to the equilibrium condition of moments:

$$\int_0^{2b} \sigma_x \, y dy = 0. \tag{5.2}$$

This graphical method is suitable for the analysis of stresses and strains in a number of cross sections, therewith allowing for plastic strains in each successive section. Given the total plastic strains, it is then possible to estimate more accurately the stresses developed in the process of heating. In this case, the residual stresses in the heat affected zone are also equal to the yield strength of the material. The calculations involved in the method can be readily made on a computer. The method is simple and illustrative and well discloses the mechanism of developing longitudinal stresses and strains. In particular, knowing the residual shrinkage strain ε_r of a plate of length l, it is easy to estimate the linear shrinkage

$$\Delta l_x = l \, \varepsilon_r. \tag{5.3}$$

The estimates obtained from equation (5.3) agree with experimental data for narrow low-carbon steel plates, i.e., the plates in which the width $2b_p$ of the plastic strain region is smaller than the plate width $2b$ by a factor of 3 or 4.

The graphical method is also applicable for the analysis of longitudinal residual stresses developed in low-carbon and corrosion-resistant steels. The values of σ_{xr} in the laser-treated zone are close to those of the yield strength so that they agree with experimental data obtained in welding, cutting, and cladding. But the values of σ_{xr} obtained by this method for titanium, aluminum, and magnesium alloys poorly agree with experimental results because the adopted assumptions described above do not hold here and the method cannot accurately allow for changes in the properties of these alloys with temperature.

In the processes of heat treatment of thick and wide parts there appear compound stresses whose analysis by the graphical method proves to be inadequate. In this case, we need to change to the methods relying on the theories of elasticity and plasticity.

Methods based on the theory of elasticity. This theory is commonly useful for the solution of heat-transfer problems if the temperature of materials varies over a relatively narrow range and their physico-mechanical properties change insignificantly. A large number of solutions on the basis of this theory are available. But since the temperature in the laser-heated zone varies over a wide range, the theory of elasticity can only give tentative values of the components of stresses and strains. The solution procedure here involves, as it were, the first stage of the solution of an elastoplastic problem.

Proceeding from the theory of elasticity, the authors of [88] have determined the distribution of stresses arising in axisymmetric surface hardening or spot welding and estimated stresses in an infinite plate heated by a moving linear source. The results of problem solutions can provide a qualitative assessment of the kinetics of stresses. Quantitative estimates can be made on the basis of the isothermal theory of plasticity.

Methods based on the theory of plasticity. These methods ensure a more accurate estimation of stresses and plastic strains that inevitably arise in laser-treated zones. A common approach in working a plasticity problem is to solve an elasticity problem and successively refine the solution under pertinent auxiliary conditions. There are a few iteration methods designed to solve an elasticity problem at each step of iteration. The most popular are the method of additional loads, the method of additional strains, and the method of variable parameters of elasticity [88].

All these methods rely on the iteration procedure of refining the problem and its solution under auxiliary conditions. The available versions of these methods can handle elastoplastic problems. However, these problems are amenable to the solution in analytical form only where they deal with rather simplified cases of analysis. Practical problems should certainly be adapted for the computer solution.

Researchers of the Bauman Moscow State Technical University have worked out the numerical methods of solving elastoplastic problems for the analysis of stresses and strains developed in metals during the processes of arc, contact, and electroslag welding. These methods have formed the basis for the methods used to solve elastoplastic problems for estimating temporal and residual stresses and strains arising in various processes of laser heat treatment.

The basic equations of elastoplastic deformation include the differential strain-equilibrium equations and equations used to define the relations between stresses and strains. The theory of plastic flow in which physical equations establish the relations between the increments of plastic strains and stresses is best suited for the solution of hot plasticity problems. An elastoplastic problem takes the form of a system of linear differential equations.

The analytical solution of a boundary-value problem involving nonlinear differential equations is only possible for specific cases of analysis, which considerably simplifies the actual process. The general method is not yet available, and so use is made of numerical computer-oriented methods.

A net-point method is applicable to second-order partial differential equations which define a nonstationary hot plasticity problem. However, this method calls for a rather fine mesh size in order to achieve a requisite approximation accuracy when replacing the second-order derivatives by a finite-difference equation. Too fine a mesh complicates the calculations. Besides, the method is poorly adapted for curvilinear regions. The functional minimized for the solution of a plane problem involving two space variables can take on a general form

$$I = \iint_S \Big[f_1(x,y,t)\left(\frac{\partial \varphi}{\partial x}\right)^2 + f_2(x,y,t)\left(\frac{\partial \varphi}{\partial y}\right)^2 + f_3(x,y,t)\left(\frac{\partial \varphi}{\partial x}\right)\left(\frac{\partial \varphi}{\partial y}\right) + f_4(x,y,t)\varphi(x,y,t)\Big] dS, \tag{5.4}$$

where $\varphi(x, y,t)$ is the desired vector function for motions, speeds, and stresses; f_1, f_2, f_3, f_4 $(x,, y, t)$ are the specified functions indicative of changes in the mechanical properties of a metal in space and time; and dS is an elemental area.

The numerical methods of minimization presuppose the replacement of the above functional by a sum that varies as the square of the finite set of values of the unknown function at the nodal points of the net:

$$I \cong I' = \sum_{i=1}^{n}\sum_{j=1}^{m} p_{i,j}\varphi_k\varphi_l, \tag{5.5}$$

where I' is an approximate functional; φ represents the nodal values of the desired function; k, l, m, and $p_{i,j}$ are the parameters and factors which depend on the specified functions and the selected method of numerical integration;

and n is the number of nodes in the net covering the region of interest.

The solution of a problem will be exact at n approaching infinity, but approximate at a limited number of nodes for which the finite sum is at a minimum.

Applying the general rules of searching for a minimum of the function of the finite number of variables, we obtain

$$\frac{\partial I'}{\partial \varphi_i} = 0, \quad i = 1, n. \tag{5.6}$$

Since I' given by equation (5.5) varies as the square of φ_i, the system of (5.6) will obviously be a set of linear algebraic equations of b_i with respect to φ_i:

$$b_i = \sum_{i=1}^{n} \left(\sum_{j=1}^{n} a_{ijk} \varphi_i \right), \tag{5.7}$$

where a_{ijk} represents coefficients in the system of equations (5.6).

A specific method employed to reduce the functional to the finite sum of (5.5) generates a corresponding algorithm for calculating the coefficients in the system of equations (5.7).

The most approved finite-element method [68, 69] commonly presupposes a linear approximation of both the desired function and its derivatives in the region of simplex elements:

$$\varphi^{(e)}(x, y) = [N]^{(e)}\{D\}, \tag{5.8}$$

$$\{g\}^{(e)} = [B]^{(e)}\{D\}, \tag{5.9}$$

where N is the form function; D represents the nodal values of the desired function; and B is the matrix of the derivatives of the function N.

Approximation of equation (5.8) within an element region along the coordinate x and y-axes dispenses with the above mentioned limitations on the mesh size, which considerably widens the field of application of the method. As follows from equation (5.9), the method represents the values of stresses and strains at the points on interelement boundaries.

The exact solution of high-order equations (5.8) and (5.9) with a wide matrix of coefficients requires a large number of computing operations and an

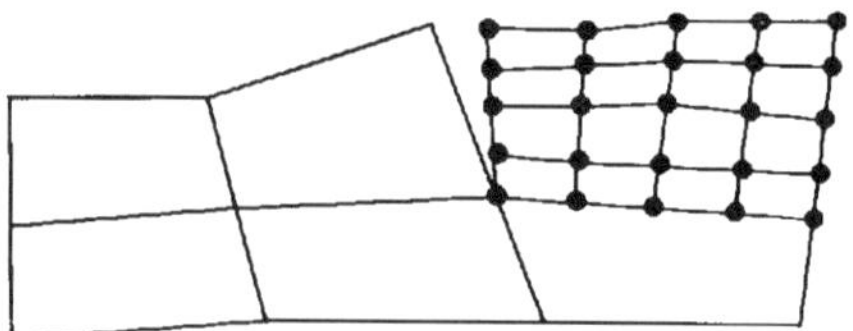

Fig. 5.4. Dividing the domain into a finite number of elements with nodal points

appreciable capacity of random access memory [90, 91]. Consider a new method designed to solve the problem of minimizing the functional of (5.4) which eliminates the above difficulties.

The region of the problem solution, i.e., integration region S in equation (5.4), consists of a set of finite elements or sections in the form of quadrangles or parallelepipeds for three-dimensional problems (Fig. 5.4). The points of intersection of the coordinate lines form a set of nodes at which the values of the unknown function are to be found.

Equation (5.4) can approximately be defined using formula (5.5) and written in the form

$$I = \int_S \int F \cong \sum_{e=1}^{n} \int_{S_e} \int F, \tag{5.10}$$

where S_e is the area of an element e and n is the number of elements.

As an example, Fig.5.5 illustrates the net formed by i and j coordinate lines for expressing derivatives by analogy with the finite element method. Let us isolate the triangles about the node of interest with coordinates x_{ij} and y_{ij}. According to expression (5.8), each triangle in any element e under examination can be defined as

$$\left\{ \frac{\partial \varphi}{\partial x}, \frac{\partial \varphi}{\partial y} \right\}_e = [B_e]\{D\}_e. \tag{5.11}$$

The derivatives at the nodal point assume the values averaged over the triangles:

$$\left\{ \frac{\partial \varphi}{\partial x}, \frac{\partial \varphi}{\partial y} \right\} = \frac{1}{4} \sum_{e=1}^{4} [B_e]\{D\}_e = [B]_{i,j} \{D\}_{ij}. \tag{5.12}$$

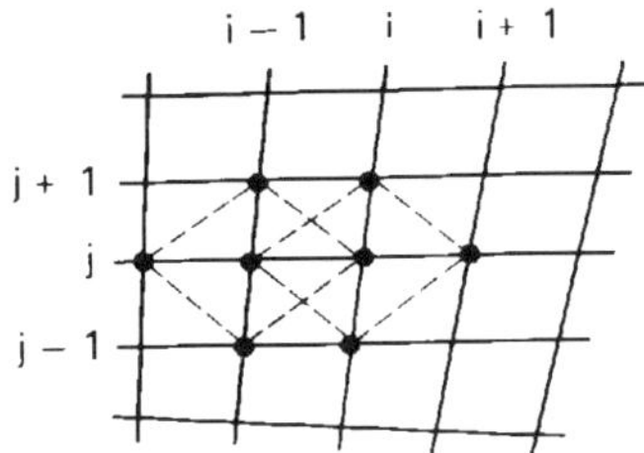

Fig. 5.5. Approximating the function for one of the internal nodes

In distinction from the finite element method, in this method the separation of an element into triangles is not permanent, but "local" with respect to a node, such as the node $(i-1, j)$ in Fig. 5.5. For a rectangular element with even sides, equation (5.12) yields the same result as that obtained from the parabolic approximation of the function along the two coordinates. So, the accuracy of approximation of the derivatives at the nodal point will be higher than that adopted in the finite element method for each calculation region. To define the node values in accordance with equation (5.12), the approximation procedure actually follows a five-point pattern

$$\{D\}_{i,j} = \{D_{i-1,j};\ \varphi_{i,j};\ \varphi_{i,j+1};\ \varphi_{i,j-1};\ \varphi_{i+1,j}\}. \tag{5.13}$$

The above described approach permits us to solve linear equations for elements by exact methods and estimate the values at interelement points by iteration methods. The parameters of elements can be chosen with regard to the optimal width of the region under examination.

The iterative solution sought for the nodes on interelement boundaries does not need to repeat the entire procedure for the element in question and involves only changes in the right-hand sides of the equations. The matrix of the equations for elements at a given time step takes a requisite form only once, and so a new solution is found comparatively rapidly after changing the right-hand sides of the equations.

The number of iterations is proportional to the number of nodes on interelement boundaries, which is much smaller than the total number of nodes in the entire region. This considerably speeds up the iterative process as against the process involved in the finite element method.

An advantage of the described method as applied to laser heating is that it enables us to separate a small high-temperature laser-heated region from a low-temperature region and assume the properties of the material of the elements to be invariable. The matrix of the system of equations for these elements then remains invariable too, and so the solution can be sought only once.

The method appreciably facilitates the procedures of developing universal programs for various processes of laser heating and offers a possibility of stating and solving the most complex and general three-dimensional problems of stresses and strains generated in bulky pieces.

This method has formed the basis for working out the algorithms and programs intended to solve the following problems: uniaxial problems involved in laser welding, cutting, cladding, and hardening of long and narrow parts; biaxial-stress problems involved in laser heating of both thin parts of various configurations and thick parts; and three-dimensional problems.

5.3. Experimental Studies of Stresses and Strains

These studies can be undertaken on original specimens or their physical models. The aim of the tests performed on physical models is to reproduce the processes of development of stresses and strains in laser heating. Scale modeling involves similarity between the specimen and its model, which gives rather authentic results when the model under test and the original piece are made from the same material [92]. The scale-model testing is quite popular for the analysis of residual stresses and strains.

Temporal stresses and strains. A technique described in work [92] for assessing temporal laser-induced stresses and strains involves tests performed on optimal models that are geometrically similar to actual metallic pieces [93]. A more popular method is to test original specimens using various strain gages and data recording sensors intended to measure and record stresses and strains. Widely used are such measuring devices as indicating heads, wire strain gages, pneumatic gages, and various electric sensors, for example, inductive and capacitive pickups.

The estimates of high-temperature strains can be made by a contactless measurement technique based on photorecording of marks inscribed on the surface to be heated. Where the process involves heating with melting, the pictures of the molten pool preliminarily covered with a metallic powder can be taken to record the positions of individual inclusions and thus estimate the strains at the stage of cooling..

The inherent strains, i.e., the sum of elastic ε_{el} and plastic ε_p strains induced by stresses, is equal to

$$\varepsilon_{el} + \varepsilon_p = \varepsilon_m - \varepsilon_{fr}, \tag{5.14}$$

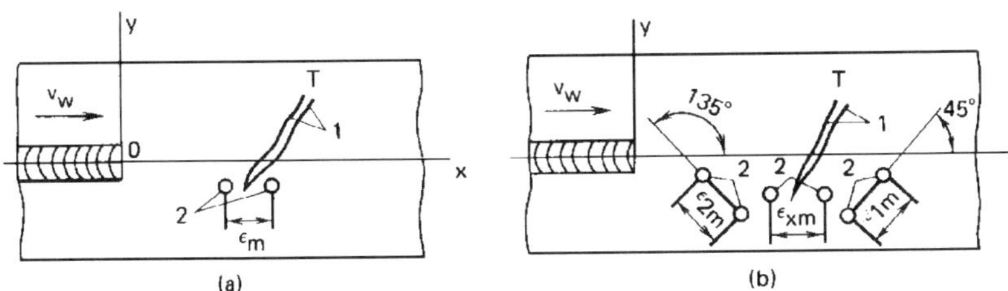

Fig. 5.6. Schematic diagrams of measurement of uniaxial strains (a) and plate strains (b): 1—thermocouple; 2—sites of measurement

where ε_m is the strain recorded by a measuring device and ε_{fr} is the free temperature strain determined from the dilatometer record that reproduces the thermal cycle in a test specimen. Figure 5.6*a* is a schematic of uniaxial strain measurement during the heat treatment. In Fig. 5.6*b* is shown a schematic diagram for measuring the thermal cycle $T(t)$ and plane strains ε_{xm} along the weld seam, strains ε_{1m} at an angle of 45° and strains ε_{2m} at an angle of 135° to the direction of travel of the laser beam.

Using Mohr's strain circle diagram, we can define the relations between the angular and the linear strains:

$$\left.\begin{aligned} \gamma_{xym} &= \varepsilon_{1m} - \varepsilon_{2m} \\ \varepsilon_{ym} &= \varepsilon_{1m} + \varepsilon_{2m} - \varepsilon_{xm} \end{aligned}\right\}. \tag{5.15}$$

So, given the experimental values of ε_{xm}, ε_{1m}, and ε_{2m}, we can estimate shear strains γ_{xym} and lateral strains ε_{ym} from the above relations.

Knowing ε_m and estimating $\varepsilon_{fr}(t)$ from the thermal cycle $T(t)$ reproduced by a dilatometer, we can use equation (5.17) to determine inherent strain components at each moment of time:

$$\begin{aligned} \varepsilon_x &= \varepsilon_{xm} - \varepsilon_{fr} \\ \varepsilon_y &= \varepsilon_{ym} - \varepsilon_{fr} \\ \gamma_{xy} &= \gamma_{xym} \end{aligned} \tag{5.16}$$

The corresponding components of internal stresses can be found from the equations of the theory of plasticity, which consider thermal plastic effects typical of laser heating. The theory of nonisothermal plastic flow is best acceptable for the purpose which considers the relationship between the infinitesimal increments of stresses and strains, thereby enabling us to follow the development of stresses at the stages of heating and cooling.

In order to get more accurate values of internal stresses, the calculations should account for the resistance of metals to strains during the laser treatment process using, for example, the thermal strain records obtained on special test machines that reproduce the thermal and strain cycles in specimens [88].

Residual stresses and strains. A method used to determine residual stresses is to cut out a test specimen to relieve stresses and to record the elastic strains induced in the specimen while it is cut. Knowing the elastic strains, we can estimate the residual stresses from appropriate equations.

The devices designed to measure strains include: electric strain gages such as wire or foil resistance sensors bonded to test specimens, photoelastic sensors from optical materials, and mechanical strainometers. The latter devices can measure strains at different sites on a test specimen, from which we can determine uniaxial (longitudinal), biaxial (plane), and triaxial residual stresses.

In order to estimate residual stresses σ_x the strains are measured at a gage site arranged along the x-axis (Fig. 5.7*a*). The length l and width b of the strain measurement site depend on the expected gradients of longitudinal stresses σ_x and lateral stresses σ_y, respectively. If the difference in σ_x is expected to be small, the length l can be chosen over a wide range. The width b can be small if the expected gradient of σ_y is large. The dimensions of the sites should be measured on both sides of the plate before and after cutting out the strip.

The results of measurement of the site give us the values of absolute deformation, $\Delta l = l_2 - l_1$ and $\Delta l' = l_2 - l_1$, on both sides of the plate, where l_1 and l_2 are the lengths of the site after welding and after cutting, respectively. An average value of absolute deformation is $\Delta l_{av} = (\Delta l + \Delta l')/2$.The relative deformation arising in the course of metal cutting is $\varepsilon = \Delta l_{av}/l_1$. Considering that in cutting the plate the residual stresses σ_x in a metal decrease in an elastic manner, it is easy to determine σ_x from the known values of Young's modulus E and strain ε:

$$\sigma_x = -E\varepsilon. \tag{5.17}$$

By measuring the deflections of the plate laser-treated along its length, we can estimate the residual stresses in the plate layers using appropriate equations of the theory of elasticity.

Biaxial stresses arise in thin laser-treated parts of various configurations and on the surfaces of thick parts along the principal x- and y-axes of which the x-axis is coincident with the direction of the weld seam (Fig. 5.7*b*). In order to determine the residual stresses σ_x and σ_y, it is sufficient to arrange the

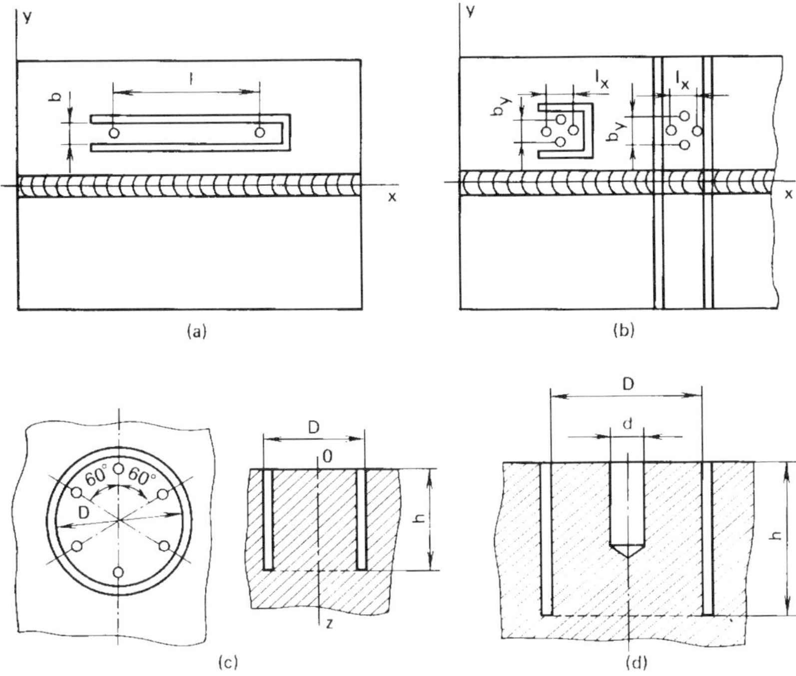

Fig. 5.7. Sites of strain measurement for estimating uniaxial (a), biaxial (b) and (c), and triaxial (d) residual stresses

measurement regions both along the *x*- and *y*-axes. In this case, the tangential stresses τ_{xy} are taken equal to zero.

The measurements should be made on both sides of the plate which can be cut into separate squares or strips. Given the strains ε_x and ε_y developed in the course of cutting, we can estimate the stresses along the two axes from the following equations:

$$\left.\begin{aligned}\sigma_x &= -E\left(\varepsilon_x + \nu\varepsilon_y\right)/\left(1-\nu^2\right)\\ \sigma_y &= -E\left(\varepsilon_y + \nu\varepsilon_x\right)/\left(1-\nu^2\right)\end{aligned}\right\}, \tag{5.18}$$

where ν is Poisson's ratio.

When the directions of the principal axes are unknown, measurements should be made in three regions with their axes being at 60° to one another for estimating the stress components easily (Fig. 5.7*c*). The same test pattern applies for the estimation of biaxial stresses formed on the surface of a thick plate. To relieve the surface layer of residual stresses, the plate can be cut to a depth $h = 0.6\ D$, where D is the distance between the edges of the circular kerf. Of course, cutting the plate in this fashion cannot remove the stresses completely, and so the calculations can only give approximate values of residual stresses.

The methods used to evaluate triaxial residual stresses are more complex. These stresses can arise in the processes of spiral uniform hardfacing of cylindrical surfaces or circumferential butt welding of cylindrical shells. The residual stresses in a cylindrical part can be found by measuring the diameter of the part when boring it. To estimate these stresses in a thick part butt-welded in the longitudinal direction, we need first to measure the residual strains at the gage sites of the plate along the three axes and then to cut the plate into strips. The elastic strains so induced enable us to calculate the stresses averaged over the plate thickness [88].

Triaxial residual stresses in the bulk of the metal can be estimated by measuring the strains, for example, with a resistance strain gage placed into a small-diameter hole pierced in a specimen (Fig. 5.7*d*). The gage glued onto a cylindrical metal pin which is put into the hole of diameter d and sealed with epoxy resion can measure strains when the specimen is cut with a milling cutter or a hollow drill [88].

In mechanical methods of destructive testing, laser holographic systems have recently gained acceptance for measuring the test regions on a specimen. A high sensitivity and increased accuracy of measurement of these systems permit estimating the residual stresses at any local points within the measurement region [92].

In use are also the physical methods of nondestructive testing for determining the residual stresses in laser-treated zones. These methods, however, yield the measurement results that cannot uniquely specify the residual stresses. The reason is that the laser beam interacting with a material initiates various physico-chemical processes that not only change the material properties but also induce elastoplastic strains in the heat affected zone. The text below briefly describes some nondestructive testing methods and their potentialities.

The magnetoelastic method depends on the change in the permeability of a magnetic metal with the residual stresses induced in the given bulk of the metal. The method can provide reliable values of the uniaxial residual stresses only outside the heat affected zone. The estimates obtained within the HAZ can considerably deviate from the actual values because the permiability of laser-treated regions varies with changes in the chemical composition, metal structure, and grain growth, let alone the changes in residual stresses.

The ultrasonic method relies on changes in the velocity of ultrasonic wave propagation in metals with residual stresses. A change in the velocity after laser heating allows us to assess the residual stresses. However, this method can lead to large errors in the estimates of residual stresses if the metal in the laser-treated zone displays inhomogeneous properties. An advantage of the method is that the test procedure does not involve much preliminary work.

It is possible to estimate the residual stresses on the surface of a specimen by measuring the hardness of the surface layer that varies uniquely with residual stresses. The estimates of the residual stresses in the base metal layers beyond the weld zone are found to agree well with the results obtained by mechanical methods. However, these measurements cannot give reliable results for determining the stresses within the heat affected zone.

The X-ray method of stress determination can measure the normal components of surface strains in the 10 μm region by comparing the positions of diffraction lines before and after subjecting the specimen to strains [93].

5.4. Distribution of Residual Stresses and Strains

A specimen laser-treated along a lengthy track undergoes both shrinkage and tensile plastic strains ε_p at the stage of heating and cooling, respectively. Since the former strains are higher in magnitude than the latter, the laser-treated part mostly experiences shrinkage strains (Fig. 5.8).

The text below presents the results of experimental studies of longitudinal residual strains developed in full-penetration laser butt welding and compares

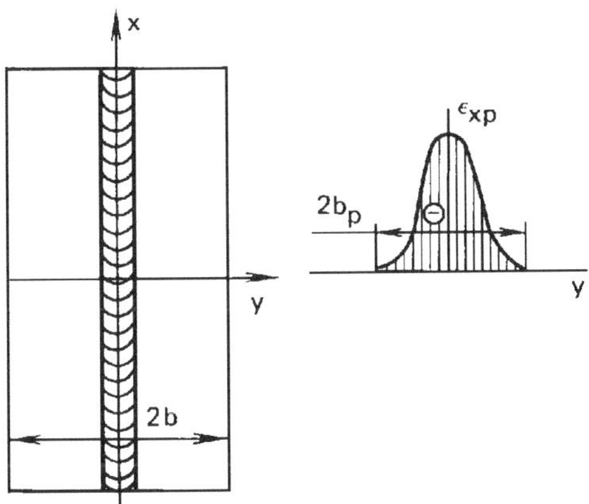

Fig. 5.8. Distribution of longitudinal plastic residual strains e_{xp} in cross sections of the welded plates

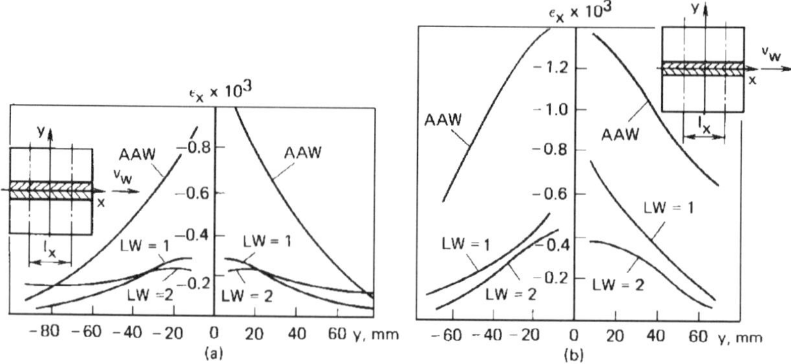

Fig. 5.9. Distribution of longitudinal residual strains in cross sections of welds between the plates made from titanium alloy 1 (a) and titanium alloy 2 (b)

them with the results obtained in the conventional processes of full-penetration argon-arc welding. In the experiments, we used a specially developed strain gage that measured reference lengths with an accuracy of ±1 μm.

In Fig. 5.9 are shown the experimental results of longitudinal strains ε_x in the cross sections of arc-welded and laser-welded butt joints between thin plates made from titanium alloys of two grades.

Table 5.1 presents the parameters of three types of welding of titanium alloy plates: argon-arc welding (AAW); laser welding in the LW-1 mode with a

Table 5.1. Parameters of Titanium Alloy Plates Welding

Material	Welding	Net power, W	Heat input, 10^{-5} J/m	Welding current, A	Arc voltage, V	Welding speed, mm/s	Laser power, kW	Net efficiency
Alloy grade 1, plate 2 mm thick	AAW	860	1.51	1.38	10	5.7	–	0.62
	LW-1	900	0.83	–	–	10.8	3.5	0.26
	LW-2	1890	0.43	–	–	44.5	4.0	0.48
Alloy grade 2, plate 3.5 mm thick	AAW	1820	2.33	280	10	7.8	–	0.65
	LW-1	1840	1.38	–	–	13.3	3.0	0.61
	LW-1	3230	0.73	–	–	44.5	5.0	0.65

Table 5.2. Parameters of Low-Carbon Steel Welding

Welding	Welding speed, mm/s	Net power, W	Heat input, 10^{-5} J/m	Current, A	Voltage, V	Laser power, kW	Efficiency
AAW	5.6	1290	2.30	180	11	–	0.65
EBW	33	1750	0.53	0.042	52×10^3	–	0.80
LW	33	2730	0.82	–	–	4.2	0.65

net power equal to about that consumed in argon-arc welding; and laser welding in the LW-2 mode with an increased net power and higher welding speed, which results in a lower energy per unit length (heat input) of the weld seam. The second mode of laser welding is more advantageous from the point of view of energy consumption since it offers a lower heat input per-unit-length.

Table 5.2 presents the parameters of argon-arc welding (AAW), electron-beam welding (EBW), and laser welding (LW) of low-carbon steel plates.

The graphs of Fig. 5.9 indicate that longitudinal strains in welded plates reach the maximum values within the weld zones. There is a significant difference between these values in argon-arc welding and laser welding. The strains $\varepsilon_{x\ max}$ in arc-welded joints are seen to be 2.5 to 3 times as high as those in the first mode of laser welding performed at the same net power. This difference is due to the fact that the laser beam of the same power can deliver a higher power density to the target, so that the weld seam can be run at 2 or 3 times the speed of argon-arc welding. In other words, the per-unit length laser energy coupled to the target to form the weld is a factor of 2 or 3 lower than that for arc welding.

The second mode of laser welding exhibits still lower values of the per-unit-length energy, and so the residual strains $\varepsilon_{x\ max}$ decrease still further.

As follows from the experimental results, laser welding ensures a stable penetration depth and can form a quality joint at a speed that is a few times the speed of argon-arc welding, with the result that the per-unit-length energy is a factor of 3 to 5 smaller than it is in argon-arc welding. The strains $\varepsilon_{x\ max}$ are approximately as many times lower.

Figure 5.10 illustrates the residual lateral strains defined in terms of transverse shrinkage Δl_y in the direction normal to the weld seam. The experimental values of shrinkage Δl_y are given for weld joints produced in keeping with the welding parameters presented in Tables 5.1 and 5.2.

As is evident from the graphs of Fig. 5.10, in comparison with laser welding, the argon-arc welding results in a more nonuniform distribution of transverse

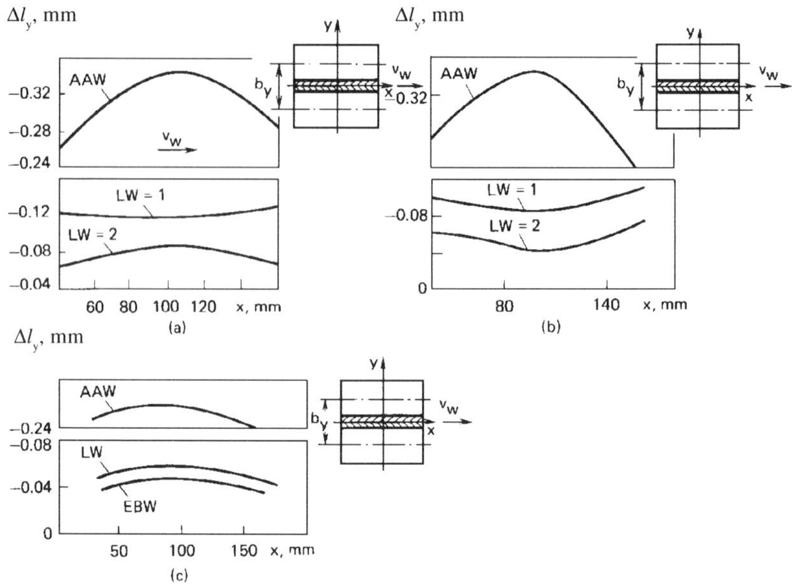

Fig. 5.10. Distribution of transverse shrinkage Δl_y over the weld lengths in alloy 1 (a), alloy 2 (b), and steel (c)

shrinkage over the length of the weld seam, which is particularly the case for the welds in titanium alloys. The value of shrinkage is the highest in the middle of a weldment.

The curves of Fig. 5.10*a* and *b* show lateral strains typical of argon-arc welding and laser welding performed at the same net power. As is seen from Fig. 5.10*a*, the transverse shrinkage in laser welding is by about a factor of 3 less than it is with argon-arc welding. The per-unit-length energy for the LW-1 mode is a factor of 1.8 lower than is the case with argon-arc welding. The shrinkage values presented in Fig. 5.10*b* for welds between the plates from titanium alloy grade 2 vary in the same manner. The shrinkage and per-unit-length energy for the LW-1 mode are 4.0 to 4.5 and 1.7 times lower, respectively, than they are for argon-arc welding. With further increase in the laser welding speed and the corresponding decrease in the heat input, the shrinkage in the butt-welded joints between the 2 mm-thick plates from titanium alloy grade 1 and between 3.5 mm-thick plates from titanium alloy grade 2 decreases by a factor of 4 or 5 and 6 or 7, respectively, as against that observed in argon-arc welding.

The same regularities also hold for welding of low-carbon steel plates (Fig. 5.10*c*). Since the electron beam features a high power density, the shrinkage values typical of electron-beam welding and laser welding are close in magnitude. The experimental values of transverse shrinkage obey the formula derived for the calculation of lateral strains in the plates butt-welded in a single pass [88]:

$$\Delta l_y = K \frac{2\alpha}{c\rho} \frac{P}{v_w \delta} . \tag{5.19}$$

Here, K is the ratio between the experimental value of Δl_y and theoretical value of $\Delta l_{y\,max} = 2\alpha P/c\rho v_w \delta$, where α is the coefficient of linear expansion; P is the net heat power; and δ is the plate thickness.

Equation (5.19) is valid for both the experimental data presented in Fig. 5.10 and the following values of thermal constants: $\alpha = 9.2 \times 10^{-6}$ K^{-1} and $c\rho = 2.7 \times 10^6$ J/m^{3}K for titanium alloys and $\alpha = 11.7 \times 10^{-6}$ K^{-1} and $c\rho = 5.1 \times 10^6$ J/m^3 K for low-carbon steel. The values of other quantities appearing in equation (5.11) are given in Table 5.3.

As is evident from Table 5.3, the factor K for argon-arc welding varies within a rather wide range. It can be taken equal to 0.7 or 0.8 for estimating the average values of transverse shrinkage from formula (5.19). This factor for laser welding is more stable due to a more uniform distribution of lateral

Table 5.3. Welding Parameters and Values of Transverse Shrinkage

Material	Welding	Welding speed, mm/s	Net power, kW	Shrinkage, mm	Factor K	K_{min}/K_{max}
Titanium alloy	AAW	5.7	0.86	0.34	0.660	0.55/0.70
grade 1,	LW–1	10.8	0.90	0.12	0.432	0.42/0.45
plate 2.0 mm thick	LW–2	44.5	1.89	0.06	0.415	0.38/0.44
Titanium alloy	AAW	7.8	1.82	0.36	0.792	0.66/0.84
grade 2,	LW–1	13.3	1.84	0.10	0.371	0.33/0.41
plate 3.5 mm thick	LW–2	44.5	3.23	0.06	0.424	0.35/0.43
Low-carbon steel,	AAW	5.6	1.29	0.28	0.800	0.72/0.85
plate 3.0 mm thick	EBW	33	1.75	0.06	0.420	0.38/0.47
	LW	33	2.73	0.10	0.370	0.33/0.42

strains and can be taken equal to 0.40–0.42 to calculate the transverse shrinkage, given the values of α, $c\rho$, and welding parameters.

The analysis of experimental data reveals that the transverse shrinkage in laser welding is approximately half that in argon-arc welding at the same heat input. Considering that the power density of the laser beam is very high and the heat input required for full penetration is less by a factor of 2 to 3.5, the shrinkage would be one fifth to one seventh as large.

As illustrated in Fig. 5.8, the weld cycles lead to compressive longitudinal plastic strains and, hence, to longitudinal residual stresses. These stresses in various weld joints considerably exceed other stress components and are prevalent in the weld joints between plates up to 10 mm thick.

Residual stresses have a considerable effect on the dimensional accuracy and serviceability of weldments [88]. Figure 5.11 shows the profiles of longitudinal residual stresses σ_{xr} estimated by the mechanical methods in the weld joints between titanium alloy plates 2.0 mm and 3.5 mm thick. The parameters of laser welding and argon-arc welding used in the experiments are given in Table 5.3. As is seen from Fig. 5.11, the profiles of stresses in laser-welded and arc-welded joints are quite similar. The tensile stresses prevail in the weld fusion zone and in the heat affected zone, which reach a maximum value along the weld axis. The compression stresses σ_x exist in the base metal. The maximum tensile stresses σ_{xr} in laser-welded joints and arc-welded joints are approximately the same and amount to 80% or 90% of the conventional yield strength $\sigma_{0.2}$.

The region width of plastic strains and tensile stresses in laser welding is much smaller than it is in argon-arc welding, for which reason the maximum compression stresses in laser-welded joints are 40 to 70% lower than those in arc-welded joints.

High compression stresses developed in arc welding result in almost nonremovable strains due to sheet buckling. On the contrary, laser-welded plate

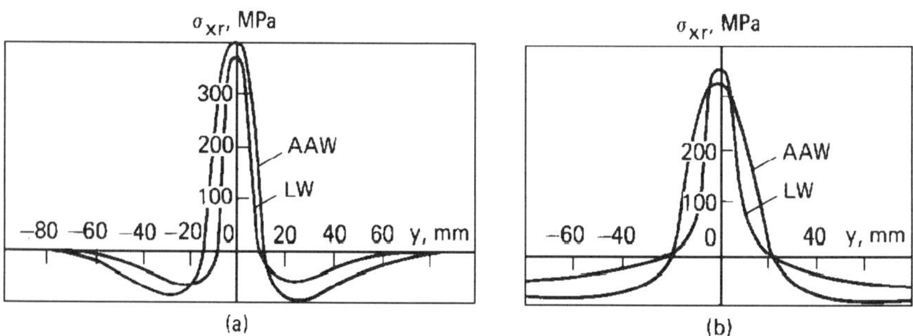

Fig. 5.11. Distribution of longitudinal residual stresses σ_{xr} in welds formed between thin plates made from alloy 1 (a) and alloy 2 (b)

joints formed at a high welding speed and low heat input are practically free of bending strains due to buckling.

Numerous experimental studies carried out at the Bauman Moscow State Technical University for estimating the residual strains and stresses in welds produced in a variety of materials confirm that the residual strains in laser-welded joints are at least by a factor of 5 smaller than they are in arc-welded joints. Laser welding is a precision process that ensures a high dimensional accuracy of weldments.

In laser surface hardening the distribution of temperature is uniform both in the bulk and on the surface of a workpiece, which results in a high gradient of residual stresses. The magnitude and sign of residual stresses have a strong effect on the wear resistance, corrosion resistance, and fatigue strength of laser-hardened parts.

The *X*-ray method can give quite reliable results in determining the residual stresses at any point on the target surface. The text below describes the results of *X*-ray tests for residual stresses on the laser-hardened surfaces of steel specimens. Surface hardening was made with a continuous CO_2 laser operating at 0.5 to 3.5 kW. A diffractometer was used to take *X*-ray photographs of specimens. The stresses were defined by the $\sin^2 \psi$ method suggested in work [93] and the experimental data were processed on a computer.

Figure 5.12 illustrates a test specimen with the regions of measurement of residual stresses σ_{yr} and σ_{xr} at different points across the laser-heat-treated track. The experimental values of σ_{yr} on the surfaces of armco-iron specimens are shown in Fig. 5.13. The profiles of stresses are nearly symmetrical with respect to the middle of the heat treated track. The distribution of stresses along the *y*-axis is nonuniform. The curves of Fig. 5.13*a* and *b* represent the stresses across the tracks heated without melting with laser powers of 0.9 and 1.3 kW, respectively. In the middle of such a track along the *x*-axis, or at $y = 0$, there are insignificant tensile stresses σ_{yr} of about 70 MPa. At the boundary of the heat affected zone, the tensile stresses increase to 270 MPa. As the distance from the middle of the track increases, the tensile stresses in the unirradiated areas gradually drop to zero and compression stresses arise. In laser heating

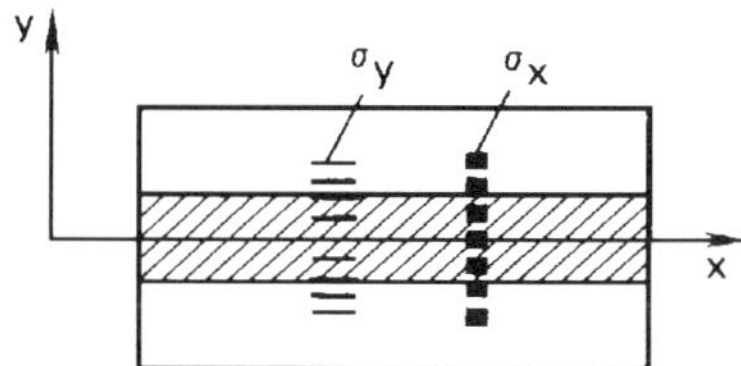

Fig. 5.12. Regions of measurement of residual stresses in a specimen by the *X*-ray method

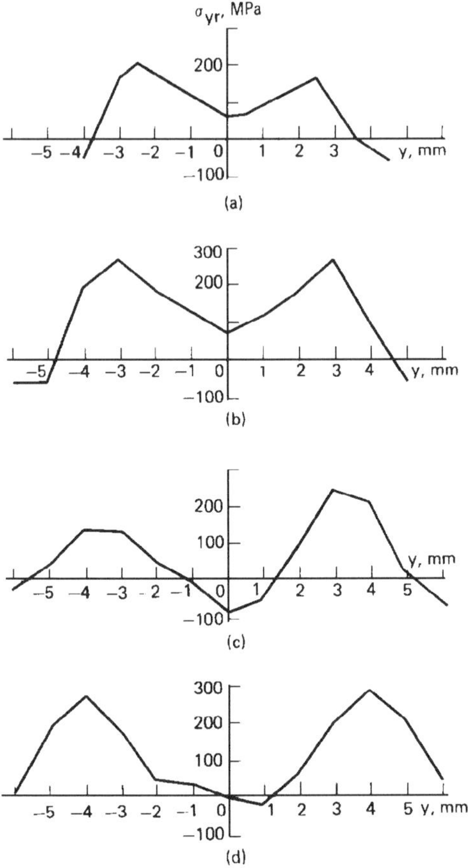

Fig.5.13. Residual stresses on the surface of an armco-iron specimen versus distance from the middle of the track hardened at a rate of 33 mm/s with a laser power of 0.9 kW (a), 2.0 kW (b), and 3.0 kW (c)

with melting, stresses in the middle of the track become compressive and reach 90 MPa at a radiation power of 2 kW (Fig. 5.13*c*) and vanish at a power of 3 kW (Fig. 5.13*d*). As laser power increases the range of tensile stresses grows together with the heat treated track.

As noted in Sec. 5.1, the mechanism of development of residual tensile stresses in the zones of laser-material interaction depends on residual plastic shrinkage strains. However, the structural and phase transformations occurring in the heat affected zone at the stage of cooling can drastically change this regularity. For example, as austenite changes into martensite, the volume of cooled metal in the laser-treated region noticeably increases and gives rise to appreciable compression stresses. This fact needs to be taken into account in the analysis of residual stresses in laser-hardened steels.

In surface hardening of carbon structural steel and chromium steel with a continuous 500-W CO_2 laser at a rate of 33 mm/s, the laser-hardened track does not exhibit any increase in hardness since the target surface does not reach the phase transformation temperature. Due to nonuniform heating and cooling plastic compression strains induce residual tensile stresses σ_{yr} on the surface of the heat treat track that range from 250 to 430 MPa. With an increase in the laser power, the surface reaches its melting temperature and the martensite transformation occurring at the cooling stage causes an increase in the metal volume, thereby initiating residual compression stresses. In the middle of the laser-treated track on a specimen either from carbon structural steel or from chromium steel, the compression stresses σ_{rr} range from –100 to –260 MPa (Fig. 5.14, curves *3* and *4*). Maximum compression stresses appear at a laser power of 1 kW because this power is only sufficient to melt a small volume of metal, which cools down quite rapidly. In surface hardening of carbon tool steel, the pattern of changes in σ_{yr} is similar (curve *1*), but the values of stresses are lower.

A higher laser power delivered to the target surface considerably changes thermal processes, for which reason the residual stresses σ_{yr} in the middle of the track and across the track change too.

Figure 5.15 gives the distribution of residual stresses on the surface of a carbon tool steel specimen heat-treated with a laser beam of 1 kW (a), 2 kW

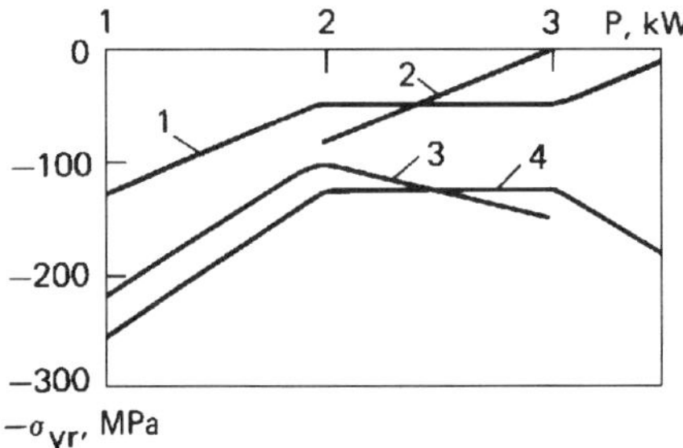

Fig. 5.14. Plots of residual stresses in the middle of the track hardened at a rate of 33 mm/s versus laser power for carbon tool steel, armco-iron, chromium steel, and carbon structural steel (curves *1*, *2*, *3*, and *4*, respectively)

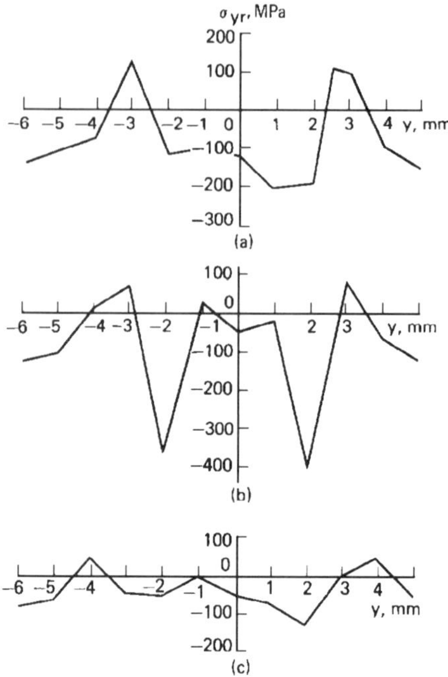

Fig.5.15. Residual stresses σ_{yr} on the surface of a carbon tool steel plate versus distance from the middle of the band hardened at a rate of 33 mm/s with a laser power of 1 kW (a), 2 kW (b), and 3 kW

(b), and 3 kW (c). At a power of 1 kW, the distribution pattern is similar to that illustrated in Fig. 5.14 by line *2* for armco-iron. At a power of 2 kW, the curve is different. The stresses are insignificant in the middle of the track, which is possibly due to an increased quantity of retained austenite. At the edges of the hardened track the compression stresses reach the values of –370 to –410 MPa, which drop to zero with distance from the edges. At the boundary of the unirradiated surface, there appear tensile stresses. At a laser power of 3 kW, the hardened track increases in width and the residual compression stresses across it decrease.

The rate of heat treatment has an appreciable effect on the distribution of residual stresses. At a rate of 8 mm/s, tensile stresses of 100 to 140 MPa appear in the middle of the track and high compression stresses of –200 to –440 MPa arise at the edges of the track in carbon structural steel and chromium

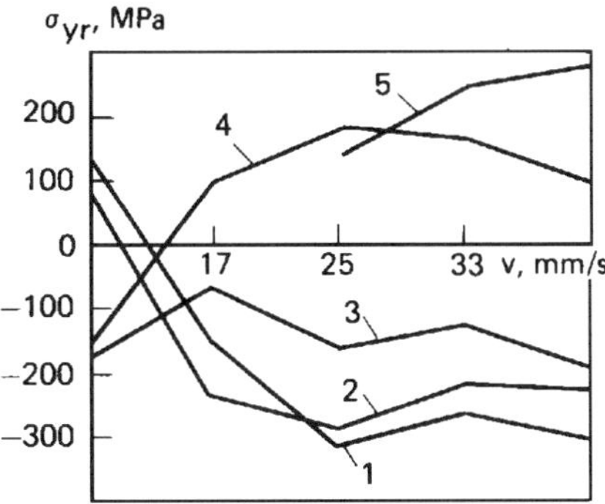

Fig. 5.16. Plots of residual stresses in the middle of the band hardened with a laser power of 1 kW versus rate of hardening. Curves *1*,*2*, *3*, *4* and *5* relate to carbon structural steel, chromium steel, carbon tool steel, chromium-tungsten-manganese steel, and boron steel, respectively

steel (Fig. 5.16, curves *1* and *2*). A further increase in the rate of heat treatment of the above steel grades gives rise to tensile stresses in the middle of the track, and the distribution of stresses across the track takes the shape of the curve of Fig. 5.15*a*.

The stresses σ_{yr} in the middle of the hardened band of a carbon tool steel specimen vary insignificantly with an increase in the rate of surface hardening (Fig. 5.16, curve *3*). For chromium-tungsten-manganese steel, the hardened track exhibits compression stresses at a low rate of heat treatment and tensile stresses at a higher rate (Fig. 5.16, curve *4*).

Heating with melting of boron steel at a rate of 25 mm/s induces tensile stresses in the track, which reach 140 MPa in the middle, gradually decrease to 20 MPa at a certain distance from the middle, and then begin to grow and reach 230 MPa at the boundary of the heat affected zone (Fig. 5.17*a*). At v = 33 mm/s, the pattern of changes in σ_{yr} is similar (Fig. 5.17*b*) and at v = 41 mm/s the distribution of stresses across the track gets smoother (Fig. 5.17*c*).

The results presented above attest to a highly nonuniform distribution of residual stresses on the surfaces of steel specimens hardened with a continuous CO_2 laser beam. The residual stress distribution varies with the grade of steel and conditions of laser heating, as is evident from the curves of Figs. 5.13 through 5.17, but gives a definite pattern of variations. The value and sign of residual stresses depend on residual plastic strains caused by the local heating

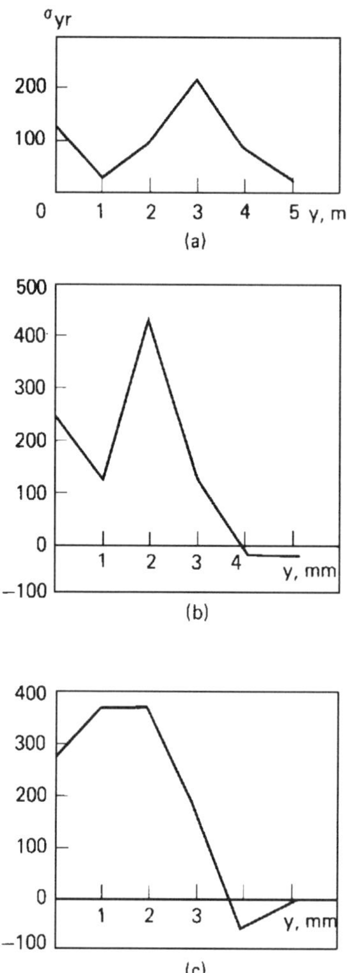

Fig. 5.17. Residual stresses on the surface of a boron steel plate versus distance from the middle of the band hardened with a laser power of 1 kW at a rate of 25 mm/s (a), 33 mm/s (b), and 41 mm/s (c)

and volume effects of structural transformations. When at the stage of cooling a sufficient amount of saturated martensite forms in carbon tool steel, chromium steel, and carbon structural steel, compression stresses appear in the middle of the hardened track. A decrease in the amount of saturated martensite weakens the volume effect and so the compression stresses decrease. A laser-treated band on the surface of chromium-tungsten-manganese steel or boron steel has a large amount of retained austenite. Since the volume effect of martensite transformation in such a structure proves to be insignificant, mainly tensile stresses arise in the hardened zone.

A knowledge of the mechanisms of residual stress development in laser surface hardening offers a means for controlling the stresses in the hardened regions for the purpose of improving the performance characteristics of laser-hardened parts.

6 Strength of Laser-Treated Metals

By the strength of laser-treated metals is meant the ability of metals to resist high-temperature effects without developing hot cracks in the process of laser heating or cold cracks at the stage of phase and structural transformations in the solid state.

6.1. Hot Cracks

In laser heating with melting, the deposited or weld metal may develop intercrystalline brittle fractures or hot cracks during its crystallization. As discussed above, a clad or weld metal undergoes heavy elastoplastic strains during its cooling. If high-temperature strains developed in the metal during its solidification are in excess of its deformability, the metal develops hot cracks. In the analysis of mechanisms conducive to the development of hot cracks, it is useful to consider the character of changes in high-temperature strains, or the rate of strains, $n_{st} = \partial\varepsilon/\partial T$, at the stage of metal solidification.

The analysis of theoretical and experimental results reveals that the rate of strains in the high-temperature range depends on thermal properties of a metal, conditions of laser processing, and rigidity of workpieces. In any process of laser heating with melting, a crystallizing metal undergoes considerable elastoplastic tensile strains responsible for the initiation of hot cracks.

The strain stability of a metal at high temperatures close to the solidus temperature T_{sol} varies with the values of instantaneous (ultimate) strength and plasticity (Fig. 6.1). The ultimate strength curves of Fig. 6.1 show two characteristic sections *AB* and *CA*, the latter section being indicative of the ductile character of rupture, i.e., dependent on the strength of intercrystalline bonds.

In accordance with the iron-carbon diagram of steels and most of the alloys, the crystallization process occurs within the liquidus-solidus temperature range (Fig. 6.2). At temperatures above the liquidus temperature T_{liq}, an alloy remains in the liquid state, but solidifies completely at temperatures below T_{sol}. In the interval between T_{liq} and T_{sol}, the alloy remains in a two-phase liquid-solid state.

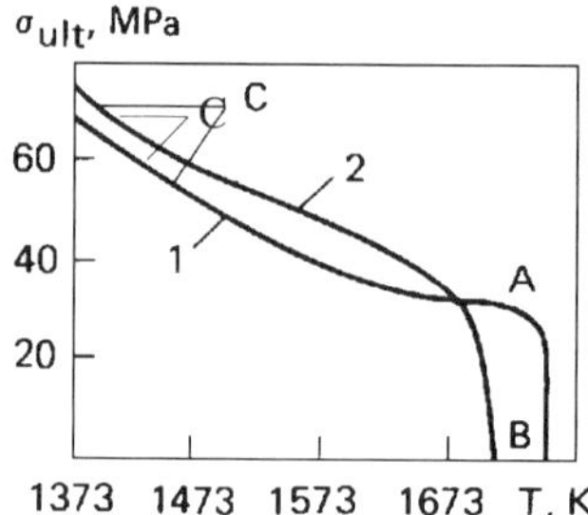

Fig. 6.1. Ultimate strength of manganese steel (curve *1*) and Cr-Si-Mn steel (curve *2*) versus temperature

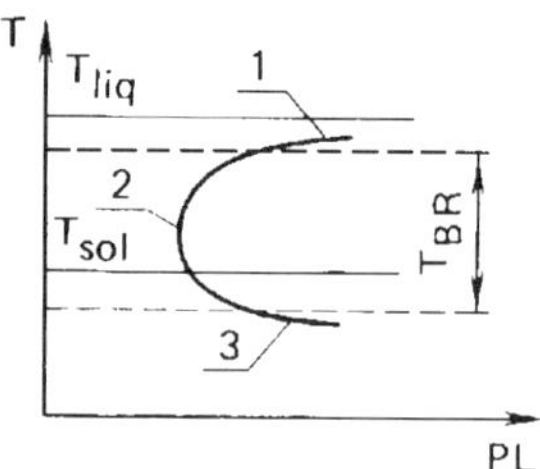

Fig. 6.2. Plasticity versus temperature

The crystallization in the liquid-solid phase entails specific changes in mechanical properties of the alloy. At the start of crystallization the volume of the liquid phase is large, and liquid interlayers separate individual crystallites. In this state, the liquid-metal system exhibits a high plasticity determined by the properties of the liquid (section *1* in Fig. 6.2).

As the temperature goes on decreasing, the volume of the solid phase grows and that of the liquid phase decreases. If the crystals form a skeleton that inhibits the liquid flow, the strains resulting from laser heat treatment may lead to brittle intercrystalline rupture because the plasticity of the two-phase system drops quite heavily (section *2* in Fig. 6.2).

With further temperature decrease, the crystals form a solid skeleton noted for an increased crystal plasticity (section *3* in Fig. 6.2). At this stage of cooling, the intercrystalline fracture gives way to intragranular plastic fracture. In the temperature crystallization interval we can thus single out a brittle range (BR) of temperatures T_{BR} within which the strength and plasticity drop considerably. Fractures tend to develop just in this range of temperatures.

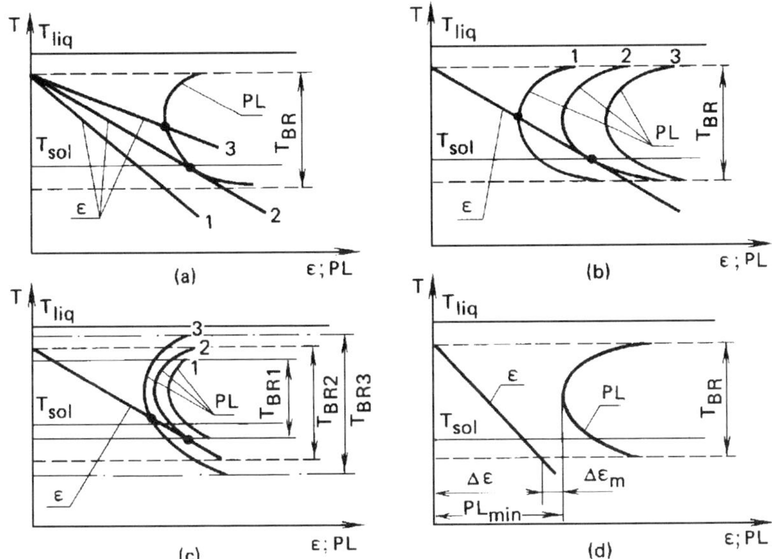

Fig. 6.3. The mechanism of hot crack formation with changes in the strain rate (a), plasticity (b), and brittle range (c), and graphs for estimating strain margin (d)

To gain insight into the mechanism of hot crack formation in welding and cladding, consider the diagrams of Fig. 6.3, which illustrate changes in metal plasticity l_p with solidification temperature and also show elastoplastic strains ε that appear below the upper limit of the brittle range [94]. To a first approximation, cooling strains increase directly with decreasing temperature, and so they change in a linear fashion.

A number of factors are responsible for the formation of hot cracks. The graphs of Fig. 6.3*a* illustrate how the rate of strains affects the metal properties. The straight line *1* does not intersect the plasticity curve l_p, which suggests that the metal is free of hot cracks since its strain capacity is still sufficient to withstand cooling strains. At higher strains the straight line *3* intersects the plasticity curve. In this case, the metal develops hot cracks because the strains prove to be beyond its plastic range. The straight line *2* tangent to the plasticity curve represents the critical rate of strains.

The graphs of Fig. 6.3*b* illustrate the effect of metal plasticity on the formation of hot cracks. An increase in plasticity leads to the formation of cracks (curve *1*). As the plasticity grows, the probability of crack formation decreases (curve *3*). The curve *2* is the critical case for the plastic range, indicative of the absence or presence of hot cracks at high or low values of plasticity, respectively.

As follows from the graphs of Fig. 6.3*c*, other conditions being equal, the probability of crack formation grows with an increase in the brittle range from T_{BR1} to T_{BR2}.

So the decisive factors in the formation of hot cracks are the rate n_s of elastoplastic strains, brittle range T_{BR}, and minimum plasticity $l_{p\,min}$ within this range. The probability of hot crack formation grows with an increase in n_{st} and T_{BR} and a decrease in $l_{p\,min}$ [95]. From the graphs of Fig. 6.3*d* we can estimate the strength margin of a metal by comparing $l_{p\,min}$ with the strain $\Delta\varepsilon$ in the brittle range. The deformability margin is

$$\Delta\varepsilon_m = l_{p\,min} - \Delta\varepsilon. \tag{6.1}$$

Dividing both sides of equation (6.1) by T_{BR} yields the strength margin

$$k = \frac{\Delta\varepsilon_m}{T_{BR}} - \frac{l_{p\,min}}{T_{BR}} - \frac{\Delta\varepsilon}{T_{BR}}. \tag{6.2}$$

Here, $l_{p\,min}/T_{BR} = n_{lim}$ is the limiting rate of strains, above which a metal develops hot cracks and $\Delta\varepsilon/T_{BR} = n_{st}$ is the rate of elastoplastic strains arising in the metal during its solidification. The strength margin can thus be written as

$$k = n_{lim} - n_{st}. \tag{6.3}$$

6.2. Test Methods for Estimating the Weld Resistance to Hot-Cracking

Since it is difficult to obtain experimental or theoretical values of n_{lim} and n_{st} for determining the strength margin k from equation (6.3), a common approach used to assess the strength of a metal in the process of its crystallization is to varify the test samples for defects and indirectly evaluate the metal strength. There are also various testing machines for estimating the resistance of metals to hot cracking.

The tests performed, for example, on welded test samples permit us to verify the tendency of alloys and filler metals to hot cracking and thus select suitable alloys and adequate conditions of welding.

Figure 6.4 illustrates a test sample with a circular full-penetration butt weld formed with a laser beam and used to estimate the metal strength in laser

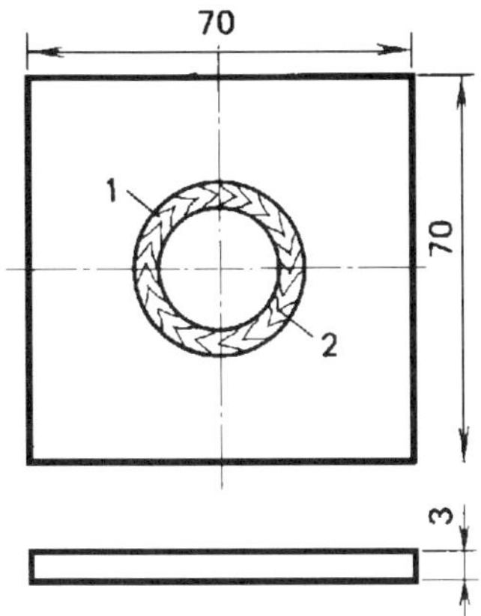

Fig. 6.4. Circular butt weld with hot cracks *1* and *2*

welding [95]. In the welding tests, circular welds of various diameters are made in the samples of different dimensions in order to check most accurately the behaviour of the weld metal and its resistance to hot cracking. After its cooling, the sample can be examined under a microscope for cracks on the weld bead surface. The total length of cracks serves as the criterion for estimating the metal strength in laser welding.

The test methods relying on testing machines can provide quantitative estimates of hot crack formation in metals. A widely used method developed at the Bauman Moscow State Technical University combines the process of welding with the concurrent stretching of a sample by a tester in which the speed of grippers determines the stretch rate of the crystallizing weld [94, 95].

The test procedure usually consists in welding together a set of samples under specified conditions and simultaneously subjecting each sample to strains, increasing the strain rate for each successive sample until it reaches a critical value above which the sample develops hot cracks. The critical rate of strains, v_{cr}, is taken as the index of strength suitable for the quantitative comparison of various alloys as regards their susceptibility to hot cracks during an invariable thermal cycle of welding.

To compare various welding methods that differ in their thermal effect exerted on metals, we can use a more universal criterion of the strain rate given as the ratio of the critical strain rate c_{cr} to the rate v_{BR} of cooling the weld metal in the brittle range:

$$n_{cr} = v_{cr}/v_{BR}. \tag{6.4}$$

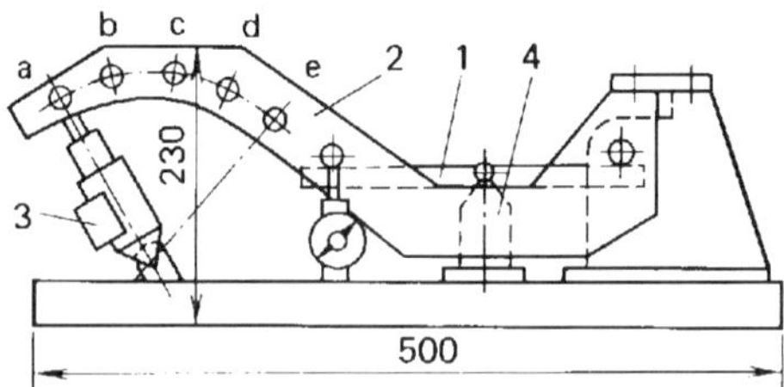

Fig. 6.5. Portable test unit

This strain rate, mm/K, is a function of weld metal plasticity in the brittle range and is independent of testing conditions.

There are a number of testers [94, 95] designed to estimate the critical rate n_{cr}, although they are only suitable for use in arc welding. A portable tester developed at the Bauman Moscow State Technical University for laser welding applications appears in Fig. 6.5. This test unit measuring 500 × 230 × 200 mm can be set on a test bed and moved relative to the stationary laser beam at a specified speed to weld the two portions of a specimen together. Specimen *1* to be bent rests on knife-edge support *4*. Lever *2* hinged to the frame and linked to the rod of drive *3* forces the specimen to bend as the rod moves downward. The tester can provide the desired force and rate of bending by joining the level to the rod at any of the points *a* through *e* and changing the actuator speed.

The test unit bends the specimen during its welding until the laser beam forms a weld 15 to 20 mm long. To determine the critical rate v_{cr} of strains, the unit must test 15 to 20 specimens under the same welding conditions, but at successively increasing rates of bending until a hot crack appears. The metal pool temperature measured with a thermocouple secured in a specimen enables us to estimate the cooling rate v_{BR} and, hence, the critical strain rate n_{cr}.

In developing laser treatment processes, it is of importance to estimate, apart from the integral criterion in terms of the critical rate n_{cr}, the plastic range and the brittle range, which determine the strain capacity of the crystallizing metal. The plasticity of a metal during its solidification within the brittle range is a measure of plastic strain determined in the tests, which does not cause the intracrystalline fracture in the weld subjected to tension at high constant rate. A series of tensile tests performed on welded specimens under similar conditions and at various temperatures can give estimates of the brittle range.

Since the rate of cooling of a laser weld in the solidification range reaches a value of 2000 K/s and higher, the tests required to determine the deformability of a metal in the process of laser welding should be made at a much higher rate of strains than is the case with the test involved in arc welding.

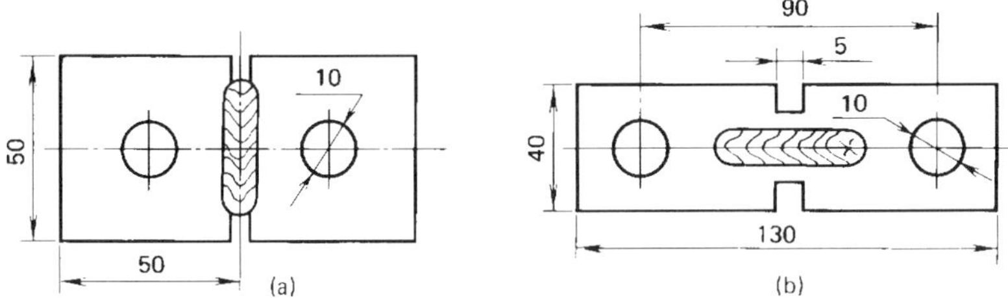

Fig. 6.6. Samples tested for longitudinal cracks (a) and transverse cracks (b) to estimate the plasticity of a crystallizing weld metal in the brittle range

Work [96] describes a test method for estimating the plasticity of a weld metal in the brittle range at a high strain rate during laser welding. A specimen to be tested for longitudinal cracks consists of two plates prepared for butt welding (Fig. 6.6*a*). In the process of welding, one of the plates is forced to move fast in the direction normal to the weld seam in order to load the entire length of the weld. In the tests for cross cracks, the specimen undergoes strains in the direction of the weld seam (Fig. 6.6*b*).

The tests are made on a set of specimens at a sequentially increased load. The value of strain above which the crack is likely to appear determines the plasticity of the weld metal in the brittle range.

The brittle range is

$$T_{BR} = v_{BR} t_{BR}, \tag{6.5}$$

where t_{BR} is the dwell time of a metal in the brittle range. This time is equal to

$$t_{BR} = l_{BR} / v_w, \tag{6.6}$$

where l_{BR} is the length of the weld portion lying within the brittle range.

Substituting equation (6.6) into (6.5) yields

$$T_{BR} = l_{BR} v_{BR} / v_w. \tag{6.7}$$

The length l_{BR} of the weld portion is the sum of the crack length l_c and the length l_{dis} of displacement of the weld portion during testing:

$$l_{BR} = l_c + l_{dis}. \tag{6.8}$$

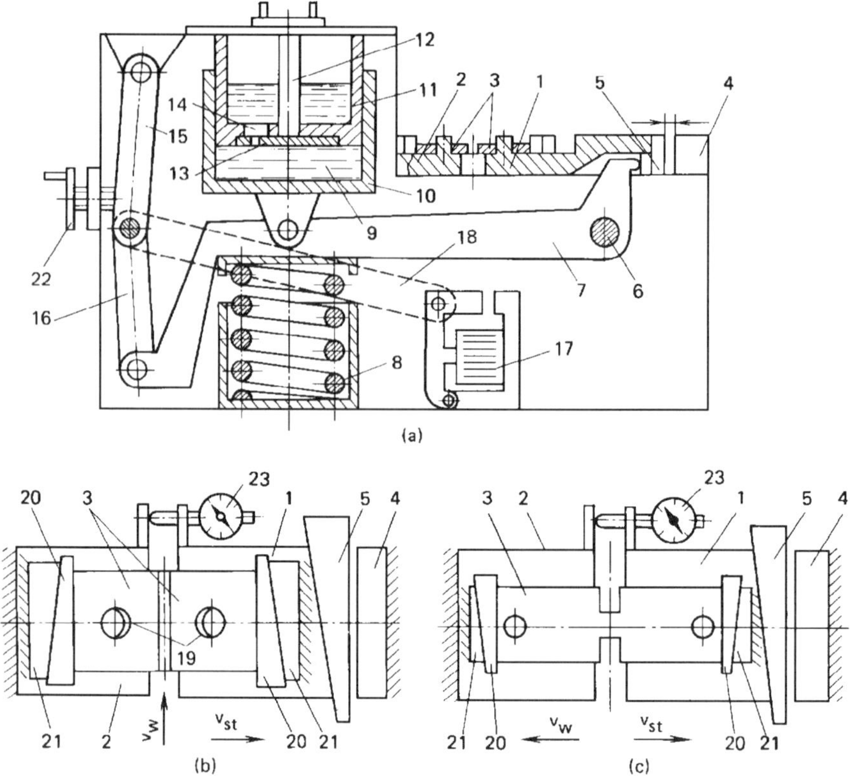

Fig. 6.7. Test unit (a) for estimating the plasticity of a crystallizing metal during welding and its fixtures (b) and (c) intended to hold samples tested for longitudinal and transverse cracks, respectively

The displacement length is

$$l_{dis} = v_w t_{st} = v_w l_p / v_{st}, \tag{6.9}$$

where t_{st} is the hold time of a test specimen in the state of strain; l_p is the plastic strain; and v_{st} is the rate of strains.

In the tests conducted at a high strain rate v_{st}, the value of l_{dis} is very small and can be neglected.

In the test procedure described above, the strain rate v_{st} should not exceed the limiting values above which the process of solidification may deviate from the general regularities and entail a decrease in the weld plasticity. Numerous tests reveal that a strain rate of 10 mm/s ensures a minimum change in the temperature of the weld during its deformation, which does not generally affect the weld plasticity.

Figure 6.7 illustrates a 550 × 200 × 450-mm weld-test unit of the spring-linkage type, which provides high tensile forces required for testing by the method described above [96]. The unit is complete with moving table *1* and stationary table *2* intended to carry specimen *3* consisting of two plates to be welded together. The specimen is fixed in position with pins *19*, wedges *20*, and stops *21*. With wedge *5* properly located to secure gap *l* equal to the expected elongation of the specimen, handle *22* is turned to move levers *15* and *16* to home position. At this stage, indicator *23* intended to register the motion of table *1* is set to zero.

At a prescribed moment of welding, electromagnet *17* is energized to actuate link *18* which abruptly shifts levers *15* and *16* from the dead point. Released lever *7* driven by spring *8* now turns on its pivot *6* and its shorter arm shifts table *1* to stationary stop *4* for the preset length *l*. The test unit controls the table speed and, hence, the strain rate with a hydraulic drive. Revolving disk *13* overlaps a portion of opening *14* in the bottom of holow piston *11* with rod *12*, thereby adjusting the flow rate of fluid *9* forced from casing *10* into the piston cavity.

In Sec. 6.3 we present the results of tests performed on specimens using the above test unit, which reveal the plastic properties of weld metals in laser welding (LW), argon-arc welding (AAW), and electron-beam welding (EBW).

6.3. Hot Crack Formation

The tests for hot crack formation in laser, electron-beam, and argon-arc welds were made on 3- to 5-mm thick specimens from carbon and alloyed steels widely used in industry. In selecting welding parameters it was essential that the welding set-up should provide a weld with full penetration in a single pass without applying a filler metal. The laser welding set-up focused beams of 3 to 4.9 kW-power with a 200-mm KCl lens and welded parts in the atmosphere of helium, argon, and carbon dioxide [22]. In argon-arc welding, the arc voltage A_{arc} and welding current varied from 9 to 16 V and 160 to 290 A, respectively. Electron-beam welds were made at a voltage of 52 kV and a current of 0.03 to 0.065 A. The critical strain rate n_{cr} estimated from equation (6.4) served as the measure of metal susceptibility to hot cracking.

Figure 6.8*a* illustrates the critical rates v_{st} of bending strains of specimens tested on the unit shown in Fig. 6.5 in the processes of argon tungsten-arc and laser welding. Since laser welding differs sharply in thermal cycles from arc welding, a correlation between the values of v_{st} can hardly reveal a true character of hot crack formation. For this, it is necessary to obtain the values of cooling rate v_{BR} of the weld metal within the brittle range and

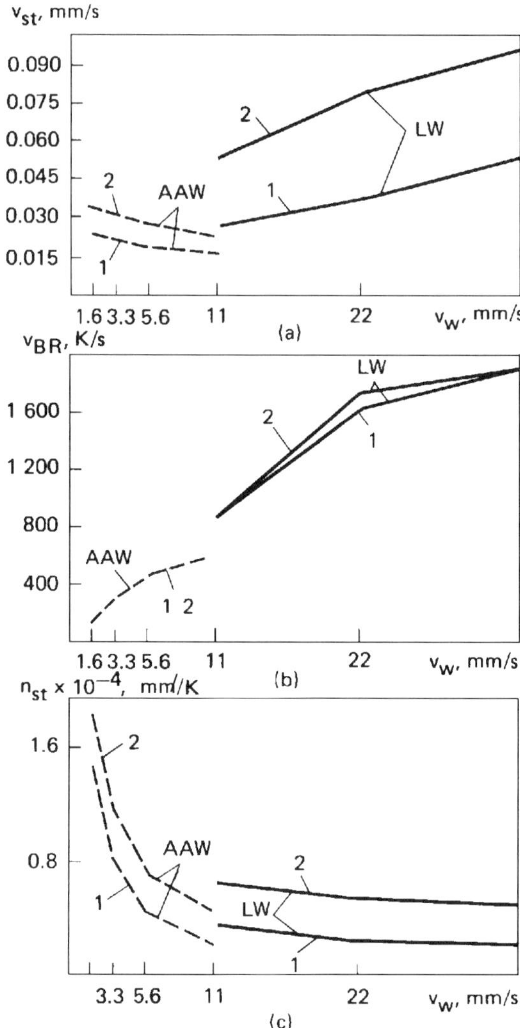

Fig. 6.8. Plots of critical rate v_{st} of bending strains in weld samples (a), cooling rate v_{BR} of weld metal in the brittle range (b), and critical strain rate n_{st} (c) as functions of speed of welding of scale-resistant steel (curves *1*) and carbon structural steel (curves *2*)

subsequently estimate the strain rate n_{st} from equation (6.4). In Fig. 6.8*b* are shown the curves of v_{BR} in the temperature range from 1673 to 1473 K, which are measured with a platinum to platinum-rhodium thermocouple located in the heat affected zone. The cooling rate in laser welding is 1.5 to 3 times that in argon-arc welding at the same welding speed. As the speed of laser welding increases, the cooling rate v_{BR} grows appreciably and reaches 2000 K/s. Figure 6.8*c* gives the values of the critical strain rate n_{st} obtained from equation (6.4) using the experimental results illustrated in Fig. 6.8*a* and *b*.

The strain rate n_{st} reaches the highest value in arc welding at a welding speed of 1.66 mm/s, but sharply decreases as the speed rises up to 11 mm/s. The increase in speed above 11 mm/s leads to an unstable formation of the weld bead.

At the same welding speed the values of n_{st} in laser welding are higher than in argon-arc welding, and so the laser weld metal shows a higher resistance to hot cracking. A further increase in the speed of laser welding does not practically decrease n_{st}.

The tests performed on samples welded with full penetration in a single pass to form circular welds 25 mm in diameter produced similar results. The speed of laser welding and that of electron-beam welding ranged from 0.83 to 33.3 mm/s and that of argon-arc welding from 0.55 to 10.4 mm/s.

Figure 6.9 demonstrates the results of experiments on samples with circular welds produced at speeds which lead to hot cracking. As is evident from the graphs, the total length of cracks in the case of laser welding in helium or carbon dioxide is smaller than that in EBW and AAW. This testifies that the

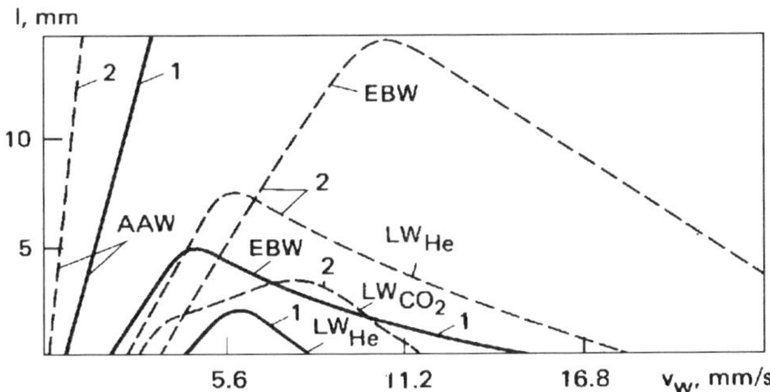

Fig. 6.9. Total length of cracks in the weld produced in scale-resistant steel (curves *1*) and carbon steel (curves *2*) as a function of welding speed

laser weld has a higher crack resistance than the electron-beam weld and argon-arc weld. The largest number of hot cracks appears at a welding speed of 4.1 to 8.2 mm/s in laser welding and 4.1 to 11 mm/s in electron-beam welding. As the welding speed increases, the cracks in laser welds gradually disappear, while the number of cracks in argon-arc welds continues growing. The laser welds formed at high speeds (above 22 mm/s) are totally free of hot cracks.

The metallographic analysis reveals that high crack resistance of metals is due to the fact that the parent metal acquires a finer structure with increasing laser and electron-beam welding speed. Also, the crack resistance varies with the content of sulfide inclusions in welds. The results of X-ray microanalysis indicate that the fusion zone of crystallites in an argon-arc weld exhibits an increased volume of inclusions whose X-ray radiation intensity grows in proportion to the sulfur content. In the central portion of the arc weld the number of inclusions is small. The central portion of a laser weld is practically free of sulfide inclusions and its fusion zone does not display any increase in the dimensions and number of inclusions.

An increased content of sulfur of the argon-arc weld may lead to the interdentritic and zonal segregation, which decreases the lower limit of the brittle region and thus renders the metal less resistant to hot cracking. The experiments carried out by the authors of [94, 95] confirm that the lower limit of the brittle region for scale-resistant steel welded with a laser beam and an electron beam is 15 percent higher than is the case with the same steel welded with an argon arc.

The welding tests performed on nickel alloys reveal a similar character of hot crack formation and clearly indicate that the crack resistance of the alloys in laser and electron-beam welding is much higher than it is in argon-arc welding. The resistance to hot cracking appreciably grows at increased laser welding speeds exceeding 36 mm/s, which leads to a high plasticity of weld metal since the metal structure becomes finer and exhibits a volume pattern of crystallization. Besides, the lower limit of the brittle region shifts to a high-temperature range because of a reduced chemical inhomogeneity of the molten metal.

To gain a deeper insight into the process of hot crack formation in laser welding and nonconsumable-electrode arc welding, a comparative analysis was made on test welds produced in 1.5- to 3.0-mm specimens from heat-resistant austenitic steels and alloys at speeds of 5.5 and 11 mm/s. The laser beam was focused by 140- to 160-mm focal length lenses and helium is delivered on the side of the laser to cover the bead and argon on the opposite side of the plate to the laser. In the process of its welding, a specimen was subjected to high-speed strains to determine the plasticity $l_{p\,min}$ of the crystallizing metal and the weld portion length l_{BR} within the brittle range. Table 6.1 lists the values of $l_{p\,min}$ and l_{BR}, given as numerators and denominators, respectively,

Table 6.1. Plastic Characteristics of Crystallizing Metals

Welding speed, mm/s	Type of welding	Plasticity $l_{p\ min}$, mm, and length l_{BR}, mm					
		Alloy grade 1, 1.5 mm thick	Steel grade 1, 3.0 mm thick	Alloy grade 2, 1.5 mm thick	Steel grade 2, 3.0 mm thick	Steel grade 3, 3.0 mm thick	Steel grade 4, 3.0 mm thick
5.5	AAW	0.05/1.60	0.12/2.00	0.04/2.00	0.06/2.20	0.05/2.80	0.11/1.80
	LW	0.06/0.80	0.12/2.00	0.04/1.00	0.06/2.20	0.05/2.80	0.11/1.80
11	AAW	0.04/1.80	0.09/2.20	0.03/2.50	0.05/2.70	0.03/3.40	0.08/2.00
	LW	0.07/0.90	0.11/0.60	0.05/1.60	0.07/0.90	0.06/1.10	0.12/0.60

for laser welds and argon-arc welds produced in heat-resistant nickel alloys (Cr-Ni-W-Mo alloy grades) and Cr-Ni steels.

From the comparison of the test results it follows that at a welding speed of 5.5 mm/s the plasticity of metals in the brittle range is practically the same. At a welding speed of 11 mm/s, laser welding affords a much higher plasticity of all the welded materials than argon-arc welding.

The length l_{BR} in laser welding is much shorter than in argon-arc welding, because the cooling rate v_{BR} of the laser weld is much higher than that of the arc weld (Fig. 6.10).

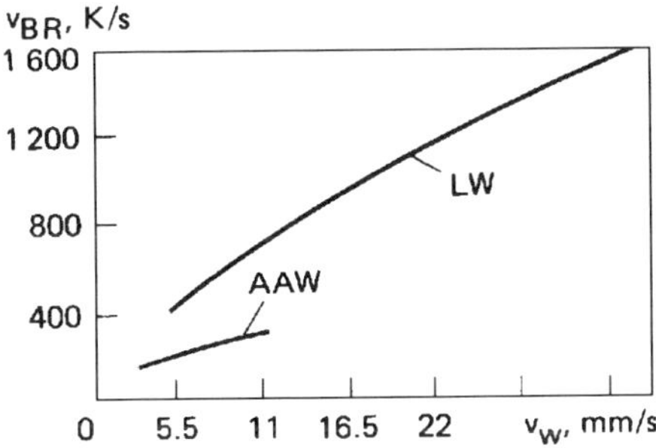

Fig. 6.10. Cooling rate versus welding speed

Table 6.2. Experimental Values of v_{BR} and T_{BR}

Welding speed, mm/s	Nickel alloy	Type of welding	v_{BR}, K/s	T_{BR}, K
5.5	Grade 1	AAW	210	61.1
		LW	420	60.6
	Grade 2	AAW	210	80.2
		LW	420	80.1
11	Grade 1	AAW	320	62.5
		LW	720	58.2
	Grade 2	AAW	320	81.8
		LW	720	78.5

The values of temperature in the brittle range are found from equation (6.7) using the experimental values of cooling rate v_{BR} (Table 6.2). As is evident from Table 6.2, the values of T_{BR} at a welding speed of 5.5 mm/s are almost the same for both types of welding, but tend to decrease at a speed of 11 mm/s in laser welding.

The experimental results of Table 6.2 allow us to estimate the critical strain rate n_{st} from equation (6.4), the values of which are given in Table 6.3.

The experimental results attest that the resistance of a weld metal to hot cracking in laser welding is higher than in argon-arc welding. With an increase in the welding speed, the advantages of laser welding become more evident. That this is indeed so is evident from the experiments carried out for estimating the plasticity $l_{p\,min}$ of the weld metal during its solidification, the brittle range,

Table 6.3. Experimental Values of n_{st}

Welding speed, mm/s	$n_{st} \times 10^3$, mm/K			
	Alloy grade 1		Alloy grade 2	
	AAW	LW	AAW	LW
5.5	0.818	0.982	0.436	0.498
11	0.642	1.270	0.366	0.637

and the critical strain rate n_{st} in laser welding of 1.5-mm chromium-nickel alloy, grade 1, and 2.0-mm steel, grade 4, at speeds ranging from 5.5 to 33 mm/s. The laser powers for welding the alloy and steel, given in the numerators and denominators respectively, are the following:

v_w, mm/s...	5.5	11	16.5	22	27.5	33
P, kW...	2.0/3.5	2.3/3.7	2.6/4.0	2.8/4.3	3.2/4.5	3.4/4.6

The experiments confirm that the plasticity of the crystallizing metal increases with welding speed (Fig. 6.11), as does the weld portion length l_{BR} (Fig. 6.12*a*), but the temperature T_{BR} slightly decreases. The values of the

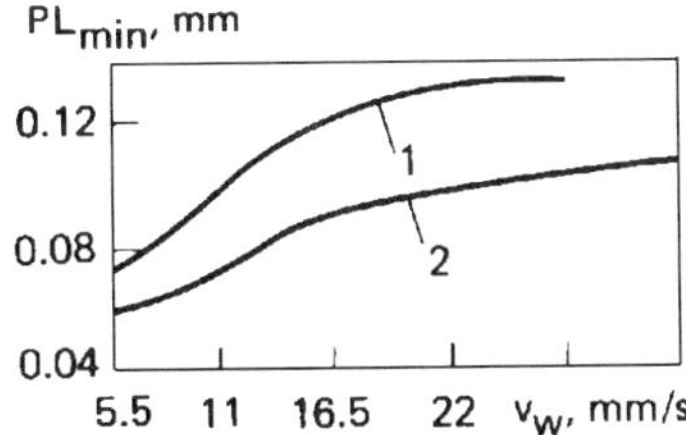

Fig. 6.11. Plasticity of Cr–Ni steel (curve *1*) and Cr–Ni–W–Mo alloy (curve *2*) versus welding speed

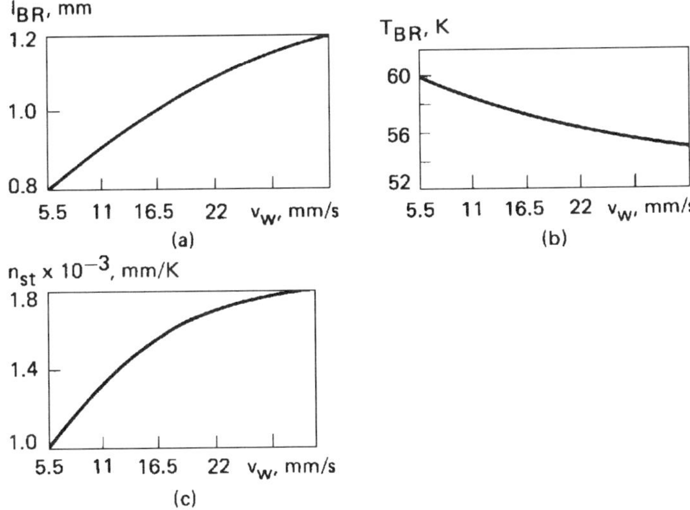

Fig. 6.12. Plots of l_{BR} (a), T_{BR} (b), and n_{st} (c) against the welding speed of the Cr–Ni–W–Mo alloy

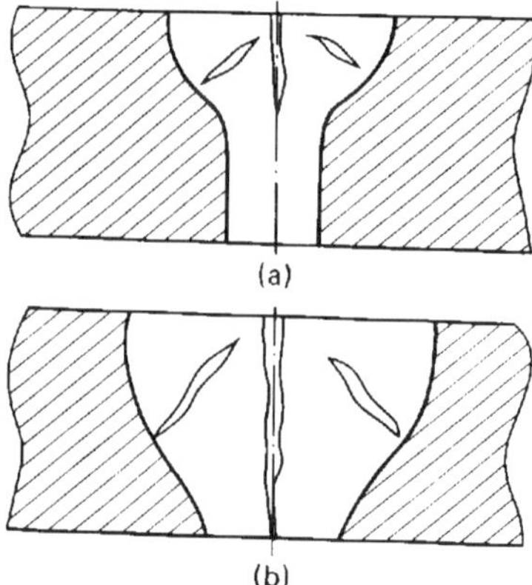

Fig. 6.13. Crack formation in a test sample during laser welding (a) and argon-arc welding (b)

critical strain rate n_{st} calculated from equation (6.4) grow with welding speed, which points to an increase in the crack resistance.

The hot plasticity has a decisive effect on weld resistance to hot cracking. What accounts for an increased plasticity of the crystallizing metal in laser welding as against the plasticity observed in arc welding is a higher degree of dispersion of the laser-welded metal structure. The laser weld has an acicular-dendritic structure that is similar to the structure of the arc weld, but the former exihibits a much higher dispersivity [97].

One more factor responsible for an increased plasticity of the laser weld is the specific character of weld metal crystallization. The high-speed strain tests performed on samples in the process of laser welding at speeds of 8 to 11 mm/s and higher reveal that the laser weld develops hot cracks only in its upper portion at a depth of not more than one-third of the penetration depth, whereas hot cracks in the arc weld propagate much deeper into the metal bulk and cause the weld rupture (Fig. 6.13). The upper and lower portions of the laser weld apparently crystallize in a different manner, for which reason they offer a different resistance to hot cracking in the brittle range.

6.4. Laser Weld Formation

The kinetics of changes in the liquid-solid interface can provide a vivid picture of weld formation. The analytical methods [94] of the analysis of weld solidification presuppose the similarity of crystallization front profiles

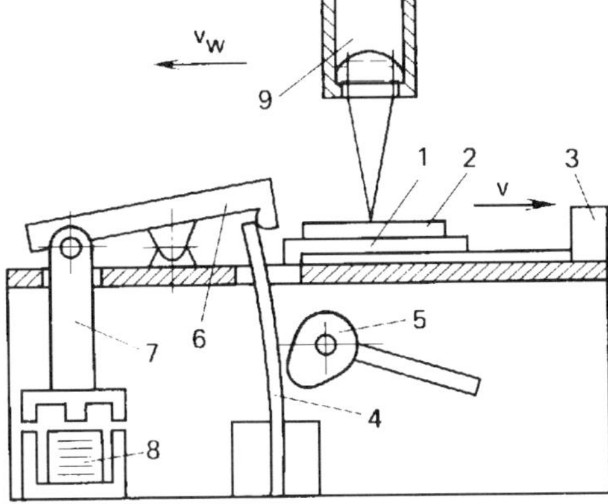

Fig. 6.14. Test unit for removing liquid metal from the metal pool during welding: (1) movable table; (2) sample; (3) stop; (4) flat spring; (5) cam with handle for spring loading; (6) catch; (7) link; (8) electromagnet; (9) laser beam

along the pool length, which is quite an appropriate condition for arc welding. In laser welding, however, the high energy density causes the phase boundary to assume a complex profile, for which reason an experimental approach must be taken to estimate the weld interface configuration.

An experimental method described in [96] involves the melt removal from the metal pool to uncover its lower surface by imparting a high acceleration (Fig. 6.14) to the sample under test. As the quasisteady temperature state of welding sets in, electromagnet *8* energizes and pulls down link *7* to raise up catch *6* which then releases spring *4*. The spring strikes table *1* with sample *2* fastened thereon, which is "forced out" from under the laser beam in the direction of the weld seam, so that the molten metal flushes out from the metal pool in the direction opposite to the table motion and reveals the phase boundary.

Each sample is then ground one 0.5-mm layer at a time across the metal seam to verify the profiles of the sections through the weld. Mapping the metal profiles in one plane gives the topograms of the phase boundary in the metal pool. The text below presents the results of experimental studies of the metal pool profiles in 4-mm thick nickel alloy, grade 1, and austenitic chromium-nickel steel, grade 4, welded with laser and argon arcs.

In Fig. 6.15 are shown the phase boundary topograms for crystallization fronts of welds produced with laser and argon arcs. The two types of weld quite differ in the shape of the metal pool. In arc welding, the profiles of

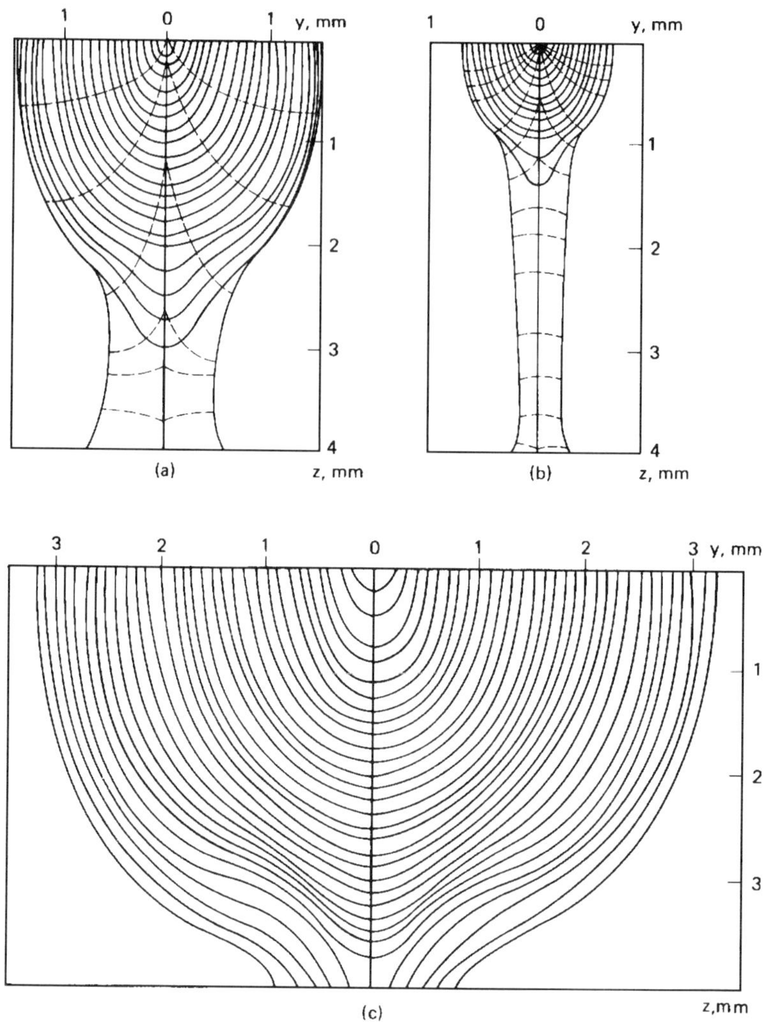

Fig. 6.15. Topograms of the crystallization fronts in laser welds (a) and (b) and arc welds (c) produced in austenitic Cr–Ni steel at a welding speed of 112, 33, and 11 mm/s, respectively

sections are fairly smooth. The longitudinal section of the laser-melted pool exhibits a solid-phase step which separates the deeper and narrower front portions of the pool from the shallower and wider back portions which extend backwards above the step in the direction opposite to that of the beam motion (Fig. 6.16). This shape of the metal pool prevails at high speeds of laser welding.

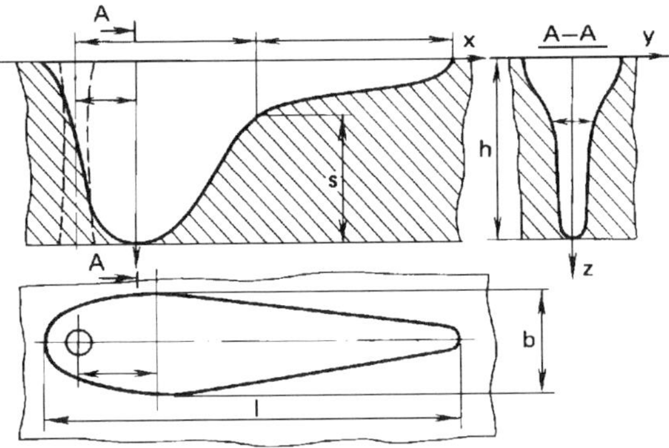

Fig. 6.16. The metal pool formed with a 3.6-kW laser at a welding speed of 22 mm/s

The longitudinal section of Fig. 6.16 is drawn using the photographs of the laser-melted pool, which were taken through quartz glass plates. In the experiments, a 1.5-mm flat sample was held in place between two 1.0-mm quartz glass plates and melted with the laser beam.

A complex shape of the molten pool results from the peculiarities of laser-induced fusion. A high-power laser beam directly falling on the front portion of the metal pool causes the metal to melt and vaporize quite rapidly. The vapors force the molten metal to flow out of the way of the beam, forming a deep cavity. In the back portion of the pool, the metal melts through mass transfer and heat conduction.

The process of melting varies with laser power, conditions of beam focusing, welding speed, and thermal properties of the target material. As noted above, the welding speed is the decisive factor responsible for the formation of a step-like pool. From the topograms obtained in the tests involving the liquid metal splashing it is evident that the step begins to form at a welding speed of up to 8 mm/s. A further increase in welding speed tends to form a clearly defined step in the pool that displays a characteristic keyhole.

It is found useful to assess the process of step shaping in terms of the step height s-to-penetration depth h ratio. The higher the value of s/h, the greater the tendency to pool formation such as that illustrated in Fig. 6.16. The weld in its cross section exhibits a narrow, lower portion tapering down and a wide, bowl-shaped, upper portion (Fig. 6.17). The depth of the tapered portion corresponds to the step height.

The graph of the s/h ratio against the laser welding speed is shown in Fig. 6.18*a*. With welding speed and the position of the focal spot relative to

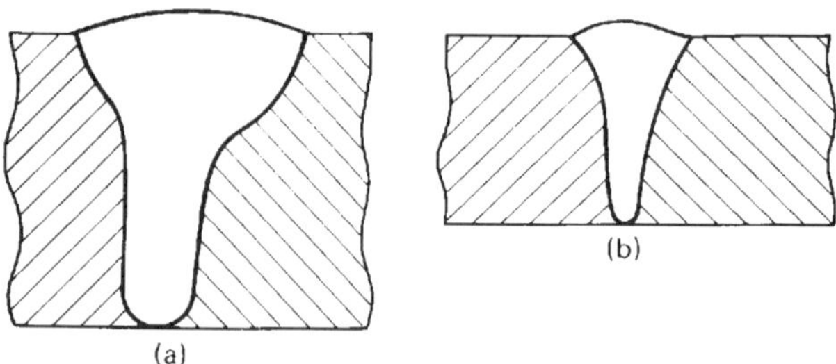

Fig. 6.17. Laser weld profiles in austenitic Cr-Ni steel welded at a welding speed of 16.5 mm/s (a) and 27.5 mm/s (b)

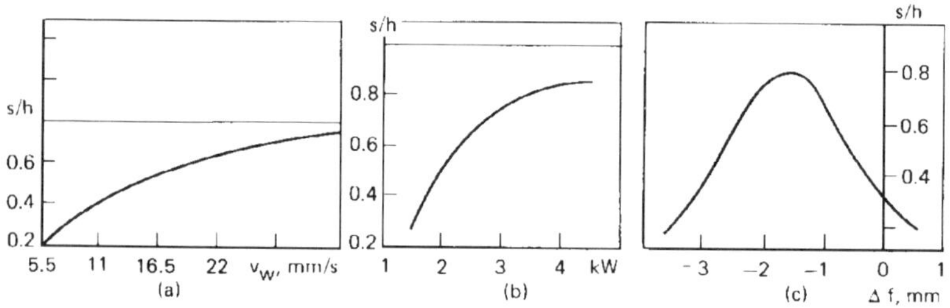

Fig. 6.18. The *s/h* ratio as a function of welding speed (a), laser power (b), and position of the focal spot relative to the target surface (c)

the target surface kept constant an increase in the beam power from 1.5 to 4.5 kW leads to a continuous increase in the values of *s/h* (Fig. 6.18*b*). The graph of Fig. 6.18*c* illustrates how the *s/h* ratio varies with the position of the focal spot on the target surface, $\Delta f = 0$, above the target surface, $\Delta f > 0$, and below the target surface, $\Delta f < 0$. At $\Delta f = -1.5$ mm, the depth of the tapered portion of the cavity is the largest and at $\Delta f = 1.0$ mm, the profile of the weld is close to the arc weld profile.

The choice of definite parameters of laser welding can ensure the conditions of weld solidification at which the crystallizing metal exhibits an increased resistance to hot cracking [98]. To verify these conditions, the researchers of [98] have carried out laser welding tests at a speed of 22 m/s on a tubular sample from a cast austenitic steel that is highly prone to hot cracking. Figure 6.19 illustrates a tubular sample with an inner diameter of 75 mm and the wall thickness varying from 1.5 to 5.5 mm.

In welding tests, a stationary 3.5-kW laser beam formed the weld around the periphery on the revolving sample without subjecting it to strains in the

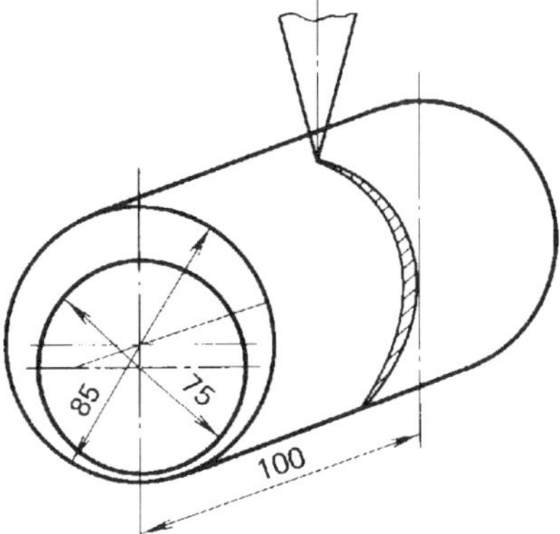

Fig. 6.19. Test sample for determining the effect of a weld on hot crack formation during laser welding

process of welding. The focal spot was held below the target surface both at a constant depth of 1.0 mm (Fig. 6.20*a*) and at a variable depth of 1.0 to 5.0 mm (Fig. 6.20*b*). In the first case, the tube was rotated about the axis of the outer surface to provide for the constant focal position. The conditions of circular weld formation without full penetration were invariable and promoted the fusion with a maximum depth of the tapered portion of the cavity. The entire circular weld was free of hot cracks.

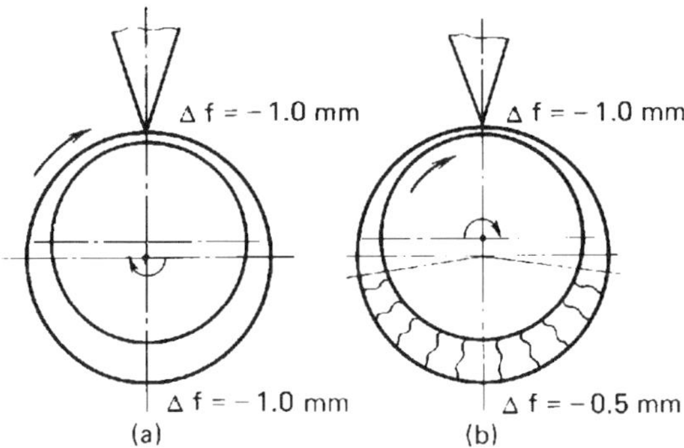

Fig. 6.20. Versions of penetration welding with a laser source

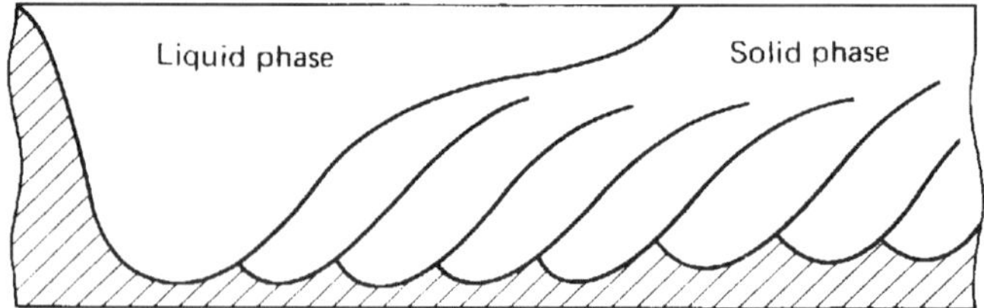

Fig. 6.21. A longitudinal section of a laser weld formed in Cr–Ni–W–Mo alloy at a speed of 16.5 mm/s

In the second case illustrated in Fig. 6.20*b*, the tube was rotated about the axes of the inner surface to vary the depth of the focus below the target surface in accordance with the variation of the tube wall thickness from 1.0 to 5.0 mm. In these conditions of laser energy coupling, the character of weld formation was quite different. In the portion of the weld seam where the focal spot was held at a depth of 3 to 5 mm, the depth of the tapered hole was minimum and the liquid-solid interface was rather smooth. It was this portion of the weld seam that exhibited hot cracks which were only in the upper section of the weld beyond the keyhole region, as in the case of the weld formed with the laser beam and simultaneously subjected to strains.

The above results suggest that the metal filling a lower narrow portion of the keyhole is much more resistant to hot cracking than the metal in the upper portion of the weld. The metallographic analysis of longitudinal sections and cross sections of the weld reveals that the keyholing process promotes the formation of the region of equiaxial crystals or the region of the disoriented close-grained dendritic structure, which adds to cracking resistance [97].

The studies of the weld metal structure in its longitudinal section attest to an intermittent character of the weld metal solidification due to a periodic transfer of the liquid metal from the front portion of the metal pool to the solidification region (Fig. 6.21). The overheated liquid metal melts the base metal. From the analysis of macrographs of the sections along the laser weld it follows that the rate f of liquid metal transfer increases with welding speed and reaches 70 Hz at a v_w of 33 m/s (Fig. 6.22).

The periodic transfer of molten metal is responsible for the intermittent solidification of the weld. The layer structure so formed augments the metal deformability. So the intermittent pattern of molten pool solidification is one of the causes of a high crack resistance of the metal in the tapered weld portion.

The analysis of solidification in various portions of the weld can disclose the conditions of hot cracking. The known models for calculating the rates of solidification are too complex for use where the metal pool displays a step-

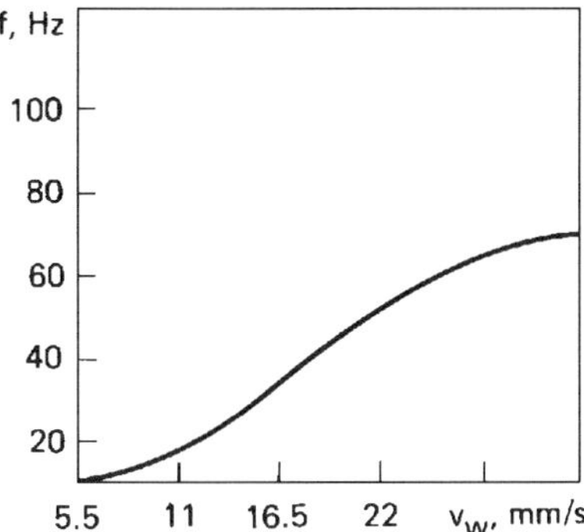

Fig. 6.22. The rate of molten metal transfer in the metal pool as a function of speed of laser welding of the Cr–Ni–W–Mo alloy

like phase boundary [94]. A more practicable approach is to calculate the solidification rate at any point of the weld seam using the topograms, such as illustrated in Fig. 6.15. The rate of solidification, v_{sol}, is defined as the solid phase growth per unit time in the plane normal to the isothermal line of the solidification front:

$$v_{sol} = \Delta l/\Delta t, \tag{6.10}$$

where Δl is an increment of the solid phase at a given point, which is defined as the shortest distance between the two neighboring isotherms; $\Delta t = \Delta x/v_w$ is the time it takes the solid phase to increase by Δl; and Δx is the distance between two cross sections of the weld or the thickness of the layer ground from the sample. The solidification rate can be rewritten in the form

$$v_{sol} = v_w \Delta l/\Delta x. \tag{6.11}$$

Thus, given an appropriate topogram, we define Δl at the point of interest and calculate v_{sol} from equation (6.11).

Figure 6.23 illustrates the plots of solidification rates calculated with the use of topograms of Fig. 6.15 for the lower portion (curve *2*) and the upper portion (curve *1*) of the laser weld in chromium-nickel alloy, grade 1, produced at a speed of 5.5 to 33 mm/s. The solidification rate in the lower portion of the weld is seen to be 3 to 4 times that in the upper portion and grows with welding speed faster than it does in the upper portion. This is because the lower portion has a small volume of the molten metal which rapidly gives

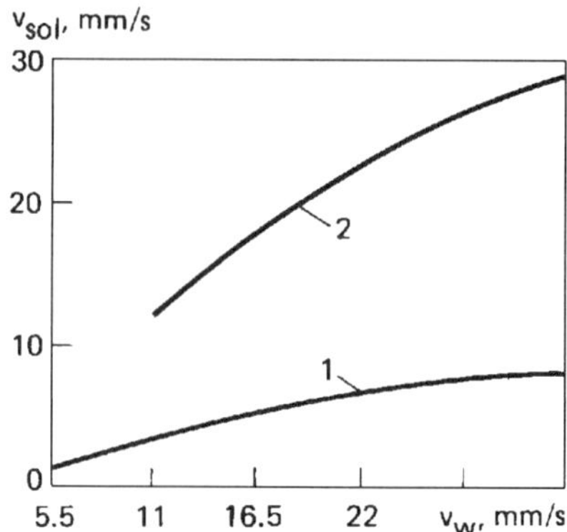

Fig. 6.23. Solidification rate of a weld in the Cr–Ni–W–Mo alloy versus welding speed

up heat to the base metal. A large variation of the solidification rate depth tends to form a solid-phase step on the phase boundary. The lower portion of the pool solidifies under the molten metal layer which replenishes the crystallizing metal with the melt, thereby appreciably increasing the resistance of the weld to thermal cracking [77, 95].

To sum up, at speeds of laser welding exceeding 11 mm/s, the solidification front exhibits a step that divides the metal pool into a deep, narrow portion and a shallow, wide portion. The step-like character of metal pool formation in the process of keyholing considerably augments the crack resistance since the upper molten layer makes up for the molten metal required in the lower layer of the crystallizing metal. Other factors that favor an increased crack resistance are the intermittent character of solidification and the formation of the region of equiaxial crystals with a fine grain structure in the tapered portion of the weld.

6.5. Cold Cracks

Cold cracks may develop in weld joints or built-up layers at the stage of cooling in the temperature range below 473 K and also after welding or cladding over a period of a few days [94, 99].

These cracks show up as a delayed fracture that occurs under the prolonged action of a steady load or a slowly varying load. In this case, the fracture strength is much lower than the short-time strength because steady stresses result in the elastoplastic shear of grains with respect to one another along the boundaries. The grain boundaries exhibit a reduced shear resistance as against the bodies of crystals in view of a highly disordered metal structure, appreciable lattice distortions, and other defects at the boundaries. These factors may promote the crack nucleation at the apexes of crystals where the sliding boundaries converge. Thus, cold cracks commonly originate at grain boundaries, which may further develop in the bodies of crystals.

Cold cracks in laser welds, laser clad deposits, and laser-treated surface portions may appear only under the effect of residual stresses which arise from nonuniform heating and cooling and also in the course of structural and phase transformations. If the strain capacity of a metal proves insufficient to withstand the residual stresses, cold cracks are likely to develop.

Numerous investigations reveal that cold cracks mostly arise from structural and phase transformations that yield martensite, namely, as a consequence of full or selective hardening and also under the effect of hydrogen.

Let us consider cold cracks formation in a laser-hardened metal where austenite transforms into martensite exhibiting high hardness, low plasticity, and increased volume density. Austenite commonly converts to martensite at temperatures of 473 to 673 K. Retained austenite can sometimes dissociate at low temperatures of from 293 to 473 K. The process of martensite formation gives rise to additional stresses due to volumetric effect of structural transformations.

Laser-treated steels and titanium alloys can develop cold cracks under the effect of hydrogen whose solubility heavily grows at elevated temperatures. The source of hydrogen that dissolves in the laser-melted metal is the moisture in shielding gases, surface coatings, and in the environment. Hydrogen can also find its way into a high-temperature laser-material interaction region from lower-temperature areas under the effect of thermal diffusion. Molten metal contains hydrogen in the monatomic form that displays a high diffusivity.

At the stage of metal cooling, the equilibrium solubility of hydrogen decreases, and so hydrogen diffuses from the molten metal into neighboring regions where it fills voids and micropores. Atomic hydrogen collected in these voids and pores changes to a molecular form. A further ingress of atomic hydrogen into the surrounding regions considerably increases the pressure of molecular hydrogen, which gives rise to high stresses in the metal. There is evidence that hydrogen adsorbed on the metal surface is responsible for embrittlement. Increasing the resistance to cold cracking is of decisive importance in widening the scope of welding and cladding processes. Preheating, high-temperature tempering, and application of austenitic filler materials are some of the ways of preventing cold cracking.

The studies of the effect of laser heating on the strength of a metal give ambiguous results. In particular, an increase in the rate of heating decreases the dwell time of the beam on any surface spot, which considerably refines the primary structure and changes the character of solidification. The refined structure is more homogeneous and contains a reduced amount of impurities at crystal boundaries. A small dwell time precludes the ingress of hydrogen from the environment. On the other hand, a high rate of cooling promotes the formation of local hardened regions that are conducive to cold cracking.

These results cannot provide the prior estimates of the effect of laser heating on cold cracking. For this, certain criteria need be used to obtain quantitative estimates of the effect of laser treatment parameters on cold crack formation.

6.6. Estimation of Cold Crack Resistance

There are various methods for estimating the cold crack resistance of metals, which commonly involve tests performed on laboratory and product samples of definite shape and size [95, 99]. The laboratory samples intended for welding or cladding are chosen so as to verify most completely the properties of metals under test by checking them for cold cracks and to select adequate parameters for the process of interest. The tests carried out on both laboratory and product samples can provide the qualitative estimates of cold crack resistance.

Product samples duplicate the design features of actual parts and serve for the final assessment of selected materials, filler compositions, and process performance. The welded sample passes the acceptance test if it does not exhibit cold cracks. This also bears witness to an adequate performance of the process worked out in the test procedure.

A major disadvantage of the above test methods is that they do not provide the quantitative estimates which help in selecting the best version of the process under examination. The quantitative methods involving the mechanical tests of specimens after their welding or cladding are free of this disadvantage.

A specimen is left to sustain a steady load for a long stretch of time in order to test it for slow rupture. A minimum tensile stress that produces a crack in the specimen loaded for 20 hours can serve as a criterion of crack resistance. An increase in the loading time does not lead to more accurate results.

A variety of specimens that differ in thickness and size are put to test for tension, cantilever bending, three-point or four-point bending, and bending at a load distributed along the perimeter of a rigidly fixed specimen. The test performed on these specimens can reproduce typical cracks that appear in weldments or any pieces subjected to heat treatment.

In estimating the cold crack resistance of laser-clad metal plates and laser-welded thin plates, it is found useful to test a flat specimen for bending under distributed load. Sheet specimens can be made by forming a bead on the sheet surface, melting the sheet in its central portion, and butt-welding two plates. A specimen rigidly fixed along its perimeter is bent under hydrostatic pressure so that its central portion can be in biaxial tension, thus imposing the same stresses along the two axes.

A measure of crack resistance of such a specimen is the critical value of pressure or stress that induces cracks in the specimen held under a definite load for 20 hours. If the thickness of the sheet metal under test is large, it is advisable to use a tee-welded specimen and subject it to bending in a testing machine. The critical value of bending stress at which the specimen loaded for 20 hours develops cracks serves as the criterion of fracture strength.

These methods determine the crack resistance under the joint effect of basic factors responsible for cold cracking. In order to estimate the effects of individual factors and to establish the mutual relation between these factors, use is made of the test methods which include the test stages intended to hydrogenate samples, simulate the thermal and stress cycles, and subject these samples to tests for slow rupture. By varying the hydrogen contents and the parameters of thermal cycles, it is possible to simulate the actual conditions of laser welding, cladding, and other heat treatment processes.

The test specimens used for the purpose are commonly the plates from 1 to 3 mm thick, 5 to 10 mm wide, and 50 to 100 mm long. The choice of a specimen of the desired size depends on the loads applied in slow crack tests. After its hydrogenation to a specified hydrogen content, the specimen is heated by passing current through it and is thus subjected to tension, bending, or torsional forces induced under the effect of the thermal cycle. After a complete cooling, the specimen is tested for slow rupture. In order to perform a thorough analysis of the conditions of cold cracking, it is necessary to plot a dilation curve for the metal or to take the anisotropic diagram of austenite disintegration and to collect data on the structure, grain size, hardness, and hydrogen content.

The tests for rupture strength should be run in two stages. The first stage involves cold cracking tests performed on simulator-specimens, hard-faced specimens, or welded specimens produced with fillers of different compositions in various welding conditions. The quantitative estimates obtained in the test procedure permit us to select optimal conditions of the process.

The second stage involves the process of welding or cladding of laboratory or product specimens under the conditions selected at the first stage. If the specimens do not develop cold cracks, the process is considered suitable for industrial applications. If this is not the case, recourse is made to various methods to control the initial composition and structure of the base metal, the chemical composition of the filler metal, the conditions of laser welding or cladding, and heat treatment after welding or cladding.

6.7. Cold Crack Formation in Laser Welding

The text below presents the comparison studies of cold crack formation involved in laser welding (LW), electron-beam welding (EBW), and argon-arc welding (AAW). Testing has been done on welds produced in various steels and alloys, the welding conditions of which are given in Table 6.4.

The critical load pressure p_l that gives rise to cracks in a weld specimen held in the stressed state for 20 hours is taken as a measure of crack resistance. Figure 6.24 shows the pattern of pressure p_l exerted on a weld specimen for estimating the crack resistance by the methods described in Sec. 6.6. Figure 6.25 illustrates the plots of the crack resistance expressed in terms of pressure p_l for high-strength steels as a function of welding speed v_w. As is seen from the curves, an increase in the speed of argon-arc welding leads to a decrease in the crack resistance of all steels, except for martensitic Cr-Si-Mo steel grade 2. The cracks develop entirely within the weld area and mostly across the weld. The longitudinal cracks only appear in martensitic steel grade 2.

The laser welds produced in martensitic steels exhibit cold cracks. The crack resistance of laser welds increases with the weld speed and exceeds that of arc welds. The values of p_l for the bainitic steel and martensitic Cr–Si–Mn steel grade 1, welded with the laser and argon arcs are practically the same. With an increase in the speed of electron-beam welding of martensitic steel

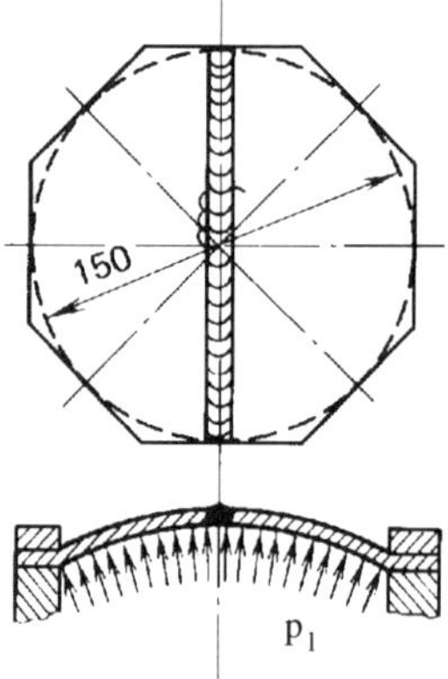

Fig. 6.24. Sample loading for determining the weld resistance to cold cracking

Table 6.4. Welding Conditions for Some Steels and Alloys

Material	Thickness, mm	LW		EBW			AAW		
		v_w, mm/s	P, kW	v_w, mm/s	I, mA	V, kV	v_w, mm/s	I, A	V_{arc}, V
Steel:									
bainitic martensitic,		11	3.5	11	20	52	3.3	110	9
grade 1	2.2	27.5	4.5	27.5	30	52	11.0	190	10
maraging		44	5.5	44	48	52	27.5	330	12
Steel:									
martensitic,		11	4.0	11	27	52	3.3	220	11
grade 2	3.3	27.5	5.5	27.5	32	52	11.0	320	12
		44	6.0	44	55	52	27.5	460	13
Steel:									
pearlitic, grade 1		11	3.0	11	30	52	3.3	110	10
	2.0	22	3.3	22	35	52	5.5	135	11
pearlitic, grade 2		33	3.8	33	40	52	11.0	170	13
Carbon steel 35		22	3.5	22	37	52	3.3	135	11
	3.0	33	4.2	33	42	52	5.5	180	13
Low-carbon									
steel St.10		55	5.2	55	55	52	11.0	210	15
Titanium alloy,		11	3.2				3.3	125	9
grade 1	2.0	22	3.5				5.5	135	10
		33	3.8				11.0	175	10
		44	4.0				22.0	240	11
Titanium alloy,		16.5	2.9	16.5	20	52	3.3	160	10
grade 2	3.5	22	4.1	22	27	52	5.5	250	11
		33	4.3	33	33	52	11.0	310	12

grade 2, the pressure p_1 decreases appreciably. Both laser welds and electron-beam welds are more resistant to cracking than arc welds.

The welds produced in titanium alloys by the electron beam and laser beam in a helium atmosphere are much more crack-resistant than arc welds (Fig. 6.26).

Figure 6.27 illustrates the crack resistance versus welding speed for welds formed in pearlitic and carbon steels. The laser welds and electron-beam welds of pearlitic steels do not develop cracks over a wide range of welding speeds. The crack resistance of arc-welded steels is much lower. All welds in low-

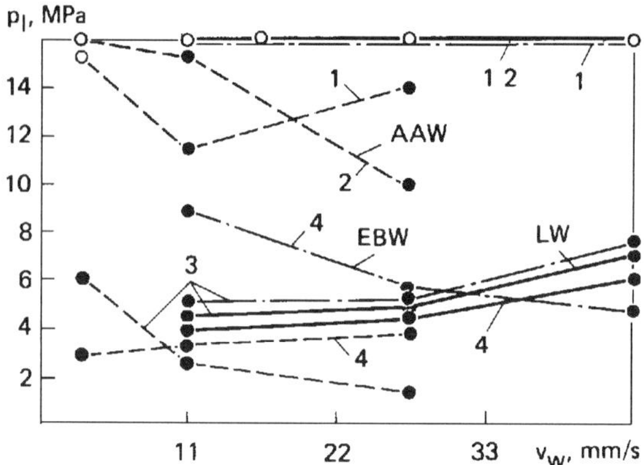

Fig. 6.25. Cold crack resistance expressed in terms of critical load pressure p_l versus welding speed for high-strength steels: (1) bainitic steel, (2) maraging steel, (3) martensitic Cr–Si–Mn steel, (4) martensitic Cr–Si–Mo steel; solid lines, dash-dot lines, and dash lines identify data obtained in laser welding, electron-beam welding, and argon-arc welding, respectively; full and open circles stand for presence and absence of cracks, respectively

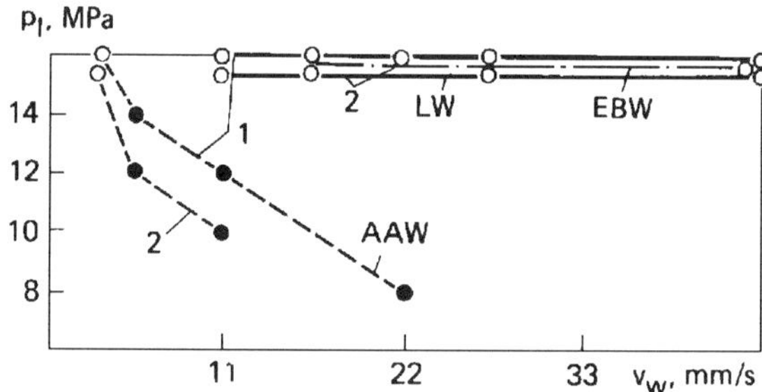

Fig. 6.26. Cold crack resistance versus welding speed for Ti alloys of two grades

carbon steel are crack resistant, but welds in carbon structural steel 35 and the welds of this steel to low-carbon steel St.10 are not very crack resistant. The arc welds produced at speeds of less than 11 mm/s are free of cracks, but the laser welds and arc welds formed between these steels exhibit cracks which extend along the weld seam and into the weld zone of carbon structural steel. Also, the crack resistance decreases with an increase in the welding speed.

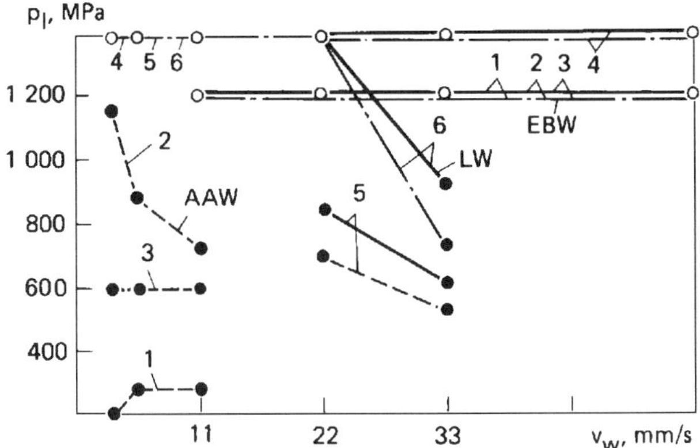

Fig. 6.27. Cold crack resistance versus welding speed: (1) pearlitic Ni–Cr steel, (2) pearlitic Cr–Mn steel, (3) Ni–Cr steel welded to Cr–Mn steel, (4) low-carbon steel St.10, (5) carbon steel 35, (6) steel St.10 welded to 35 steel

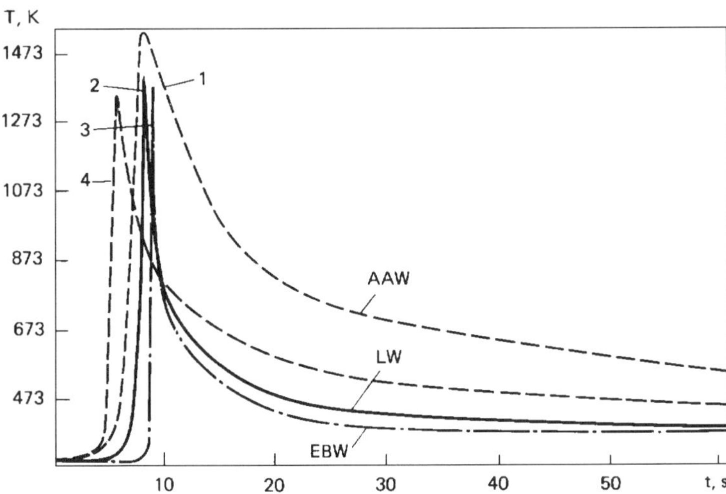

Fig. 6.28. Thermal cycles in butt welds of 3.3-mm martensitic steel at welding speeds of 12, 14, and 27.5 mm/s (curves *1*, *2* and *3*, *4*, respectively)

The crack resistance varies with the type and rate of welding on account of changes in the thermal cycles that modify the structure of the heat affected zone. The plots of thermal cycles against the rates of welding with the laser beam, electron beam, and arc are shown in Fig. 6.28. The thermal cycles appreciably differ from one another at the stage of weld cooling. The cooling rate of laser welds and electron-beam welds is much higher than that of arc welds at increased welding speeds, which is evident from the graphs illustrated in Figs 6.29 and 6.30. The approximate values of the cooling rate v_c in the

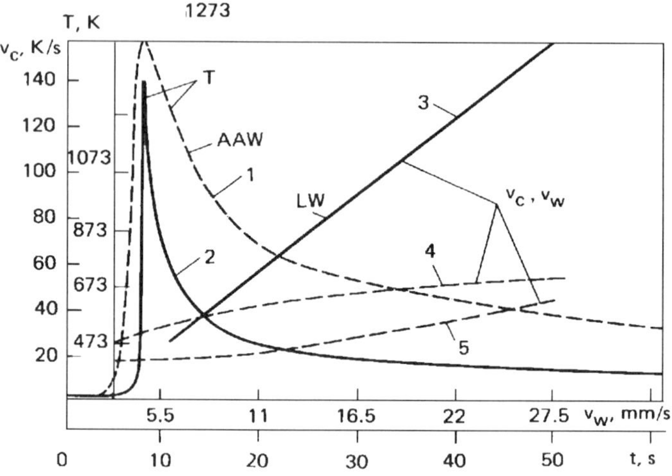

Fig. 6.29. Thermal cycles and cooling rates against butt welding speed: (1) and (2) martensitic steel, at 12 and 28 mm/s, respectively, (3) 4-mm steel St. 3, (4) 3.3-mm bainitic steel, (5) 3.3-mm martensitic steel

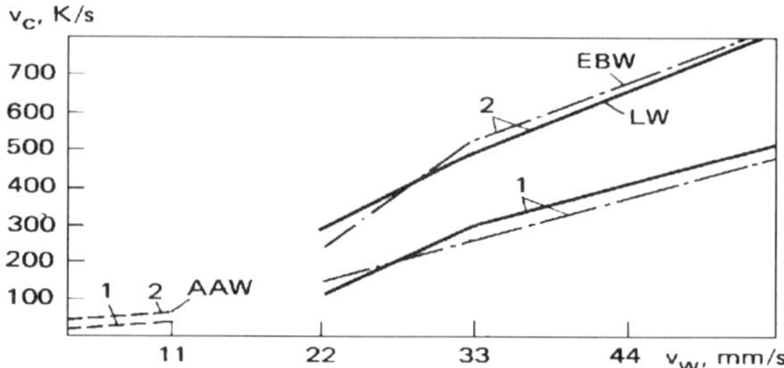

Fig. 6.30. Cooling rates in the temperature range of 873–773 K against welding speed: (1) CR–Ni and Cr–Mn steels, (2) 35 steel and steel St.10

heat-affected zone for martensitic steel grade 2 in the temperature range from 873 to 773 K are the following: 25 K/s at v_w = 12 mm/s and 66 K/s at v_w = 27 mm/s in argon-arc welds; 100 K/s at v_w = 14 mm/s and 280 K/s at v_w = 27 mm/s in laser welds.

The analysis of changes in the weld metal structure is easy to perform using continuous cooling transformation diagrams taken from a fast-speed dilatometer (Fig. 6.31). As is seen from the diagrams of Fig. 6.31 for the bainitic steel, the initial transformation temperature decreases steadily from 668 to 638 K with an increase in the cooling rate from 8 to 50 K/s. A further increase in the cooling rate does not change the initial temperature of austenite-to-martensite transformation. This points to the fact that at v_c higher than 50 K/s the steel undergoes a purely martensite transformation rather than the bainite transformation. The final temperature of gamma-to-alpha transformation decreases from 488 to 453 K with the cooling rate increasing from 8 to 14 K/s, so that the steel undergoes bainite transformation. At a cooling rate exceeding 14 K/s the final transformation temperature does not change, which points to bainite-martensite transformation.

The graphs of Fig. 6.29 show that an increase in the arc welding speed from 2.8 to 11 mm/s leads to a higher cooling rate, and so the thermal cycle

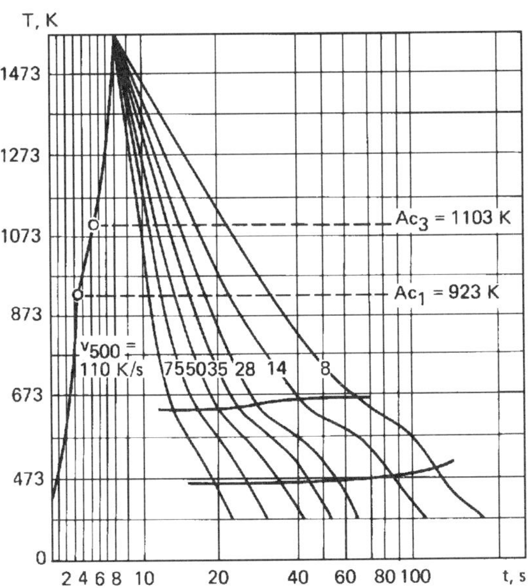

Fig. 6.31. Cooling transformation diagram for bainitic steel

changes more abruptly. In the temperature range from 873 to 773 K, the cooling rate rises from 25 to 45 K/s and intensifies the gamma-to-alpha transformation process, thereby reducing the crack resistance, which is clear from Fig. 6.25. Here, the minimum value of crack resistance points to the exhausted potentialities of the bainite and bainite-martensite transformations at a cooling rate v_c of less than 45 K/s. A further increase in the crack resistance can be accounted for by the refinement of the original structure and the formation of the low-alloy martensite that is highly stable to cold cracking.

The metallographic examinations confirm the above findings. Figure 6.32 gives the profiles of welds produced in the bainitic steel. The argon-arc weld of Fig. 6.32*a* has the shape of a bowl with an aspect ratio *h*/*b* (depth-to-width ratio) of about 0.3. The laser weld of Fig. 6.32*b* takes the shape close to that of a triangle at a low welding speed, and has an aspect ratio of about 1.0. With an increase in the welding speed, the laser weld assumes the keyhole profile with an aspect ratio *h*/*b* approaching 2.0 (Fig. 6.32*c*). The argon-arc weld formed at a speed of 28 mm/s becomes too wide and exhibits undercuts and root porosity.

The arc weld made in the bainitic steel at a speed v_w of 2.8 mm/s shows a cellullar-dendritic structure near the fusion line [97]. As the middle of the weld is approached, the metal gradually acquires the dendritic structure with developed secondary branches and with the dendrites oriented in the direction of welding.

The laser welds are arc welds produced at a speed of 11 mm/s and have a columnar-dendritic structure [97]. Unlike the arc weld, the laser weld has a region of very thin dendrites near the fusion junction, which points to a high degree of overcooling during crystallization. The main dendrites in the laser weld are shorter and thinner and feature developed secondary branches. Such a structure is more resistant to cold cracking than that of the arc weld.

Increasing the speed of argon-arc welding up to 27.5 mm/s changes the crystallization process, so that the weld metal structure becomes cellular-

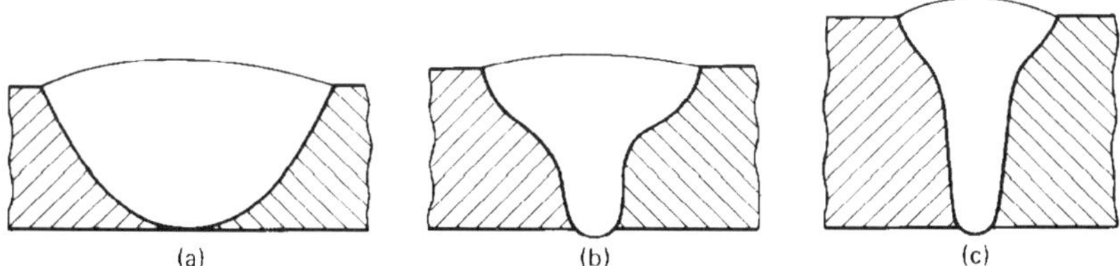

Fig. 6.32. Weld profiles at argon-arc welding speed of 12 mm/s (a) and at laser welding speeds of 14 mm/s (b) and 28 mm/s (c)

dendritic. The laser weld produced at the same speed displays a columnar-dendritic structure, as does the weld formed at a speed of 11 mm/s. However, this structure is still finer and consists of shorter and thinner dendrites. In the middle of the weld there appear regions with a polyhedral equiaxial structure. The size of austenite grains in the laser weld formed at v_w = 11 mm/s is approximately one-half or one-third that of the grains in the arc weld produced at the same speed. Increasing the speed of laser welding to 27.5 mm/s leads to an appreciable grain refining.

The degree of structure inhomogeneity of a weld metal can be judged by measuring the microhardness H in the cross section of the weld. Figure 6.33 illustrates the measured values of hardness H of laser welds (curve *1*) and argon-arc welds (curve *2*) produced at a welding speed of 11 mm/s. From the comparison of the regression curves it follows that the hardness of the laser weld remains on the average constant and reaches 6300 MPa over its entire width b, whereas the hardness of the arc weld increases from 6500 MPa in the middle of the weld to 7000 MPa at the fusion junction, which attests to the structure inhomogeneity of the arc weld metal due to different conditions of cooling of the weld portions. A lower increase in hardness of the fusion area of the laser weld points to specific conditions of crystallization in laser welding.

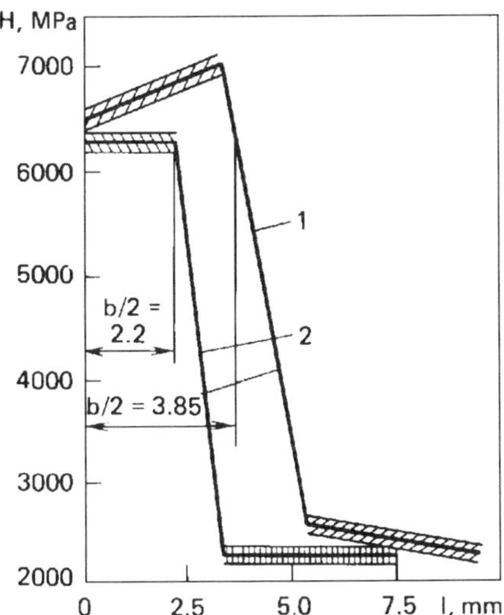

Fig. 6.33. Hardness of welds in martensitic steel versus distance from the weld axis

The regions of hardening and tempering in the laser weld are much narrower than they are in the arc weld, which is particularly the case for welding at a high speed of 27.5 mm/s. Laser welding at this speed affords a lower degree of softening in the tempering region by virtue of a high temperature gradient and a small length of time over which the heat affected zone is held at hardening and tempering temperatures.

The crack resistance of laser welds and electron-beam welds in most of the steels under study increases with welding speed, as is evident from Figs. 6.25 and 6.27, because of a more refined structure and increased homogeneity of the weld metal. However, the results of Fig. 6.27 indicate that the crack resistance of laser welds and electron-beam welds in carbon structure steel and between this steel and low-carbon steel is lower than that of an arc weld and decreases with increasing welding speed. It is more expedient to consider this fact from the viewpoint of specific effects of thermal cycles on the structures of weld metals. For this, let us compare the results obtained for steel 35 and pearlitic steel grade 1 which differ in their tendency to crack formation in laser welding and in argon-arc welding.

At a rate of cooling less than 50 K/s, steel 35 undergoes pearlitic transformation and its hardness increases to 2400 MPa. As the cooling rate grows, the structure becomes bainitic and, starting from a cooling rate of 80 K/s, the weld metal acquires a bainitic-martensitic structure whose hardness increases to 4500 MPa. At a cooling rate exceeding 140 K/s the martensite transformation prevails, yielding a course-needled martensitic structure whose hardness reaches 6800 MPa. The plots of Fig. 6.30 show that laser welding and electron-beam welding can ensure cooling rates above 140 K/s at high welding speeds within the temperature range from 873 to 773 K. As seen from Fig. 6.27, the weld in steel 35 exhibits cold cracks which are inherent in the bainitic-martensitic or the martensitic structure.

In pearlitic Cr-Ni steel, the bainite-martensite transformation takes place at slow cooling rates of 12 to 30 K/s that are typical of argon-arc welding, as is obvious from Fig. 6.30. At cooling rates above 30 K/s that are specific to laser welding and electron-beam welding, this steel undergoes martensite transformation.

The weld structures of steel 35 and pearlitic Cr-Ni steel differ in degree of dispersion. The laser welds and electron-beam welds formed in the pearlitic steel at high speeds feature a considerable grain refinement as against the argon welds. The size of austenite grains in the weld-metal zone and the heat affected zone of the arc weld produced at a speed of 5.5 mm/s is 120 μm and 80 μm, respectively. In the laser weld formed at a speed higher than 33 mm/s the grain size is much smaller and equals 50 μm in the weld-metal zone and 35 μm in the heat affected zone. With further increase in the laser welding speed, the grain size decreases to 20 μm.

The laser welds and electron-beam welds made in steel 35 at speeds of from 2.7 to 55 mm/s have austenite grains that are slightly smaller than those in arc welds. At high speeds of laser welding and electron-beam welding, the steel acquires a coarse-needled martensitic structure at the cooling stage, which is highly prone to cold cracking. The arc weld has a pearlitic structure of low hardness and thus displays a higher crack resistance.

Where the process of laser welding involves favorable changes in the structure, the weld proves less sensitive to cold cracking despite the sharply varying thermal cycle. The comparison studies of the structures of welds in Ti alloys support the above conclusion. The structure of an arc weld formed at a speed of 7.7 mm/s is inhomogeneous and represents a combination of α'-, α-, and β-phases, where the martensite-like α'-phase predominates. The martensite-like phase consists of thickened and elongated needles that appear at a slow cooling rate in the martensite transformation range.

In laser welds and in electron-beam welds the metal structure becomes more homogeneous and consists mainly of a martensite-like phase with a small amount of the equilibrium alpha phase. The degree of dispersion of these phases is much higher than is the case for arc welds. This explains why laser welds and electron-beam welds in titanium alloys are more resistant to cold cracking than argon-arc welds, as is obvious from Fig. 6.26.

To sum up, laser welding ensures a more homogeneous structure of the weld than arc welding despite the abruptly varying thermal cycles during laser heating. A laser beam or an electron beam is able to form a weld with a fine-grained structure of high plasticity, which offers a higher resistance to cold cracking as compared to an arc weld.

Laser welding and electron-beam welding performed under optimal conditions afford a more favorable process of crystallization than arc welding, so that the weld has a smaller width, shorter zones of hardening and tempering, and a lower degree of softening in the heat affected zone. The welds produced with a laser source at a low heat input and at a speed, higher than 27.5 mm/s are commonly more resistant to cold cracking than those produced with arc welds.

Part 2
LASER TECHNOLOGY

7 Welding

7.1. Weld Shaping

In its ability to deliver a laser power to a small area, the laser is by far superior to conventional welding sources, except for electron-beam devices now widely used for welding critical parts. However, the electron beam requires a high vacuum in the welding chamber for the process to proceed steadily. In contrast, laser welding can be done in an atmosphere of shielding gases such as argon, helium, and carbon dioxide. The lens-mirror system can direct the laser beam to any elements of large-size members and to hard-to-reach locations [100]. The welding process is easy to control by adjusting the beam energy parameters. Unlike the electron beam and arc, the laser beam is uneffected by the magnetic field of a workpiece, so it can yield a quality weld throughout its length.

Both solid-state and gas lasers operating in the continuous or pulse mode can function as welding sources. Because of the high energy density of a laser during welding, the bulk of the molten metal is small, the dimensions of the metal effected by the heat are insignificant, the rate of heating is high and the rate of cooling near the weld is small. A small heat input results in a very slight distortion of the weldment, and a specific thermal process secures a high-strength weld.

Laser welding covers a large variety of techniques capable of producing welds in various metals ranging from a few micrometers to tens of millimeters in thickness. Figure 7.1 illustrates the laser welding methods classified by the beam energy characteristics, performance parameters, and process characteristics.

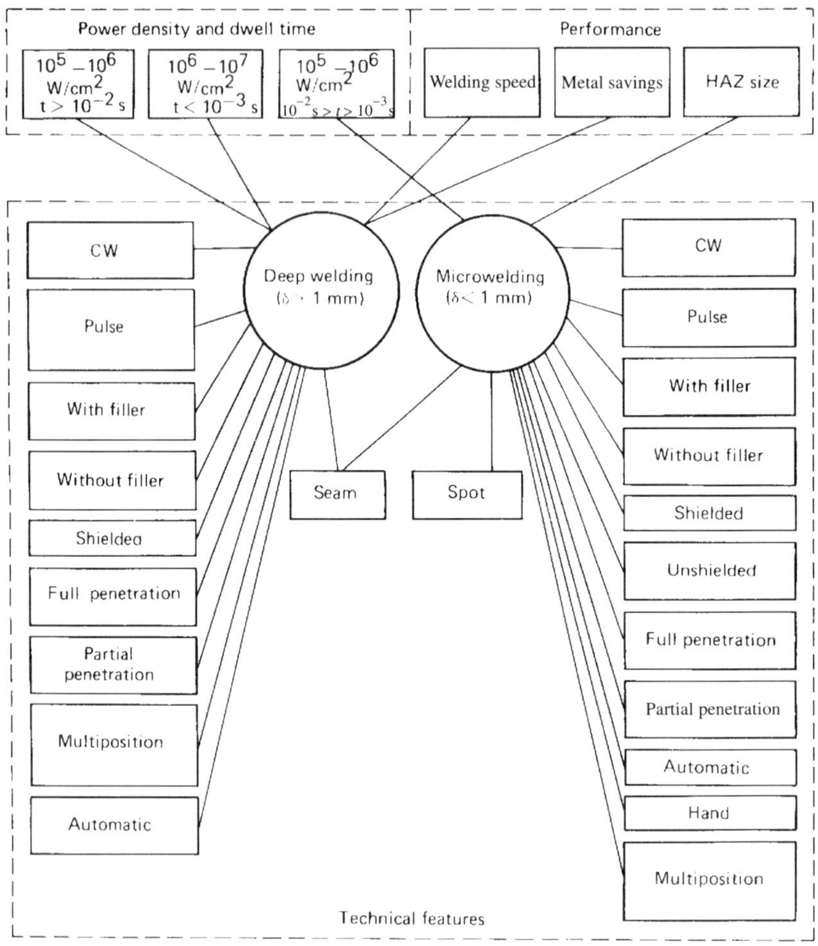

Fig. 7.1 Classification of welding methods

Energy characteristics. These are the laser power density q and the action time t, or dwell time, which depends on the welding speed in CW laser welding and on the pulse length in pulse welding.

As noted in Sec. 4.4, laser welding presupposes an upper limit to the power density, defined as the threshold power density q_{th} above which the metal vigorously vaporizes, so that the material ejection impairs the weld joint. In practice, the welds are made at power densities in the range from 10^5 to 10^7 W/cm^2. The power density below 10^5 W/cm^2 is insufficient to provide a quality of weld. In this case, a traditional technique of welding with melting is a better choice.

A feature specific to laser welding is that it requires a definite combination of the power density and dwell time. Three combinations of q and t are practicable in laser welding.

1. First combination: $q = 10^5$ to 10^6 W/cm^2 and $t > 10^{-2}$ s. This combination is typical of continuous laser welding. Varying q and t in the specified limits allows us to carry out the fusion welding of various materials in a wide range of thicknesses. The dwell time can be defined as the ratio of the diameter of the focal spot to the welding speed v_w:

$$t = d/v_w. \tag{7.1}$$

2. Second combination: $q = 10^6$ to 10^7 W/cm^2 and $t < 10^{-3}$ s. This combination is specific to pulse welding with a short-time action of recurrent pulses whose rate reaches tens and hundreds of hertz. The pulse length t_p is much shorter than the dwell time defined by equation (7.1), but the total action time of a few overlapping pulses is sufficient to ensure deep penetration. Pulse welding under these conditions can yield welds in a wide range of depths and requires a much lower heat input than CW laser welding.

3. Third combination: $q = 10^5$ to 10^6 W/cm^2 and $10^{-3} < t < 10^{-2}$. Here, the dwell time is approximately the same as it is in the second case, but the pulse length is larger. A train of pulses can yield a spot weld of a requisite depth which is comparatively small.

Performance parameters. These figures of merit determine the effectiveness of a welding process and include the welding speed, the factor used to account for metal saving, and the spot size or HAZ size.

Continuous laser welding can be carried out at speeds several times exceeding those of conventional welding processes. A high welding speed raises the process capacity at a reduced heat input, i.e., the power needed per unit weld length. But the speed of pulse welding is much lower and generally comparable with that of the conventional processes.

The process of continuous laser welding of plates 15 to 20 mm thick effects savings in metal since, unlike arc butt welding, it dispenses with plate beveling and filler wire and can form a weld in a single pass. In laser welding of thin sheets, savings in metal are insignificant.

A laser beam focused to a spot 0.1 mm or less in diameter provides a narrow heat-affected zone which rapidly cools without inducing large plastic strains. So, laser welding can serve as an effective tool for the fabrication of precision weldments [101]. Highly accurate, narrow, and precisely located welds do not require subsequent dressing or machining and allow for quite a compact layout of the components of devices and assemblies. The use of laser welding, for example, in the fabrication of electronic circuits leads to a considerable increase in the packing density of the components on the board, which reduces the overall dimensions and mass of the devices. The precise locality of welds manifests itself most vividly in pulse welding.

Process features. A convenient way to illustrate some features of laser welding is to consider two types of the process that provide penetration depths above and below 1 mm, respectively.

Both CW lasers and repetitively pulsed lasers are suitable for penetration welding and microwelding at small heating depths. In the latter case, the laser operates under less rigid conditions at which the heat source only melts the base metal without raising the metal temperature to its vaporization point.

Where it is necessary to reduce the requirement for the assembly accuracy or to alloy the weld metal in the process of welding, use can be made of filler materials in the form of a powder or thin wire 1.0 to 1.5 mm in diameter or even less, which must be precisely fed into the interaction region.

Deep-penetration welding commonly calls for protection of the molten metal against oxidation and contamination by shielding gases. A simpler process of welding thin parts produced from low-carbon steels and some other materials can do without shielding gases.

Full-penetration welding mostly applies for critical load-bearing units. Partial penetration welding serves to form an air-tight joint between thin parts or to join them to thick parts.

It should be noted that the optical system can direct the beam to any location to produce the weld in any position in space.

Welding under small-penetration conditions relies on the thermal effect produced by the beam as it interacts with an opaque material. For the metal to be melted up efficiently, the pulse duration must correspond to the thermal time constant t for a given metal, an approximate expression for which takes the form

$$t = \delta^2/4k,$$

where δ is the thickness of a part; $k = K/c\rho$ is the thermal diffusivity; c is the specific heat; and K is the thermal conductivity.

The values of t' for thin parts, 0.1 or 0.2 mm thick, are comparable with the free-running laser pulse length of the order of a few milliseconds. For parts thicker than 1.0 mm, the thermal time constant appreciably exceeds the highest pulse length and presents difficulties in implementing the process of pulse welding.

The mechanism of shaping a deep-penetration weld with a multikilowatt laser involves the phenomenon of keyholing described briefly above. The laser power of a density above the critical value heats up the target at a velocity much higher than that of heat removal by conduction. The metal vaporizes rapidly at the focused spot and the hole so produced allows the beam to penetrate unimpeded into the part and form a deep cavity. The vapor pressure of the molten material covering the cavity walls proves sufficient to overcome the hydrostatic pressure and surface tension forces, so the molten material does not fill the hole until the beam passes it.

At an appropriate welding speed, the cavity profile acquires a dynamic stability. The beam melts the front wall of the cavity and the molten metal flows to the rear wall and back into the cavity where it solidifies as the beam moves away in the direction of the weld seam. Since laser energy penetrates deep into the target and efficiently couples to the cavity walls, the beam forms a narrow weld with a large aspect ratio.

The process of keyholing gives rise to a luminous plume above the target, which consists of the vapors, ejection material, and particles of condensed vapor. The plume and the plasma arising from an optical breakdown in the gaseous atmosphere absorb laser radiation and considerably affect the melting efficiency. At atmospheric pressure, a proper choice of a shielding gas for laser welding can improve the melting parameters illustrated in Table 3.1.

The studies of the dynamics of keyholing play an important role in optimizing the parameters of welding and evolving an efficient process. Experimental methods involving the welding tests on specimens with simultaneous photorecording of keyhole profiles offer quite reliable results. The specimens used for the purpose are simulator models fabricated from transparent materials such as quartz or acrylic glass, or glass ceramic, and also composite models prepared from the above glasses and metals. There are individual reports on the analysis of keyholing by X-ray examination of metal specimens in the course of their welding.

Specimens used to model the process of keyholing appear in Fig. 7.2. The weld profile constructed on the basis of experimental results is shown in Fig. 7.3. The metal pool has a specific form elongated in the direction opposite to that of welding.

The front portion of the pool exhibits hole *3* filled with metal vapors. This is the region of the highest glow. The molten metal flows down from the curved front wall and moves upward along the solidified step. A major portion of the molten metal moves to the back of the pool along the cavity walls. At the back portion, the molten metal swirls upward to the target surface. A bright cloud of plasma plume *2* that appears above the surface periodically changes in size and brightness at a frequency of several hundred hertz. At a high speed of welding, the plume declines in the direction opposite to that of welding by 20 to 60°.

The process of molten metal transfer in the pool under the action of reacting forces of vapors has an appreciable effect on the weld quality. The reacting

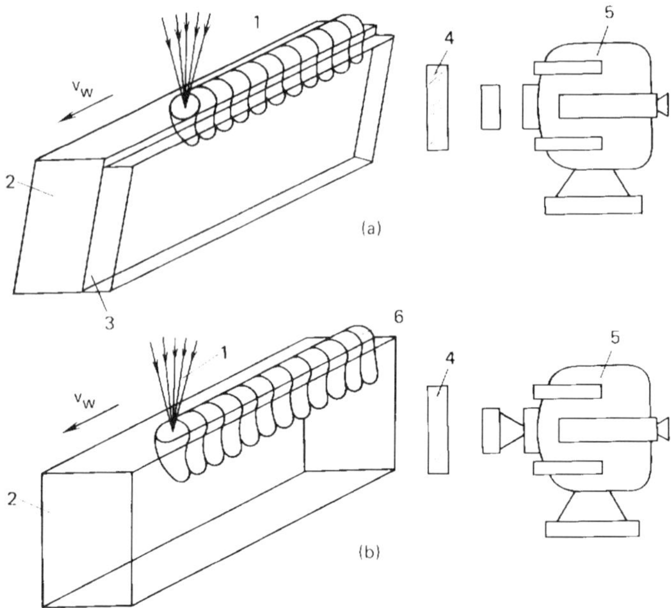

Fig.7.2. Physical simulation of dynamic processes in the metal pool: (*a*) metal-transparent material specimen; (*b*) transparent material specimen; (*1*) laser beam; (*2*) metal; (*3*) transparent material; (*4*) light filter; (*5*) cine camera; (*6*) remelted material

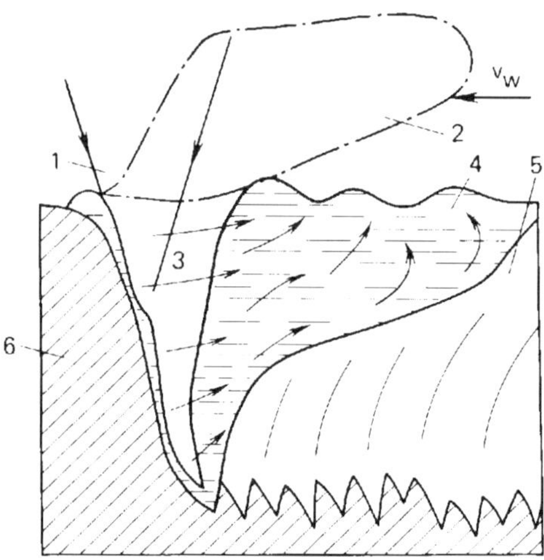

Fig. 7.3. Metal pool in the longitudinal section: (*1*) laser beam; (*2*) plasma plume; (*3*) vapor-gas crater; (*4*) back portion; (*5*) solidified metal; (*6*) weld material

forces cause the molten metal to move both down along the front wall of the cavity and along its side walls. The uncovered front wall now melts down and the reacting forces initiate the next cycle of molten metal transfer. Experiments reveal that the speed of the molten metal transfer reaches 1000 to 2000 mm/s at a welding speed of 2 to 5 mm/s. The recurrence rate of mass transfer varies in proportion to the welding speed and ranges from 10 to 50 Hz.

It is found that the process of welding entails self-excited oscillations [48]. When a time-constant flux of radiant energy above a certain critical value heats up the target, the surface temperature continuously varies with time, which points to a resonant character of the material heating. This effect can be used to advantage in developing more efficient methods of welding on the basis of dynamic focusing of the laser beam and pulsating delivery of shielding gas [63].

Overlapping pulse welding relies on the use of pulses 10^{-3} to 10^{-6} s long at a power density of 10^6 to 10^7 W/cm^2 and a repetition rate of 100 to 1000 Hz. As with CW laser welding, spot welding under deep-penetration conditions depends on the keyholing process in which the hole does not clog as the next pulse decays.

At a mean power of laser radiation of 1 kW, the power in a pulse may reach 100 kW. The metal rapidly heats up to the boiling point over a short time. The reacting forces of vapors periodically displace the molten metal from the front wall of the cavity to the back one at a pulse repetition rate [35]. The cavity wall has no time to cool down at a high pulse repetition rate, and so the penetration depth can be even larger than when welding with a CW laser.

Welding with a repetitively pulsed laser leads to a periodic generation and relaxation of the plasma above the target surface. The plasma plume builds up after some time delay t_1 from the instant the pulse arrives and then abates for time t_2 once the pulse ceases. The basic parameters of a repetitive waveform are the pulse length t_p and space width t_{sp} between the successive pulses:

$$
\begin{aligned}
t_p &= 1/gf_p \\
t_{sp} &= (1/f_p) - t_p,
\end{aligned}
\tag{7.3}
$$

where g is the pulse period-to-pulse duration ratio and f_p is the pulse repetition rate.

By selecting a waveform for which $t_p < t_1$ and $t_{sp} > t_2$, the plasma effect on the melting efficiency can be done away with to a large extent.

7.2. Microwelding

Welding of thin parts can be performed both with CW lasers and with repetitively pulsed lasers. The main parameters of the process of pulse welding are the pulse energy E_p, pulse duration, pulse repetition rate, focused spot radius r, and position of the focal point relative to the target surface.

An approximate expression for the pulse energy required to melt a metal without flushing assumes the form

$$E_p = 0.885 T_m K t_p \pi r^2 / \sqrt{k t_p}. \tag{7.4}$$

The pulse duration determines the time of melting of a metal depending on its properties and thickness. Approximate values of pulse duration needed to melt copper, aluminum, and steel are $10^{-4} < t_p < 5 \times 10^{-4}$ s, $5 \times 10^{-4} < t_p < 2 \times 10^{-3}$ s, and $5 \times 10^{-3} < t_p < 8 \times 10^{-3}$ s, respectively. More accurate values of t_p can be found from experiments.

The trapezoidal or triangular waveform with a steep leading edge and a gently sloping trailing edge is the best wave for welding.

The focused spot diameter d determines the area of heating and laser power density

$$q = 4E_p / \pi d^2 t_p. \tag{7.5}$$

The diameter d is chosen in the range from 0.05 to 1.0 mm so as to ensure a required power density of 10^5 to 10^6 W/cm^2. The simplest way to adjust the spot diameter and thus the power density q is to defocus the beam, therewith placing the focal spot of the least diameter above or below the target surface.

Welds are made by overlapping weld spots at a definite overlap ratio of 0.3 to 0.9, which varies with the type of weld, its strength, and air-tightness. The speed of seam spot welding depends on the weld spot diameter equal to about the diameter d of the focused beam, overlap ratio a, and pulse repetition rate f_p:

$$v_w = d f_p (1 - a). \tag{7.6}$$

Solid-state laser welding set-ups can produce a weld seam at a speed of up to 5 mm/s and f_p of up to 20 Hz. An increase in f_p can lead to a higher value of v_w.

A spot weld features quite a small fusion zone in cross section. A low dwell time results in a high cooling rate of 10^5 to 10^6 K/s in the fusion zone which

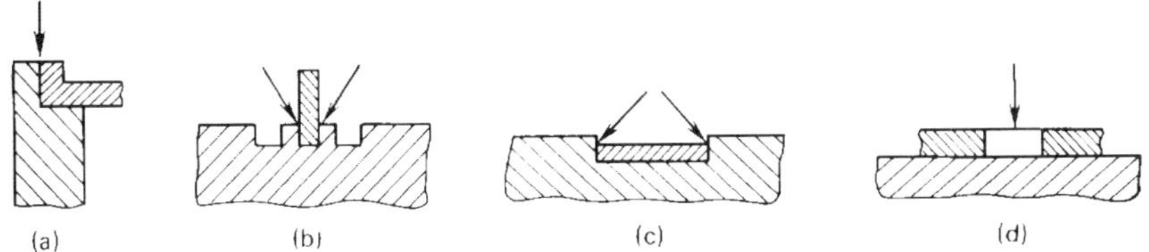

Fig. 7.4. Designs of weld joints between thin and thick parts. Arrows show the beam direction

displays a fine-grained dendritic structure with interdendritic inhomogeneity. The size of the heat-affected zone that has an inhomogeneous structure is generally small and ranges from 100 to 150 μm in cross section.

A filler material introduced into the weld pool in the form of a powder, wire, or strip serves to change the chemical composition of the weld metal, adjust its structure, and increase the cross-sectional area of the weld to eliminate the weld concavity. The filler can be sprayed or smeared on the part edges to be welded.

Spot welding enjoys the widest uses in electronic engineering for assembling components with butt, lap, and corner joints, and especially for joining small pieces to large parts.

Figure 7.4 illustrates joint designs for welding small parts to massive components, with arrows showing the beam direction. In the joint design of Fig. 7.4*a*, a thicker body of revolution is cut to provide a ledge for a thinner part. If the two parts differ considerably in thickness, the beam falls on a thicker part to equalize the temperature field and uniformly melt both parts. In the tee joint design of Fig. 7.4*b*, the joint preparation is more complex and calls for grooving the thicker part so as to form flanges raised on both sides of the thinner part. The groove cut in the massive part of Fig. 7.4*c* accommodates a thin part to be seam welded along the edges of the groove or spot welded at individual points to a penetration depth below the thin part. In the joint design of Fig. 7.4*d*, a thin part has a hole to allow the beam to melt over the hole edges and the surface of the thick part along the circumference.

An optimal joint design is one that ensures the most efficient melting and solidification of the metal. In preparing a butt joint it is essential that the gap and the edge skewing should be held as small as possible.

Both solid-state and gas lasers operating in the CW mode at powers of up to 1 kW are suitable for microwelding.

7.3. Deep Penetration Welding

As mentioned above, penetration welding is a keyholing process producing a specific profile of the metal pool such as illustrated in Fig. 7.5. and described in Chaps. 4 and 6 and in Sec. 7.1. The basic parameters of this type of welding is the laser power and welding speed, including the parameters of a focusing system.

The laser power determines the melting efficiency and the character of weld shaping. Along with the laser power, such characteristics as the mode pattern, polarization, power density distribution, and beam divergence need be taken into account since the same welding process may require different values of power in changing from one laser model to another. At a specified value of power, the welding speed can range from a minimum value at which no keyholing occurs to a maximum one at which the beam forms a poor weld with undercuts, pores, faulty fusion, and other defects.

The optical system focuses the beam to a spot 0.5 to 1.00 mm in diameter. At a smaller spot size, the beam of increased power density overheats the molten metal and causes it to vaporize intensively, thus producing a poor weld. At a spot size in excess of 1.00 mm the process efficiency decreases.

The optimal shape of the metal pool cross section is a keyhole profile that ensures an aspect ratio above unity. The aspect ratio depends on the position of the focal plane relative to the target surface. Locating the focal point below the surface affords the largest penetration depth. The depth of the focal point below the surface varies with metal properties, parts thickness, and welding conditions.

The pool profile in Fig. 7.5 shows that deep penetration involves the process of keyholing coupled with the process of surface melting due to heat conduction through the metal. It is the welding conditions that primarily determine which of the processes prevails and secures a definite profile of the metal pool.

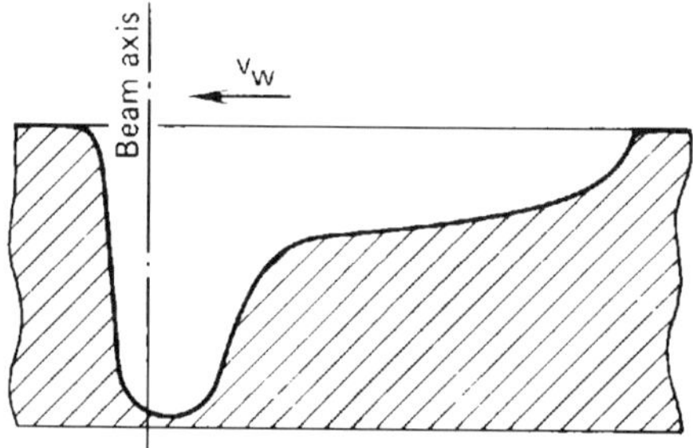

Fig. 7.5. Laser metal pool profile in the longitudinal section

The melting efficiency depends on the conditions of beam focusing. Along with ensuring a maximum power density at the spot of a minimum size, the focusing system must afford a requisite angle of beam convergence.

The procedure of calculating the welding parameters with the aim to achieve high efficiency of the process that would provide quality welds is quite laborious because of the intricate interrelation between the parameters. A practical approach is to use experimental and reference data for deriving empirical formulas from the regressive analysis of the results. In a first approximation, the penetration depth *h* as a a function of laser power *P* and welding speed can be given by the formula [64]

$$h = \beta \sqrt{P}\, v_w^{-\gamma}, \tag{7.7}$$

where β and γ are the factors used to account for the laser source applied, type of focusing system, and the properties of metal welded.

In [64] the interrelation between welding parameters is given in terms of normalized power P/hT_mK and normalized speed $v_w d/k$, where T_m is the melting temperature, K is the thermal conductivity, and k is the thermal diffusivity.

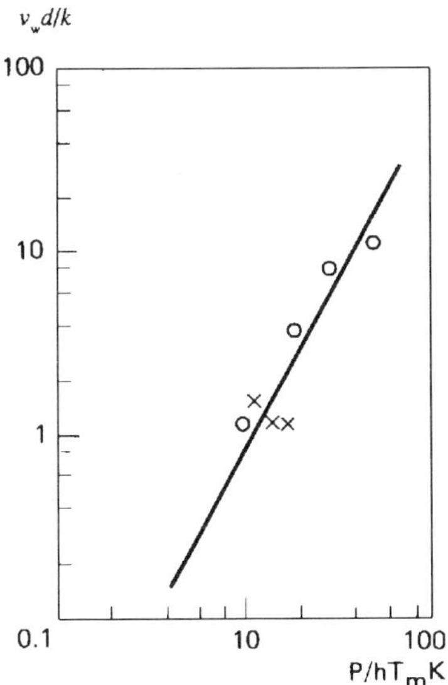

Fig. 7.6. Normalized welding speed versus normalized power for welding steel (light circle) and aluminum (cross)

The results of experimental data confirm that the normalized speed varies in a linear fashion with normalized power (Fig. 7.6).

An increased power density of the laser source as against the arc-welding source enables the weld to be made at a higher speed and, hence, a lower heat input, so that the weld metal heats up and cools quite rapidly. These conditions are of decisive importance for the weld strength, i.e., the weld metal resistance to hot and cold cracking. Laser welding carried out at speeds of 30 mm/s or more greatly adds to the weld strength and makes many structural materials not suitable to arc welding more weldable.

Residual strains in laser welds formed at speeds of 25 to 30 mm/s are 3 to 10 times those in arc welds. Because of a narrow plastic strain region in a laser weld, the compression stresses induced in laser welding are by 40 to 70percent lower than those observed in arc welding, with the result that the residual strains do not distort the shape of a sheet material. In arc welding, high strains can disturb the buckling stability of welded sheets and cause the shape distortion.

Laser welding uses both fluxes and shielding gases to protect the fusion zone against oxidation and assist in more efficient melting of the weld metal. Gas is supplied directly to the interaction region through a nozzle just as it is done in arc welding. Examples of the shielding gas arrangements appear in Fig. 7.7. In welding highly reactive metals under full penetration conditions, it is necessary to provide protection for the weld root as well.

The shielding gases used in arc welding are also suitable for laser welding. In selecting a shielding gas for laser welding, preference is given to the gas that does not interact with the metal vapor plasma and raises the melting efficiency. As noted earlier, the gas with a high ionization potential and increased thermal conductivity greatly promotes the fusion process. The weld quality depends on a gas flow rate; a low flow rate renders weld protection ineffective and excessive flow rate leads to unjustifiable expenses. A gas flow rate established in experiments that can provide adequate protection against oxidation is as follows: helium, (50–60) × 10^{-5} m^3/s; argon, (15–20) × 10^{-5} m^3/s; and mixture of 50% He and 50% Ar, (45–50) × 10^{-5} m^3/s.

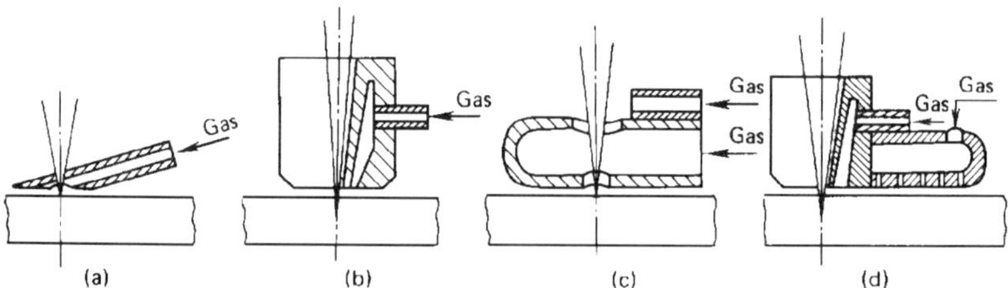

Fig. 7.7. Shielding gas arrangements

Laser welding can use the same fluxes as those applied in arc welding, preferably spreaded over to give a coat to the part to be welded.

Additives serve to achieve the following: adjust the chemical composition of the metal to obtain the desired properties of the weld joint; obviate such defects as pores, cracks, and fusion irregularities in depth; and ease the requirements for fit-up accuracy. A thin filler wire less than 1.0 mm in diameter is commonly fed into the interaction region with special precision mechanisms.

The defects inherent in penetration welding are fusion irregularities at the weld root and cavities extending deep into the weld metal. Irregular fusion in depth decreases at a higher welding speed. One more way suggested in [102] to reduce nonuniform fusion and porosity is to deflect the laser beam from its vertical line by 15 to 17° in the direction of welding. Welding at a penetration depth down to the permanent or removable backings can obviate the defects resulting from irregular fusion.

Welding with repetitively pulsed lasers at a pulse repetition rate of 0.4 to 1 kHz and a pulse length of 20 to 50 ms can afford the penetration depth three or four times that achieved with CW lasers. Spot welding offers two or three times the thermal efficiency of continuous laser welding. However, spot welding calls for quite a precise guiding of the beam on to the target and highly accurate edge preparation. Besides, the speed of spot welding is a few times slower than that of CW welding.

Shifting periodically the focal point of the CW beam relative to the target surface, as illustrated in Fig. 4.13, appreciably increases the process efficiency. The thing is that the beam can penetrate deeper into the material with an increase in the speed of travel of the fusion and vaporization fronts into the weld pool [63]. This speed increases with the power density applied to the front wall of the crater. At a fixed position of the beam focus the power density drops with depth. The power density can be held constant in depth by vibrating the lens, for example, with a piezoelectric mechanism that can change the focal point position at a rate of up to 150 Hz to a peak value equal to the thickness of the welded part.

This method of focal point oscillation can increase the penetration depth in steels, aluminum alloys, and titanium alloys by 40 percent. Although the weld width increases by about 30 percent, the aspect ratio grows by 10 to 15 percent. The weld exhibits lower fusion irregularities at the root and has an improved appearance. The thermal efficiency is by 60 to 80 percent higher than it is in welding with the fixed position of the focused spot.

The stream of a gas such as argon or helium applied at a definite pressure directly to the interaction region can promote the melting efficiency [43]. The gas flow reduces the plasma effect and controls the hydrodynamic processes in interaction regions. A properly adjusted gas flow rate at an optimal pressure provides for a deeper penetration. An increase in the flow rate above the optimal value impairs the quality of the weld which can develop large pores and cavities. A still higher flow rate may force the molten metal out of the pool and result in metal cutting.

Researchers of the Bauman Moscow State Technical University have developed an efficient laser welding process using the pulsating feed of a gas. A pulsating flow rate causes the plasma to move from the surface into the cavity and the constant component of the flow rate limits the vapor plasma above the target surface. The optimal flow rate varies with the laser power density, welding speed, and properties of the welded metal. This method can increase the penetration depth by 30 to 40 percent and appreciably stabilizes the process of fusion, which is of much importance in partial penetration welding.

Combining the laser source with a cheaper energy source can greatly improve the welding efficiency. Thus, arc-augmented laser welding offers the combined effect greater than the sum of the effects produced by each source, one acting independently of the other. The process is most efficient when the arc power is comparable with the laser power. In particular, the process can be run at a speed a few times the speed of the conventional process. Also, arc-laser welding is a cost-effective process since the arc-welding source is comparatively cheap. The addition of a conventional low-cost energy source can increase the net power coupled to the target and improve the performance characteristics, apart from retaining the advantages of the laser source.

Properly prepared surfaces and edges of parts to be welded can increase the melting efficiency [103]. The laser power coupling efficiency can be raised by coating the parts to be welded with such chemical elements as potassium and sodium of low ionization potentials, which can effectively suppress ionization and thus reduce the shielding effect of the plume [36].

7.4. Welding of Structural Materials

The present-day laser welding technology can offer a number of well developed processes tailored to weld metals up to 10 mm thick. What limits the field of laser welding application is the cost of an industrial laser that is still rather high, and so the decision to embark on laser welding may be taken if the improved part quality more than offsets high capital investment or if the conventional processes yield poor welds or are unable to weld at all. It will be found advisable to apply laser welding in order to:

1. To produce a precision weldment whose shape and size should not change in the welding process.

2. To simplify the production process in which the last operation is welding that dispenses with the operations of part straightening or machining.

3. To increase the throughput which is easy to do since the laser can produce welds at a speed of 35 mm/s and over.

4. To weld low-rigidity large-size assemblies in areas difficult to reach with conventional welding techniques.

5. To weld materials difficult to weld with other techniques, including dissimilar metals.

Low-carbon and low-alloy steels. These steels find wide uses for the fabrication of weldments. A low-carbon steel contains up to 0.25 percent carbon and a low-alloy steel contains 0.25 percent carbon and has a total content of alloying elements of up to 4 percent.

Low-carbon steels are quite weldable. The adopted techniques of welding of these steels must comply with a number of requirements, the basic ones being that the weld joint should be free of defects and equivalent in strength and other mechanical properties to the base metal. The weld joint should be resistant to embrittlement. The selected process must be efficient, cost-effective, and reliable.

Experience gained in arc welding enables us to single out the basic features of a welding process that secure a high quality of the weld.

A high rate of weld metal cooling at a certain decrease in plasticity and impact elasticity of the metal favors an increase in the weld strength. A rigid thermal cycle in deep-penetration laser welding encourages a maximum possible rate of up to 3000 K/s, thus changing the primary and the secondary structure of the metal. In the range of welding speeds from 80 to 120 m/h the original structure of steel St.3 becomes dendritic-columnar, with individual regions converting to a polyhedral-equiaxial structure because the temperature gradient in the solidification front rises slower than the solidification rate with the speed of welding.

The secondary structure in a laser weld is built up of acicular ferrite. In an arc weld, it is ferrite-pearlite. Acicular ferrite has a hardness of 1920 to 2160 MPa, while the hardness of ferrite in the base metal reaches merely 1200 MPa. This ferrite results from an intermediate or bainite transformation occurring at a high cooling rate.

During the thermal cycle, the metal structure in the heat affected zone changes appreciably. The arc weld produced in a low-carbon steel or a low-alloy steel displays the regions of hardening, incomplete hardening, and tempering. In the general case, the lengths of these regions, their structure, and the size of grains depend on the beam action time.

The thermal cycle in laser welding ensures a minimum time of overheating of the region where the metal can completely convert to austenite, thereby precluding the grain growth and impairment of the alloy properties. The heat affected zone in laser welding is smaller in size than that in arc welding by a factor of 3 to 5. This feature is of particular importance in welding heat-hardened steels. Because of a small HAZ and a rapidly varying temperature, the tempering region has no time to soften.

The properties of low-alloy steels when welded with a laser change more heavily. Arc welding of a hot-rolled steel fosters a hardened structure in the regions of overheating and normalizing. The steels subjected to hardening so that they still retain high plasticity are more difficult to weld. In the regions of recrystallization and blue brittle these steels soften under the effect of high-temperature tempering, yielding temper troostite and sorbite. The higher the strength of the hardened steel, the higher the loss of hardening. The cooling rate and mostly the heat input have a decisive effect on weld properties. An increase in the heat input leads to a lower strength and wider region of softening.

Laser welding can master the above difficulties largely due to a high cooling rate and low heat input. The heat input required to weld metals 2 to 5 mm thick with the laser is lower than that in arc welding by a factor of 3 to 5. For this reason, the weld produced in the steel containing 0.14 % C, 1.2 % Mn, 0.34 % Si, 0.16 % Cr, 0.08 % Ni, and 0.094 % V exhibits a martensite-bainite structure, with martensite formed at the fusion junction. The partial recrystallization region has a ferrite-pearlite structure, as does the base metal.

All low-carbon and low-alloy steels weld readily by any fusion welding technique. The hardened regions in the HAZ that are conducive to cold cracking do not pose too severe a problem.

Laser welds in low-carbon and low-alloy steels have good mechanical properties.

For example, welds formed in low-carbon steel 3 to 6 mm thick with a CW CO laser exhibit a strength equal to that of the base metal, impact strength of 1.2 MJ/cm, and bend angle of 180°. The spectral analysis reveals that the manganese and silicon contents in these welds decrease by 0.02 percent. The single-pass welds in low-alloy steel 15 to 20 mm thick have a strength equal to that of the base metal and a rather high impact strength.

Medium-carbon, high-carbon, and alloy steels. Medium-carbon steels contain 0.26 to 0.45 percent carbon and find wide applications for weld assemblies. High-carbon alloys contain 0.46 to 0.75 percent carbon, which are poorly weldable and rarely used for weldments. The total content of alloying elements of an alloy steel ranges from 2.5 to 10 percent.

The welds in these steels have the tendency to form highly brittle regions, hot and cold cracks, and pores. Experience shows that these steels require preheating before welding and heat treatment after welding, although these operations complicate the process.

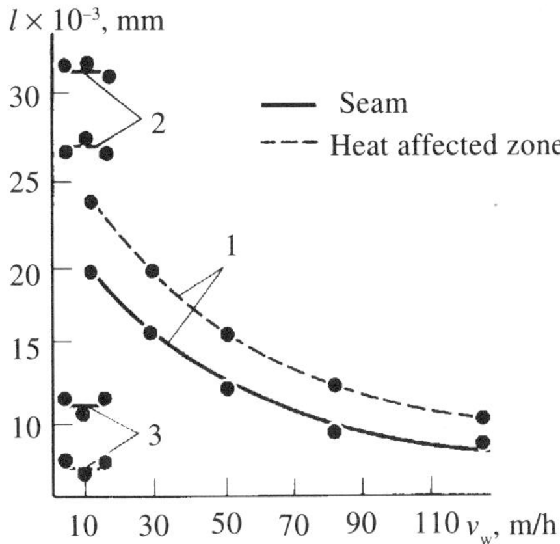

Fig. 7.8. Austenitic grain size versus welding speed and welding techniques: (*1*) CW welding; (*2*) argon-arc welding; (*3*) pulse welding

The factors responsible for cold cracking in the welds of the above steels are the following: (1) the state of the weld metal structure; (2) hydrogen content; and (3) level of principal tensile stresses. The size of an actual austenitic grain and its shape and also the distribution of chemical elements determine the structure of the weld and largely govern its properties. A decrease in the grain size adds to the plasticity and cold crack resistance.

Laser welding can appreciably refine austenitic grains. The experiments carried out on an automatic argon-arc welder reveal that the arc weld displays larger grains than the laser weld (Fig. 7.8).

The grain size in laser welding is smaller than in argon welding even at comparable welding speeds. A decrease in the heat input due to a higher welding speed refines the grains whose size decreases from 20 μm at 40 m/h to 8.5 μm at 150 m/h, which is indicative of the obvious advantage of laser welding over arc welding. The laser weld displays a strongly disoriented fine-grain structure which increases the cold crack resistance.

The segregation of alloying elements in the process of crystallization renders the weld metal chemically inhomogeneous, which impairs its mechanical properties and cold crack resistance. The main cause responsible for the formation of a chemically inhomogeneous weld is a different solubility of mixtures in the liquid and the solid phase. Since the solid phase contains a smaller amount of solute than the molten metal, the solute surplus accumulates at the solidified front. In the general case, the alloying elements diffuse slower at a higher freezing rate. It would seem that the laser weld should then be less homogeneous than the arc weld. However, the reverse is true because a high level of grain dispersion enables the elements to diffuse more uniformly over the bulk of the solidified metal.

A more homogeneous structure of the heat affected zone displays a higher stability to cold cracking. The process of phase transformations yields the final structure whose composition depends on the effect of the heat source on the target.

The welds produced in carbon and alloy steels by any fusion welding method generally display a brittle martensite structure noted for high hardness and increased strength. The hardness grows with carbon content, but then the metal becomes more brittle. Martensite containing more than 0.25 percent carbon has a decreased resistance to cracking and is highly prone to brittle failure.

The martensite structure of the laser weld metal is finer than that of the arc weld metal, with the martensitic needles being 5 or 6 times smaller in thickness. Such a structure features a high strength and adequate elasticity and offers an increased resistance to cold cracking.

Gases such as oxygen, nitrogen, and particularly hydrogen contained in steel strongly affect the weld joint and can lead to its embrittlement. Atomic hydrogen can accumulate in micropores, convert to molecular hydrogen, and raise the pressure that promotes the development of cracks. Diffusible hydrogen is conducive to the metal embrittlement to a greater extent than residual hydrogen.

The hydrogen content of the weld strongly depends on the shielding gas used and the method of welding. Given the same gas shielding against oxidation, the amount of diffusible hydrogen in the welds produced in Cr–Si–Mn steel with a CW laser is found to be 2 to 2.5 times smaller than for arc welds.

The matter is that a small amount of the laser-melted metal solidifies at a much higher rate than the arc-melted metal. Also, as the laser beam penetrates deep into the target, the gas intensively pushes the molten metal away from the keyhole, which rapidly solidifies so that hydrogen has much less time to diffuse into the metal. The weld produced with a repetitively pulsed laser has a still less amount of diffusible hydrogen.

A lower hydrogen content of the laser welded metal leads to a high resistance of the weld to cold cracking. Comparison between the values of cold crack resistance of the welds produced in Cr–Si–Mn steel with a laser and arc reveals that the laser weld has 3 to 3.5 times higher crack resistance than the arc weld (Fig. 7.9).

The factors responsible for hot crack resistance of welds in carbon and alloy steels are the following: (1) the chemical composition of the weld metal that determines the intercrystalline strength and plasticity in the brittle temperature range; (2) the value and the rate of growth of the tensile stress

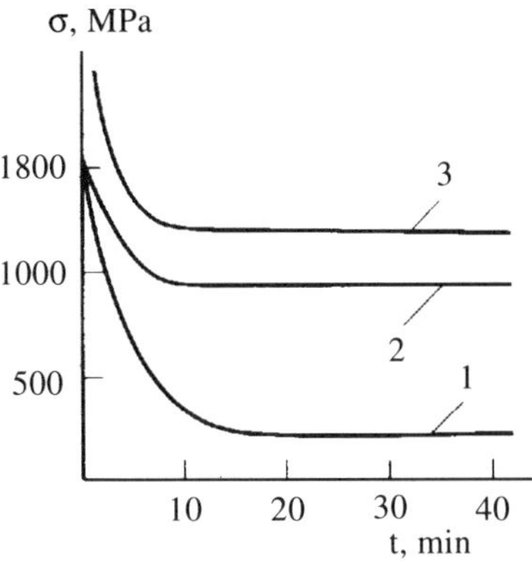

Fig. 7.9. Tensile strength versus loading time for the welds produced in Cr-Si-Mn steel with arc (*1*), CW laser (*2*), and repetitively pulsed laser (*3*)

and the corresponding strain in the brittle range; (3) the size of primary crystals; and (4) the weld pool shape which determines the direction of growth of columnar crystallites, the character of their intergrowth, the degree of zonal liquation, and the location of crystallite axes relative to the direction of tensile stresses.

The criterion of hot crack resistance is commonly taken to be a maximum strain rate of the crystallizing weld metal at which the crack does not appear. Since the thermal cycles in laser welding differ essentially from those in arc welding, the estimates of the critical strain rate defined by equation (6.4) are commonly made to judge the resistance of laser welds and arc welds to hot cracking. As seen from Fig. 6.8*c*, laser welds are more stable to hot cracking than arc welds formed at the same speed.

The weld in a hardenable steel exhibits a zone of cast structure that differs in chemical composition from the base metal, a hardening zone with a partially overheated course-grain structure, and a tempering zone. The mechanical properties of these zones may vary within wide limits for the same metal depending on its original structure, the chemical composition of the filler wire, the technique of welding and its parameters, and the conditions of subsequent heat treatment. A lower limit of weld strength depends on the strength of the base metal annealed before or hardened after welding and the strength of the tempering zone of the metal hardened before welding.

What accounts for a higher strength of the laser weld in a heat-hardened steel is that the metal in the heat affected zone softens insignificantly. For example, the ultimate strength of Cr–Ni steel, welded with a laser is equal to that of the base metal, whereas this strength in the arc weld amounts only to 87 percent of the base metal strength. The ultimate strength of chromium-manganese steel welded with a laser is 12 to 15 percent higher than that of the same steel welded with an arc. The test specimens of laser welded Cr–Ni steel develop fractures in the base metal.

A high strength of laser welded heat-treated steels can also be accounted for by the effect of "soft layer" hardening that can arise when the width of the softened zone is rather small. The strength of the softened zone in a laser weld is 10 to 15 percent higher than that in an arc weld, the microhardness is 375 MPa, and the length is 1.01 mm as against 340 MPa and 2.82 mm of this zone in the arc weld.

The soft layer in the laser weld becomes harder because of its small size. The soft, annealed zone is backed up, as it were, by the harder base metal and by the strong zone of the weld metal, for which reason the weld under test develops cracks in the base metal.

The impact strength of the laser weld in Cr–Ni steel is much higher than that of the arc weld at the weld junction, in the hardened zone and in the parent metal zone. The impact strength in the tempering zone is approximately the same for the laser weld and the arc weld.

A higher impact strength of the laser weld results from a much finer secondary structure of the weld metal and a cleaner and degased weld after its solidification. Unlike the arc weld, the laser weld is free of electrode material contaminants and has a smaller amount of such harmful impurities as sulfur, oxygen, and nitrogen since they absorb laser radiation more readily than metal and evaporate rapidly.

High-alloy steels. These steels contain more than 10 percent of alloying elements. Wide use for the fabrication of weld assemblies is made of austenitic steels of two classes which commonly contain no more than 18 % Cr and 10 % Ni as the basic elements, respectively. The total content of alloying elements of the austenitic steels can be as high as 55 percent. The welds in high-alloy steels are prone to hot cracking because the heat affected zone acquires a course-grain structure.

For the welds in these steels to be resistant to hot cracking, the conditions of welding should be such as to refine crystals and eliminate the columnar structure; to produce a certain amount of delta ferrite; and to reduce the amount of impurities which form a low-melting point eutectic.

Laser welding can in the main answer the needs. The laser weld formed in a high-alloy steel exhibits a fine-grain structure, contains 10 to 20 percent delta ferrite, and has a small amount of harmful impurities. The strength of the weld metal is comparable to that of the base metal and the plasticity is somewhat higher because of a decreased amount of nonmetallic inclusions.

Maraging corrosion-resistant steels find wide uses for critical weldments. A high strength combined with good elasticity of these steels results from the formation of high-alloy low-carbon martensitic matrix that subsequently undergoes precipitation (aging) hardening.

The arc welds in these steels are prone to corrosion cracking and intercrystalline corrosion in the atmospheric environment because the tensile residual stresses affect the area where chromium carbide precipitates along the grain boundaries and the metal in the heat affected zone experiences secondary hardening.

Experiments reveal that the weld zones heated to critical temperatures of 713 to 753 K, 838 to 853 K, and 863 to 88 K are highly prone to corrosion cracking because chromium partially redistributed in the weld adds to the electrochemical activity of the metal. The electrochemical activity begins to grow heavily after holding the metal at the above temperatures for 0.7 second and reaches a maximum after 2 seconds elapse. The tendency for intercrystalline corrosion accelerates accordingly. A further increase in the holding time does not generally change the corrosion resistance.

In arc welding, the thermal cycle has a large time of holding the metal at critical temperatures, which can amount to 10 s and above. For this reason the arc weld exhibits a high electrochemical activity and shows a tendency to intercrystalline corrosion. Laser welding carried out at a speed higher than 35 mm/s cuts down the holding time to one second and thus appreciably reduces the electrochemical activity in the heat affected zone, with the result that the corrosion resistance of the weld increases two or three times.

Maraging steels are also prone to cold cracking. Laser welding increases the cold crack resistance of these steels, and so the laser weld displays better mechanical properties than the arc weld. For comparison, Table 7.1 presents the mechanical properties of 2.0-mm corrosion-resistant steel welded with laser and arc.

When subjected to tests for tensile strength, the laser welds and arc welds exhibit fractures within the heat affected zone which softens somewhat in the

Table 7.1

Welding	Property		
	Ultimate strength, σ, MPa	Impact strength, MJ/m^2	Bend angle α, degree
Arc	1106	0.58	154
Laser	1216	0.59	158

process of welding. The laser weld has an increased strength because the original austenite grains are much smaller in size and the processes of secondary hardening and reaging have no time to terminate during the thermal cycle. The plasticity of laser welds increases insignificantly.

The tests for fatigue strength disclose the advantages of laser welding over arc welding. Laser welds offer a higher resistance to dynamic loads because they have a more homogeneous structure and lower tensile residual stresses.

For laser welds of requisite quality and reliability to be made, the welding technique developed for the purpose must cover a number of preparatory stages:

1. To plane, cut, or turn the edges to be welded in an effort to fit up and secure the parts in position most accurately.

2. To clean the metal surfaces, where the joint is to be made, of scale, rust, and moisture to guard against the formation of pores, oxides, and cracks due to the ingress of hydrogen. To use a wire brush to clean the sides and faces of plates and degrease the abutting surfaces. Where the process involves hardening of the part to be welded, the operations of edge beveling, grooving and cleaning should certainly be done after heat treatment.

3. To use assembly jigs and fixtures to adjust the edges along their entire length with a minimum gap and fit up the parts to the desired accuracy. Avoid tacking the parts where possible or, otherwise, tack them with a laser beam.

4. To adapt the process preferably to butt welding, for lap joints or vee joints formed in carbon steels are rather sensitive to stress concentrators.

5. To allow 0.5 to 0.8 mm of the plate thickness for machining the weld face and the back side of the weld to remove undercuts and poor fusion regions, if any.

6. To protect the weld bead against oxidation by shielding gases or gas mixtures, for example, a 2:1 mixture of helium and argon or 3:1 mixture of CO_2 and Ar, delivered to the weld pool through a nozzle. It is advisable to protect the weld on the back side with argon. Some metals, for example, low-carbon steels, do not need protection.

In deciding on the process of laser welding, the specific features of steel grades must be taken into account. For example, most of the high-alloy steels exhibit a low thermal conductivity and a high coefficient of linear expansion. Assuming the heat input and other parameters to be the same, the welds produced in these steels may differ in quality from those in low-carbon steels because the fusion zone and high-temperature regions in high-alloy steels tend to get wider and the total plastic strain in the heat affected zone tends to increase at a low welding speed. For this reason, high-alloy steels need be welded at a higher speed. Besides, consideration should be given to the type of weld, the conditions in which the weldment must operate, strain requirements, and to other factors.

Table 7.2

Steel grade	Steel thickness, mm	Laser power, kW	Welding, speed, m/h	Focal length, mm	Focal point depth, mm
Low-carbon, low-alloy	3.0	3.1	110	120	1.5
Medium-carbon,	2.0	2.8	100	120	1.5
alloyed	3.0	3.2	100	120	1.5
High-alloy,	3.0	3.3	100	160	1.0
austenitic	5.0	5.0	75	150	1.0
High-alloy,	2.0	2.5	100	160	1.0
maraging	3.0	3.5	80	500	1.5

Steels just like other materials, obey the general regularities related to changes in the weld geometry with welding conditions. Table 7.2 lists the values of some parameters of CW welding of certain steel grades. The process of welding at the values of the parameters presented in Table 7.2 ensures the best geometry of the weld and high strength of the weld metal. The optimal conditions of welding cover the range of welding speeds from 80 to 120 m/h, at which the laser power required to weld 1-mm steel reaches 1.0 to 1.3 kW. The depth of the focal point below the target surface varies with the design of a welded structure and the focal length. The above parameters relate to butt welding, but lend themselves to adjustment so as to produce slot, corner, tee, and other welds.

Aluminum alloys. These alloys are light, strong, and corrosion-resistant, for which reason they find use in a number of branches of industry.

The feature specific to aluminum alloys is that they intensively oxidize when heated to the melting temperature and above. The oxide film has a high melting temperature (above 2273 K) and strongly adsorbs gases and water vapors on its surface, which can render the metal pool nonuniform in density. The oxide particles penetrating into the metal pool impair the weld quality, and so special measures should be taken to destroy and remove the film and to protect the metal from secondary oxidation [104].

The welds in aluminum alloys may develop pores filled with hydrogen which well dissolves in aluminum at the melting temperature. Aluminum-magnesium alloys are strongly prone to porosity since magnesium promotes the solubility of hydrogen in aluminum. A way of reducing the porosity is to pretreat the target surface so as to remove moisture contained in the oxide film as the hydrated oxidc.

In the welds of aluminum and its alloys not subjected to heat treatment, the heat affected zone displays a coarser grain structure and softens to a certain extent because of the relief of work hardening. In the welds of heat-treated alloys, the heat affected zone also loses strength.

Aluminum alloys readily conduct heat and thus require a larger amount of energy than steels for their welding. Because aluminum alloys have a high coefficient of linear expansion and low modulus of elasticity, welds in these alloys develop much higher residual strains than welds in steels.

The use of such welding sources of high power density as a laser beam and an electron beam can obviate many difficulties mentioned above.

The text below presents the results of studies on the laser welding of aluminum alloys, primarily Al-Mg alloys that are widely employed in the manufacture of a variety of products.

The stage of weld preparation involves steps taken to remove coats of grease from workpieces, to groove the joint edges, to etch the oxide film, to clarify the parts in a 30 percent solution of nitric acid, and then to wash the parts with hot water. The oxide film should be etched with a 20 percent solution of sodium hydrate to a width of 25 to 30 mm along the entire length of the joint. The treated surface retains its properties for 3 or 4 days. The edges should be scraped bright just before welding. The surface so treated permits avoiding welding defects, primarily pores and oxide inclusions.

Shielding gases provide the most efficient protection of welds produced in aluminum alloys from oxidation. The use of conventional fluxes for the purpose proves ineffective. The weld requires protection on both sides. Helium provides the best protection and ensures a maximum penetration depth. Savings can be obtained if helium is brought directly onto the bead with argon applied onto the opposite side to protect the root.

The weld quality depends on an adequate choice of welding parameters. A feature typical of the process of welding aluminum alloys is that the laser beam provides the metal fusion only at a definite threshold power. For example, a CO_2 laser power of 2 to 2.2 kW is still sufficient to melt the target to a depth of 1.5 to 2.0 mm. However, at a lower power, the alloy does not melt at all. The thing is that at an emission wavelength of 10.6 μm, aluminum displays a high reflectivity (0.97) combined with high values of thermal conductivity and heat capacity. At temperatures close to the melting point, the reflectivity of aluminum sharply falls and the beam rapidly forces its way deep into the material. A further increase in power generally leads to a linear increase in weld penetration. The power threshold depends on the beam properties, focusing parameters, part thickness, state of a part surface, and welding speed.

The beam provides an adequate form of the weld with a minimum fusion zone under optimal welding conditions such that the weld geometry displays the specified values of depression k, sagging k_1, face width b, and root width b_1 (Fig. 7.10). The values of k and k_1 must not exceed 10 percent of the part thickness.

Laser welding of an Al-Mg alloy plate 2.0 mm thick provides, at a speed above 22 mm/s, the weld form similar to that shown in Fig. 7.10. The width of the weld with almost parallel walls in the cross section is about 2 mm and the values of sagging and depression are held within the permissible limits. The weld width increases insignificantly with laser power and penetration depth. Arc welding cannot provide such a shape of the weld. The optimal parameters of welding Al-Mg alloy with a CO_2 laser are given in Table 7.3.

The porosity of welds in Al-Mg alloy is within the limits prescribed in specifications for critical weldments. The microstructure of the laser-welded

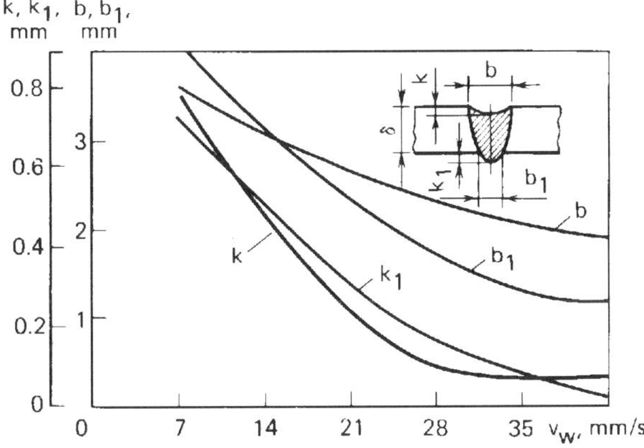

Fig. 7.10. Dimensions of the weld in the cross section as a function of welding speed

Table 7.3. Welding Parameters of Al-Mg Alloy

δ, mm	*P*, kW	v_w, mm/s	*f*, mm
2.0	2.1	25	120
	2.3	33	120
3.0	2.3	25	140
	2.8	33	140
4.0	2.8	25	140
	3.1	33	140

metal differs from that of the arc-welded metal. A laser weld exhibits a fine-grain structure with columnar dendrites that are smaller in size than those observed in the arc weld. In close vicinity to the HAZ the metal is free of a segregated eutectic and fused boundaries of grains. The region where structural transformations occur is five or six times smaller than that in the arc weld. The size of grains in this region increases only slightly. This structure improves the mechanical properties of the weld and increases its resistance to hot cracking.

Magnesium alloys. These alloys have a high strength at a low specific density. The welding of magnesium alloys presents difficulties primarily because they readily oxidize to form the oxide film of an increased melting temperature and a high density that is nearly twice the density of the metal. The oxide inhibits the edge melting and oxide particles entrapped in the molten pool impair the weld quality. For this reason, the weld bead requires a thorough protection from oxidation.

In the process of welding Mg alloys, the molten metal acquires an increased fluidity and the metal may lose the strength almost completely at a solid-liquid state temperature. To obviate the difficulty, arc welding uses copper or corrosion-resistant steel pads.

Laser welding ensures a good weld joint without backup plates, which appreciably simplifies the process of fabrication of large-size assemblies. The practice of welding magnesium alloys does not in principle differ from that used to weld aluminum alloys. The procedure of weld preparation involves the same operations of degreasing, etching, and scraping.

The optimal welding speeds range from 16 to 35 mm/s, and the laser power required for full penetration welding is slightly lower than that needed to weld aluminum alloys (Table 7.4).

The shielding gases used to protect the weld from oxidation are helium delivered to the laser-beam interaction region and argon forced to flow on the opposite side of the plate.

The microstructure formed in the heat affected zone is free of coarse grains. The two-phase structure of the base metal at the fusion line gradually changes to an acicular structure of the parent metal in the weld-metal zone. This

Table 7.4. Welding Parameters of Mg Alloy

δ, mm	P, kW	v_w, mm/s	f, mm
1.8	2.0	27.7	112
1.2	2.3	36.0	112

structure offers a high resistance to hot cracking. The strength of the parent metal is comparable to that of the base metal.

Titanium alloys. These light, strong, and corrosion-resistant alloys find wide uses in the manufacture of welded structures. The main difficulty involved in welding titanium alloys is that the molten metal strongly reacts with oxygen and hydrogen. An increased content of interstitial impurities, primarily hydrogen impurities, reduces the weld resistance to cold cracking. Titanium alloys are prone to grain coarsening at elevated temperatures.

A laser beam or electron beam can weld at a minimum heat input, thus obviating the above difficulties. At the stage of weld preparation, the edges to be welded are milled or turned, the oxide film is removed by shot-blasting or sand-blasting, and the surface is then cleaned by subjecting it to chemical etching, clarification, and washing. When fitting up the weld assembly, the parts should be well aligned and matched to each other to provide a requisite gap between the edges to be welded together.

In welding titanium alloys, it is necessary to provide protection on both sides of the weld. The shielding gas supplied through the nozzle with a shank should also shield the weld bead cooled to a temperature of 773 to 673 K. Both high-purity inert gases, such as helium and argon, and oxygen-free fluoride-chloride fluxes are suitable for protection of the metal pool. Helium serves to suppress the plasma and reduce the beam interaction and argon can protect the cooling bead and the weld root.

Welding should preferably be done at speeds above 25 mm/s because at lower speeds the weld gets wider, the structure becomes less favorable, and the metal can absorb more harmful gases which reduce the cold crack resistance. Table 7.5 presents the parameters of laser welding of some titanium alloys.

At high speeds of laser welding, the cooling rate increases and leads to an appreciable refining of grains in the heat affected zone, with the result that

Table 7.5. Welding Parameters of Ti Alloys

Alloy	δ, mm	P, kW	v_w, mm/s	f, mm
Grade 1	3.0	3.0	22	300
	5.0	4.0	22	300
Grade 2	2.0	4.0	44	230
	3.0	3.3	27	150
Grade 3	4.0	4.0	44	500
	5.0	4.0	27	150

laser welds feature better mechanical properties than the welds produced by other techniques.

Nickel alloys. These alloys contain 55 % Ni and above and are indispensable materials in chemical, oil, electric-power, and electronic industries for their corrosion resistance, high-temperature resistance, and heat resistance.

Heat-resistant Ni-Cr alloys (nichromes) offer an increased corrosion resistance in gaseous atmospheres at elevated temperatures. The composition of these alloys consists of an alpha-solid solution as the base with additions of such alloying elements as W, Mo, Al, and Ti, which increase the thermal stability of the solid solution and only slightly strengthen it through age hardening. The alloys are readily weldable.

High-temperature nickel alloys contain metallides and at elevated temperatures exhibit a higher strength than iron-base alloys and even cobalt-base alloys. Alloys containing titanium or titanium and aluminum with or without additions of refractory metals receive an added strength owing to such metallides as $Ni_3(Ti, Al)$ and $(Ni, Fe)_2$.

Age-hardenable alloys heat treated in the process of hardening and aging are moist stable to high temperatures. These alloys are poorly weldable and some of the alloys containing more than 4 % Ti and Al are unweldable by conventional techniques.

The thing is that the strengthening phases disintegrate at the heating stage of the welding process and separate out at the cooling stage, with the alloying elements segregating at grain boundaries. As a result, the regions of a decreased resistance to high temperatures appear, which can develop cracks under the action of elastoplastic strains. A complex welding technique including the stages of preweld and postweld treatment is helpful in alleviating the problem, but cannot yield the weld metal comparable in high-temperature strength to the base metal.

A decrease in the heat input or increase in the welding speed can improve the weldability. In this case, the strengthening phase has no time to dissolve and precipitate during cooling. In arc welding that commonly applies to high-temperature alloys, the above approach is inadmissible because the three-dimensional pattern of crystallization converts to the plane or linear pattern, which renders the metal quite sensitive to crystallization cracks. It is thus essential to work out a more elaborate techniques of welding that would promote the formation of a three-dimensional or equiaxial structure of the weld metal at a minimum heat input.

Experiments reveal that three types of cracks appear in the process of welding. The first type of cracks of the crystallization-segregation nature develops in the direction of the weld seam.

The second type of cracks develops across the weld and primarily occurs in alloys containing more than 5 % Ti and Al.

The third type of cracks appears in the heat affected zone of the weld during its heat hardening. The number of cracks of this type increases with the total content of Ti and Al, strain energy, and welded structure rigidity. These cracks are typical of rigid units subjected to dispersion hardening, where the welds sustain internal stresses close to the yield point and have highly softened heat-affected zones due to the coagulation and dissolution of the strengthening phase. This explains why the welds in these structures display a reduced creep resistance during heat treatment.

The common cause responsible for the formation of cracks of all types is the diffusion process that develops during welding in the liquid or the liquid-solid phase. Suppressing this process to a certain extent offers a substantial increase in the crack resistance during welding and in heat treatment.

Laser welding carried out at speeds of 70 to 100 m/h can improve the crystallization process, yield a more homogeneous structure, and decrease the dispersion hardening of the weld.

The crack resistance during crystallization increases with the speed of laser welding, whereas the opposite is true for argon-arc welding (Fig. 7.11). The factors accounting for an increase in the crack resistance of the laser weld are a higher deformability of the weld metal due to the enhanced refinement of the structure, the shift of the lower limit of the brittle range to a high-temperature region, and the conversion from the two-dimensional to the three-dimensional pattern of crystallization at speeds above 70–100 m/h.

The hardness of the laser weld decreases with increasing speed because the strengthening phases precipitate slower, thus raising the weld deformability at the stage of cooling.

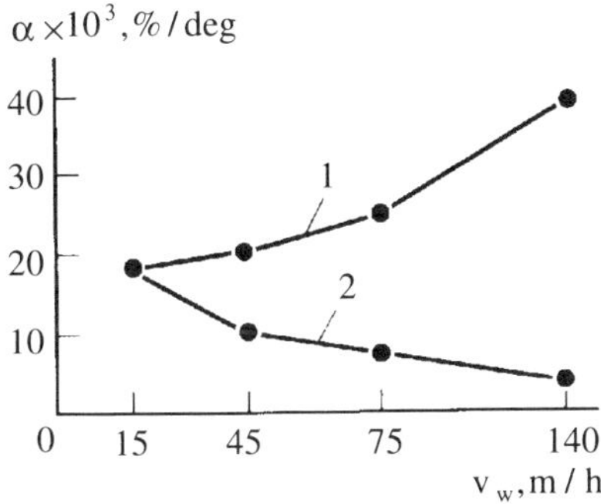

Fig. 7.11. Crack resistance of the solidifying weld in 1.5-mm Ni alloy as a function of speed of laser welding (1) and argon-arc welding (2)

In arc welding of nickel alloys, the heat affected zone gets weaker since the strengthening phases such as Ni_3(Ri, Al) disintegrate. In the process of heat treatment, the strengthening phases precipitate little, which localizes the strains as a result of the residual stress relief that occurs at the time of weld embrittlement. The loss of deformability leads to weld cracking.

Laser welding at high speeds can decrease the time of high-temperature heating by a factor of 10, thus retaining the strengthening phase properties and, hence, canceling the strain localization. This increases the crack resistance of welds in the subsequent process of heat treatment.

Laser welding carried out at speeds of 120 to 150 m/h can ensure an intensive heat hardening of the weld metal whose long-term high-temperature strength can almost be equal to that of the weld metal subjected to heat treatment (Fig. 7.12). The strength properties of the weld metal are 10 percent and 23 percent lower than those of the base metal for the laser weld and arc weld, respectively.

The short-term strength of heat-treated laser welds at a temperature of 20 °C is higher than that of unhardened welds and is about the strength of the base metal.

The tests carried out on heat-treated and untreated welds for a long-term strength at a temperature of 650 °C indicate that welds differ little in strength because the untreated welds undergo aging at the test temperature which corresponds to the aging temperature of nickel alloys.

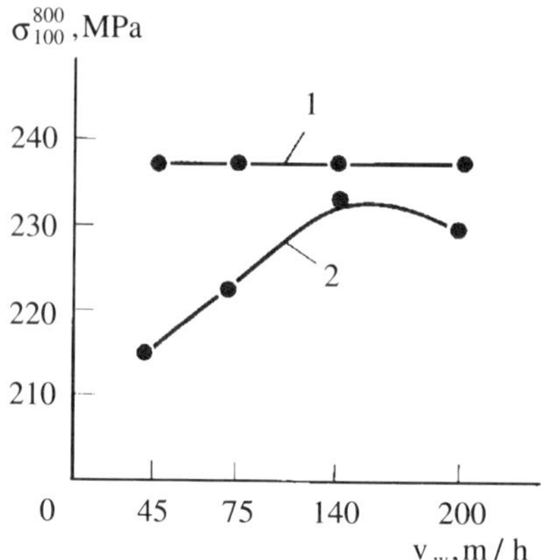

Fig. 7.12. Long-term high-temperature strength of laser welds in nickel alloys versus welding speed: (*1*) after heat hardening and aging; (2) without heat treatment

The tests for the fatigue strength measured within the fatigue range of 10^7 cycles show that the fatigue limit for the heat-treated laser welds in high-temperature nickel alloys is 10 percent higher than that for similar welds produced with an arc. The untreated laser welds are comparable in fatigue strength to heat-treated arc welds.

Considering that high-temperature nickel alloy parts operate under long-term load conditions at temperatures of 700 to 900 °C, it is advisable to use untreated laser welds for short-time loaded structures and heat-treated age-hardened welds for heavy-duty long-time loaded structures.

Nickel alloy parts stored for a long time become covered with sulfur-containing films which cannot be removed during degreasing. Before welding, the part should be ground along the section 20 to 30 mm wide and cleaned with acetone, white spirit, or purified gasoline.

To guard against weld cracking, the molten metal should be out of contact with free air. The molten metal can heavily absorb nitrogen, hydrogen, and oxygen, which form pores in the solidifying metal. The alloying elements such as titanium, chromium, and vanadium can reduce the porosity, while cerium, manganese, carbon, niobium, silicon, and iron tend to increase the porosity.

To protect the laser welds formed in nickel alloys, use is made of helium or a 1:3 mixture of helium and argon, or helium with 20 percent hydrogen. Helium and its mixtures are suitable for protection of the weld face and argon can provide shielding to the root. Hydrogen strongly dissolves in nickel as the latter melts and solidifies. Solidifying nickel can contain two times and three times the amount of hydrogen in austenitic steel and carbon steel, respectively.

Hydrogen exerts a positive effect on the molten pool because the atmospheric oxygen reacts with hydrogen, thus preventing the formation of the nickel oxide and, hence, pores.

Nickel and nickel alloys are more viscous in the molten state than steels, and so they require a higher power of the beam to melt them to an adequate depth.

For welds of high serviceability to be made, laser welding should be performed at speeds of 70 to 150 m/h, which ensure optimal dimensions of the weld, good structure, and appropriate mechanical properties.

In fitting up the weld joints, care should be taken to reduce the gaps, eliminate edge skewness, and prepare the butt joints without flanges for, otherwise, the pockets formed on the opposite side of the weld would promote crevice corrosion.

7.5. Welding of Composite Metals

Last years have seen an ever increasing use of composite metallic materials for their high strength and rigidity, low sensitivity to stress concentrators, high resistance to fatigue failure, and enhanced wear resistance and corrosion resistance.

This section deals with the technique of fusion welding of both composite materials and these materials with homogeneous materials. Proceeding from the results of welding composite and dissimilar materials, the composites can be broken down into five classes according to the interphase interaction of their components during welding.

The first class includes composite metals whose components form a homogeneous melt, but generally do not dissolve in each other in the solid state. These are Fe–Cu, Al–Be, Ti–Mg, and other alloys. After melting, these alloys solidify to form a heterogeneous structure with alternating particles of the matrix and the reinforcing agent. The fusion welding of these metals may yield the weld without softening its structure.

The second class includes the materials whose components display a limited or an unlimited solubility in the course of melting and crystallization. These are Nb-W, Ni-W, Al-Si, Co-Cr, and other alloys. The weld in such an alloy consists of a solid solution where the concentration of components smoothly varies from the fusion line to the weld axis. The strength of the weld metal is comparable to that of the alloy. The weld in the metal with a limited solubility of its components can also contain a eutectic or a peritectic alloy depending on the constitution diagram. The eutectic can be made to crystallize in the oriented fashion so as to reduce the weld zone softening to a minimum.

The third class covers the materials of fibrous structure, such as Al-B, Ti-B, and Al-steel. The fusion welding of these composites causes the fibers in the weld zone to disintegrate, yielding a large number of intermetallic compounds which render the weld brittle.

The fourth class embraces the materials whose components do not interact with each other because of a definite relation between the melting temperature of a refractory component and the boiling temperature of a fusible component. The two liquid phases cannot be in equilibrium and remain mutually insoluble at any level of overheating.

Examples of these composites are W-Cu, W-Ag, Mg-Be, and Nb-Ga. The melting point of the filler material used for their welding should be lower than the boiling point of a fusible component.

The fifth class includes such materials as Fe-Pb and Al-Pb, whose components are mutually insoluble in the liquid state. The weld in such a composite is prone to rupture quite readily within the fusion zone.

Table 7.6. Classification of Composite Metals by the Type of Interaction of Their Components in Melting

Class of composites consisting of	Group of composites		
	Fibrous	Laminated	Dispersion-hardened and reinforced with particles
1. Insoluble components in solid state	Al-Be, Fe-Cu Pb-Sn, Ag-Pb	Steel-copper	Al-Be, Fe-Cu
2. Soluble components in solid state	Ti-Mo, Al-Si Al-Zn, Ni-Cr Cu-Cr, Ni-W	Steel-steel	Co-Cr, Nb-W Ni-W
3. Components forming intermetallic compounds	Al-B, Ti-B Ti-Be, Ni-Mo Co-W, Al-Mo	Steel-brass Steel-titanium Steel-aluminum	Al-C Fe-brass
4. Noninteracting components	Mg-C, Mg-B		
			Cu-W, W-Ag
	Cu-W		Fe-Mg, W-Ga
			Mo-Ga
5. Insoluble components in liquid state	Mg-Ti, Mg-Ta Mg-steel	Fe-Pb, Ag-Fe	Fe-Pb, Cu-Mo Cu-Pb, Al-Pb Fe-Ag Fe(Cu-Pb) Fe(Cu-Pb-Sn)

This classification is rather conventional because one and the same material may belong to a few classes depending on the effect of the thermal welding cycle. Nevertheless, it allows us to single out the basic groups of composites and the criteria by which we can evaluate their weldability (Table 7.6).

The most promising materials are the composites reinforced with particles and based on mutually insoluble components, one supplementing the other with the required properties. It is difficult to produce these materials by fusion because the melt separates into two liquids which strongly segregate and crystallize in a wide temperature range since they greatly differ in melting temperature. These composite metals are generally made by sintering the components in the solid or the solid-liquid state, by contact alloying, and other methods.

The text below presents the experimental results of laser welding of Fe-Cu, Fe(Cu-Pb), and Fe-(Cu-Pb-Sn) systems based on immiscible components.

The Fe-Cu system is a class 1 composite whose components are totally soluble in the liquid state and practically insoluble in the solid state.

The Fe-(Cu-Pb) system is a class 5 material composed of insoluble components in the liquid and the solid state. The constitution diagram displays a wide range of insolubility of liquid lead in iron. In this system, lead and copper form a monotectic.

The Fe-(Cu-Pb-Sn) system belongs to the 5th class and also to the 3d class since it interacts with iron and copper, yielding intermediate phases. The constitution diagram of Fe-Pb components is similar to that of class 4 materials in which the liquid phases do not interact with each other.

Solid-phase welding is suitable for joining the composite metal parts together or these parts to steel, aluminum, and titanium parts. Bolted joints and rivetted joints are also practicable. Solid-phase welding of the diffusion and pressure types lends itself to control so as to mix up the components that tend to separate in the liquid state. However, this welding technique cannot provide butt welds between thin-wall parts or welds in the assemblies with bolted and rivetted joints. Also, it is difficult to join together the parts in complex-shaped assemblies.

Fusion welding offers indisputable advantages over solid-phase welding. This is a highly efficient process that can weld complex structures and is suitable for use at the assembly site.

In order to retain the basic properties of composite metals in the weld zone, preference should be given to laser welding which ensures a small HAZ size, minimum dwell time at a temperature above the specified value, and high rates of crystallization and cooling.

In the process of fusion welding, the components of a composite metal heat up to high temperatures at which the weldability of the components primarily depends on their metallurgical compatibility defined from equilibrium diagrams. These diagrams cannot afford quite exact estimates, but allow us to judge qualitatively how the components interact with each other during melting and solidification.

However, since there is not yet a sufficient number of well studied ternary and quarternary diagrams, recourse has often to be made to binary diagrams for predicting the behavior of complex systems.

The constitution diagram of the Fe-Cu system indicates that the system components are compatible, totally soluble in the liquid state, and do not form intermetallic phases, nor generally interact in the solid state. The welding process should only slightly change the structure of this material in order to avoid defects.

Lead and iron of the Fe-(Cu-Pb) system cannot yield a uniform structure because they separate out and segregate in the liquid state. Copper and lead also separate in the temperature range from 1227 to 1263 K. The metallurgical compatibility of these components in the ternary alloy, i.e., their miscibility in the liquid state, is rather problematic.

In the Fe-(Cu-Pb-Sn) system, tin can interact with copper and iron to form intermetallic phases in melting. The welds produced in Fe-(Cu-Pb) and Fe-(Cu-Pb-Sn) can obviously develop defects that are not typical of the defects observed in the welds formed between homogeneous materials.

In an effort to investigate the weldability of the above mentioned composites, clarify the defects, and work out the method for eliminating these defects, we have carried out welding experiments on composites with CO_2 lasers of up to 5-kW power in the atmosphere of helium or the helium-argon mixture to protect the weld pool and weld root.

The analysis of welds in Fe-Cu reveals that the resolidified zone has a sharply defined boundary with the base metal. A high thermal conductivity of the material promotes a rapid removal of heat from the solidifying metal to the base one. The metal crystallizes in a step-like manner, forming a heterogeneous but quite a regular structure with the morphology of the components similar to that inherent in the base metal. But the solidified metal differs from the base metal in dispersity because the rates of heating and cooling in laser welding are much higher than they are in the manufacture of the Fe-Cu material.

It is of interest to compare the process of laser welding of Fe-(Cu-Pb) and Fe-(Cu-Pb-Sn) with that of electron-beam welding. The EB welding set-up maintained a vacuum at 10^{-4} to 10^{-5} mm Hg and welded at a speed of 0.5×10^{-2} to 2×10^{-2} m/s with a beam of 1.0 to 2.5 kW focused on the target surface. In all the conditions of welding the electron beam action caused the molten metal to flush out of the pool, thus forming a cavity-type defect.

It is found that lead is mainly responsible for metal flushing. The vaporization temperature of lead at a pressure of 10^5 N/m^2 is 2023 K, while the melting temperature of the alloys under study reaches 1725 K. A slight overheating of the bath causes lead to vaporize and splash the liquid out of the pool. The boiling of lead in the bulk of the molten metal tends to form lead vapor bubbles which rapidly grow when the pool temperature rises above the vaporization point of lead. The analysis of the critical content of lead in the alloys of interest and experimental findings provide support for the boiling lead effect.

The experimental results confirm that one of the ways to eliminate the cavity-type defect is to increase the external pressure, i.e., to change from electron beam welding in vacuum to laser welding at the atmospheric pressure.

In the experiments, the laser beam power, the focused spot size, and the welding speed were chosen to be the same as in electron-beam welding, which resulted in cavity-type defects. In laser welding of 50Fe-(Cu-Pb) and 50Fe-(Cu-Pb-Sn), which contained from 5 to 20 percent lead by volume, the cavity defects in weld joints did not appear. The critical content of lead that exerted its effect on metal flushing was 5 percent and 5.2 percent for 50Fe-(Cu-Pb) and 50Fe-(Cu-Pb-Sn), respectively.

In the Fe-(Cu-Pb) and Fe-(Cu-Pb-Sn) alloys, the components are totally soluble in the liquid state only at temperatures above the critical "mixing" temperature T_{mix}. When the melting temperature drops below T_{mix}, the finest droplets begin to form in the homogeneous solution and rapidly settle down under gravity until the melt laminates into two layers. The liquid metal can have time to solidify without segregation only at a high cooling rate that is typical of laser welding.

Numerous experiments of laser welding of 50Fe-(Cu-Pb) and 50Fe-(Cu-Pb-Sn) indicate that the lamination of their components in the metal pool takes place when the former and the latter alloy contain 2.5 percent and 3 percent lead by volume, respectively. When the base metal contains more than 3 percent lead, the lamination occurs under any welding conditions and at any solidification rate since it is impossible to raise the metal to the critical temperature T_{mix} without causing lead to vaporize.

In laser welding of Fe-(Cu-Pb) and Fe-(Cu-Pb-Sn), the molten metal proves inhomogeneous. Even at the initial stage of melting, the particles immiscible in the liquid state begin to coalesce on the front wall of the crater because of nonuniform temperature distribution through the bulk of the liquid layer and local whirling of the liquid (Fig. 7.13).

The capillary forces and reactive vapor forces then rapidly transfer the liquid to the back wall of the cavity, where it solidifies. The copper-lead solution turns out to be under the molten iron as it flows along the side walls under the action of centrifugal forces. The pressure on the back wall does not allow this solution to flow down. The molten metal solidifies in layers. First, the iron begins to crystallize and forms, as it were, a rigid frame within which the copper-lead phase solidifies. Such a mechanism of crystallization must involve the process of squeezing out some amount of the copper-lead solution to the weld surface. The experiments confirm this conclusion. The surface layer of the weld consists of the Cu-Pb eutectic.

Let us note that high rates of solidification over the range of a few thousand degrees per second do not reduce the segregation because the copper-lead and iron particles begin to coalesce just on the front wall as a result of thermal-gradient-induced flows in the liquid phase. The liquid separated into the layers flows to the back wall and rapidly solidifies.

The above described mechanism of lamination and segregation of the alloy components clearly shows that the additional ultrasonic or electromagnetic action intended to mix up the molten metal cannot afford the desired result and can only increase the rate of coalescence of single-component particles. To uniformly distribute the components in the weld pool, use should be made

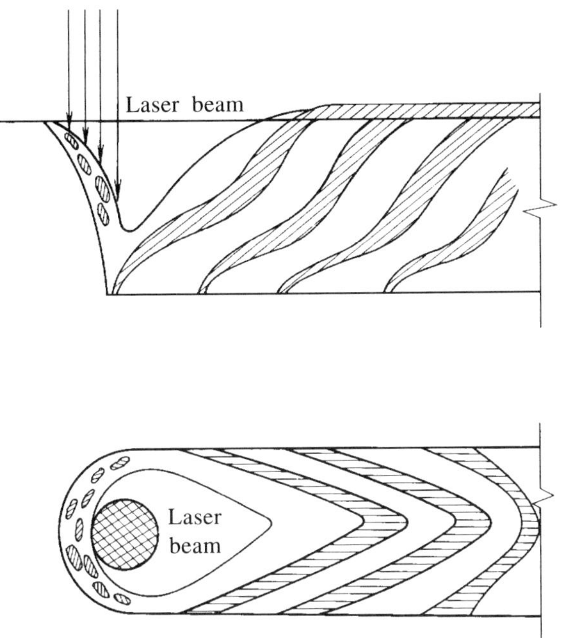

Fig. 7.13. Lamination of liquid components of Fe-(Cu-Pb) and Fe-(Cu-Pb-Sn) in laser welding

of alloying elements which reduce the critical temperature T_{mix} in the Fe-(Cu-Pb) and Fe-(Cu-Pb-Sn) alloys and suppress the segregation of their components. The method of introducing the alloying elements should certainly be suitable for industrial use.

In our experiments we used aluminum and nickel as alloying elements. An aluminum or a nickel strip served as a spacing placed between the butt-joint specimens which were brought in contact with the strip to eliminate the gap (Fig. 7.14). The strip thickness varied from 0.05 to 0.55 mm.

The studies of the effect of the lead content on the requisite thickness of an aluminum spacing reveal that the relation between the required aluminum content and the lead content takes a linear form:

$$V_{Pb} = a + KV_{Al}, \tag{7.8}$$

where $a = 2.5\%$ by volume and $K = 0.25$.

The straight line in Fig. 7.15 shows how the lead content defined by relation (7.8) varies with the aluminum content by volume. The plot locates the

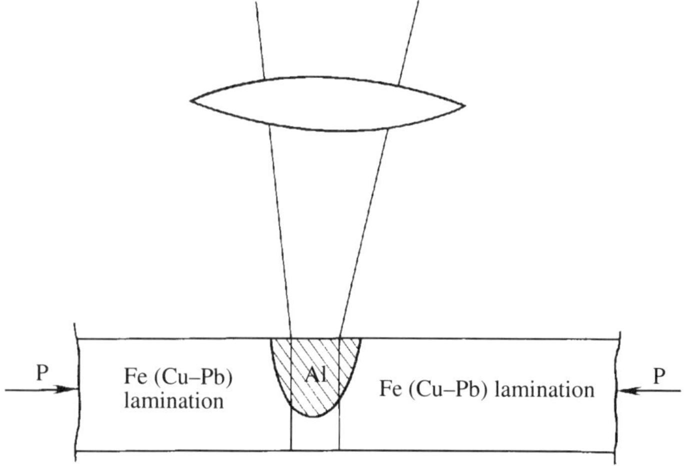

Fig. 7.14. Design of the butt joint with an aluminum spacer placed in the gap between the plates welded together

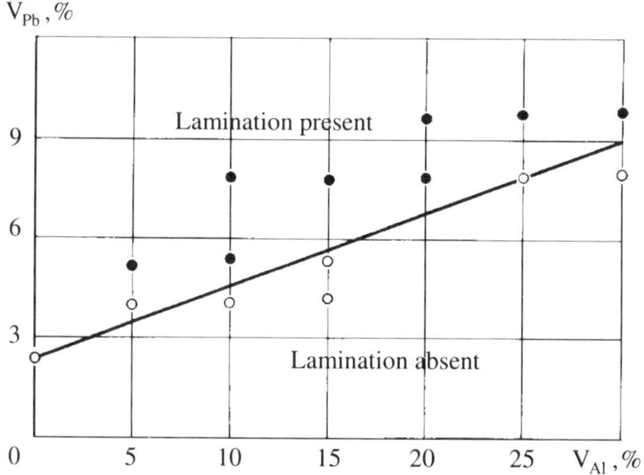

Fig. 7.15. Lead content of 50Fe-(Cu-Pb) versus content of aluminum used to suppress the segregation in laser welding at $P = 3$ kW and $v_w = 2.4 \times 10^{-2}$ m/s. Dark and light circles denote the presence and absence of the component lamination in the weld, respectively

miscibility region within which the weld shows uniform distribution of the components and aluminum.

The required thickness b_{sp} of a spacing can be found from the equation

$$b_{sp} = 0.4\, b_w (V_{Pb} - 2.5), \tag{7.9}$$

where V_{Pb} is the lead content of a composite and b_w is the weld width.

However, aluminum reacts with the components to yield intermetallic compounds in the weld metal which becomes rather inferior in strength to the base metal. For example, the weld in 50Fe-(40Cu-10Pb) alloyed with aluminum, which suppresses the lamination of the components, has a strength merely 30 percent that of the base metal.

Nickel used as an alloying element in the form of a spacer between laser-welded specimens does not form intermetallic compounds with the components of the composite. The experimental studies of the effect of nickel on the miscibility of the components in the weld pool have enabled us to determine the optimal nickel content that effectively suppresses the segregation depending on the lead content of the 50Fe-(Cu-Pb) composite (Fig. 7.16).

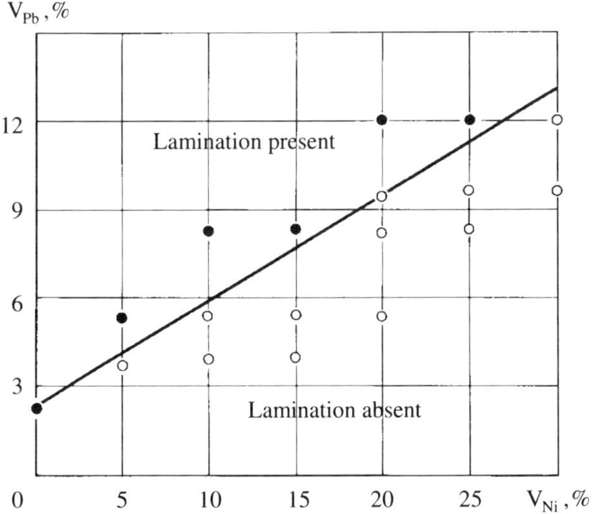

Fig. 7.16. Lead content of 50Fe-(Cu-Pb) versus content of nickel intended to suppress the segregation in laser welding at $P = 2.5$ kW and $v_w = 2.4 \times 10^{-2}$ m/s. Dark and light circles denote the presence and absence of component lamination, respectively

The thickness of a nickel spacing varies with the lead content and can be found from the formula

$$b_{sp} = 0.27\ b_w(V_{Pb} - 2.5). \tag{7.10}$$

The weld structure consists of nickel dissolved in iron and copper and particles of the copper-lead eutectic dispersed in the bulk of the weld metal. The strength of the weld in 50Fe-(40 Cu-10 Pb) alloyed with nickel reaches 80 percent of the base metal strength.

Aluminum and nickel used for alloying the 50Fe-(Cu-Pb-Sn) material suppress the component separation but do not cancel the formation of cracks in the weld. The weld formed in Fe-(Cu-Pb-Sn) containing 10 to 20 percent Sn by volume develop both longitudinal and transverse cracks. These cracks appear because of the material low plasticity in the brittle range and because of low-melting elements and eutectics. Alloying the weld metal with aluminum and nickel leads to a higher probability of crack formation in the heat affected zone, and for this reason it should not be used for the above composite material.

From the results of experimental studies we have worked out recommendations for the manufacture of parts and units from Fe-Cu, Fe-(Cu-Pb), and Fe-(Cu-Pb-Sn) composites. These materials have good damping characteristics, increased wear resistance, high corrosion resistance, and improved radiation protection properties. They are suitable for the fabrication of parts intended to carry impact loads, casings of devices, parts for friction units, and workpieces of welded structures.

There are two techniques of manufacturing composite metals, namely, contact surface alloying and bulk alloying. The processes of welding the composites differ with the technique of their manufacture.

As noted earlier, the procedure of butt joint preparation includes operations called upon to mill or grind the edges and to file off the burrs or grind them by an abrasive wheel.

In machining surface-alloyed composites such as Fe-Cu, Fe-(Cu-Pb), and Fe-(Cu-Pb-Sn), the surface alloyed with Pb and Sn should be given a softer cover such as a plastic plate. Before welding, it is necessary to scrape the Fe-Cu surface with a metal brush and the alloyed layer with thin emery paper, to keep it clean of fatty contaminants, draw off the dust with a solvent, and degrease the edges with ethyl alcohol.

Butt welds have been made with different gaps to determine how they affect the weld strength. The permissible gaps that do not reduce the strength of welds under static loading are 0.1×10^{-3} m for plates 1×10^{-3} or 2×10^{-3} m thick and range from 0.15×10^{-3} to 0.20×10^{-3} m for plates 3×10^{-3} to 5×10^{-3} m thick.

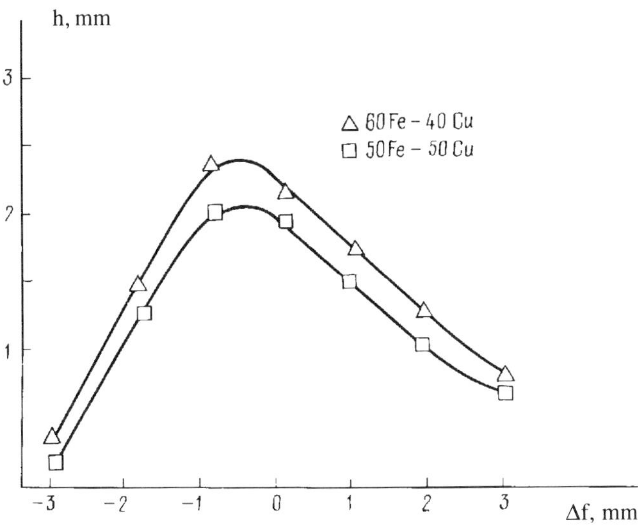

Fig. 7.17. Weld penetration depth h versus focal point penetration Δf relative to the target surface for laser welding of the Fe-Cu composite at $P = 3000$ W and $v_w = 1.4 \times 10^{-2}$ m/s

The shielding gases suitable for welding include pure helium, a mixture of helium and nitrogen, and also small additions of argon, of up to 20 percent.

As described in Sec. 4.2, the laser beam ensures a maximum penetration depth if the optimal system puts the focal point deeper into the target surface. Figure 7.17 illustrates the plots of the penetration depth h against the focal point penetration Δf relative to the surface in laser welding of 50Fe-50Cu and 60Fe-40Cu composites using the lens with a focal length of 140×10^{-3} m. The curves of Fig. 7.17 enable us to estimate an optimal location of the focal point relative to the target surface.

The welds formed in 50Fe-50Cu and 60Fe-40Cu composites may develop both longitudinal and transverse hot cracks because of too high a content of low-melting lead in the base material. A number of tests performed on laser-welded specimens for the tendency to hot cracking suggest that the Fe-Cu composite must contain not more than 15 percent lead by weight. In laser welding noted for high cooling rates, this amount of lead is able to dissolve in the copper component.

As noted above, the laser weld in Fe-(Cu-Pb) should be made with a nickel spacer placed between the plates with their edges butted up against the spacer. The thickness of a spacer can be found experimentally by producing butt welds with and without spacers. The widths of the welds produced with and without

Table 7.7. Laser Welding of 50Fe-(Cu-Pb) Composite

Type of part welded	Material	Part size	Ultimate strength	Welding conditions
Surface-alloyed 50Fe-(Cu-Pb) 50Fe-50Cu	1st layer: 50Fe(Cu-Pb) up to 10% Pb 2nd layer: 50Fe-50Cu	σ = 5 to 10 mm x = 0.2 to 0.5δ b = 2 or 3 mm	300 MPa	P = 5 to 8 kW $v_w = 2 \times 10^{-2}$ m/s
Surface-alloyed Ni 50Fe-(Cu-Pb) 50Fe-50Cu	1st layer: 50Fe-(Cu-Pb) up to 15% Pb 2nd layer: 50Fe-50Cu	δ = 3 to 10 mm x = 0.1 to 0.5δ $b_1 = 0.27b_2(V_{Pb} - 2.5)$ b_2 = 1.5 to 3.0 mm b = 2.0 mm	300 MPa	Two-sided welding P_1 = 2 to 4 kW $v_w = 3 \times 10^{-2}$ m/s P_2 = 5 kW $v_w = 1.4 \times 10^{-2}$ m/s
Bulk-alloyed b_1 Ni 50Fe-(Cu-Pb) b_2	50Fe-(Cu-Pb) up to 15% Pb Ni spacer	δ = 1.5 to 5 mm b_2 = 1.5 to 3 mm $b_1 = 0.27b_2(V_{Pb} - 2.5)$	250 to 300 MPa	P = 2 to 5 kW $v_w = 3 \times 10^{-2}$ m/s

spacers up to 0.5 mm thick are practically the same. Given the weld width and the material composition, we can estimate the thickness of a nickel spacer from expression (7.10).

The welds in 50Fe-(Cu-Pb) are prone to form voids at the end of the laser beam action, because a thin layer of the molten metal rapidly solidifies on the back wall of the cavity and has no time to fill the crater. Whenever possible, removable straps can be used to eliminate this defect. For workpieces with circular welds which do not allow the use of straps, the crater can be welded up by reducing the beam power.

Table 7.7 gives examples of laser welds which can be made in the 50Fe-(Cu-Pb) composite of the surface-alloyed and bulk-alloyed types. This material surface-alloyed with up to 10 percent lead by volume can be welded without spacers if the thickness of the alloying layer does not exceed two-tenths the thickness of the material. The material components will not separate because the total content of lead in the weld is not more than 10 percent by volume. If the material contains more than 10 percent lead, the weld should be made with a nickel spacer. Welding should be done on both sides where possible, first on the side of the 50Fe-50Cu layer of a higher weldability and then on the side of the alloyed 50Fe-(Cu-Pb) layer.

8 Cutting of Nonmetals

8.1. General

Nonmetallic materials are cut by a number of conventional methods such as machine, ultrasonic, plasma arc, and water jet cutting. Laser cutting that has recently gained wide acceptance offers a number of advantages over conventional methods, which include: a narrow, high-precision kerf; lack of tool contact; ability to cut materials of any hardness and reach hard-to-get-at areas; small heat affected zone; ease of automation of the process; and high production rate. The share of nonmetals roughly accounts for 70 percent of the stock of materials cut with industrial lasers.

The laser cutter can readily cut such high-strength materials as cermets, vitreous carbon, and composites based on carbon and boron fibers [67, 105]. The transition to automatic laser cutting can drastically increase the production rate [106]. Laser cutting of glass cloth provides a high-quality kerf and yields a much smaller amount of glass dust than machine cutting, thereby improving service conditions.

The use of laser cutters in wood-working industry brings forth considerable advantages. For example, the process of plywood cutting with a laser is free of sawdust and secures a narrow kerf, smooth edges, and a high accuracy of complex cut-out patterns [107]. The use of lasers in papermaking for cutting paper and binder board can appreciably raise the process efficiency [66].

Let us consider some physico-chemical features of laser cutting of nonmetals. Because nonmetals exhibit a lower thermal diffusivity than metals, a surface layer more than 0.5 mm thick can be considered as a semi-infinite body when cut at speeds v_c exceeding 10 mm/s and with a focused spot diameter not larger than 0.5 mm [52]. This means that the threshold power density q_{th} required for cutting is generally independent of the target thickness.

The absorptivity of CO_2 laser radiation at a wavelength of 10.6 μm for most of the nonmetals can be as high as 0.9. Since the laser energy effectively couples to the target, the laser beam can afford a large cutting depth. For example, a laser power of 500 kW is enough to cut 25-mm thick polymethyl methacrylate (acrylic plastic).

In its interaction with organic compounds, the laser beam initiates various chemical reactions of decomposition, vaporization, sublimation, and thermal dissociation with evolution of gases [108]. The process of cutting wood, rubber, paper, and some plastic materials results in the formation of free carbon (soot). In the interaction of the laser beam with some nonmetals, the damaged target produces aerosol in the form of fine particles of the vaporized material.

If a material rapidly evaporates, even a low power density of the order of 10^3 to 10^4 W/cm^2 is sufficient to form an erosion plume that shields the target surface, absorbs the laser energy, and reduces the heat input rate. The level of power that favors shielding depends on radiation parameters, material properties, and the environment.

The mechanisms of shielding are different in the processes of cutting metals and nonmetals. In the interaction with metals, the beam of a 10^5 to 10^7 W/cm^2 power density initiates an optical breakdown in the vapor-gas medium, and the plasma plume so formed shields the target surface. At the same power density delivered to a nonmetal target, the flow of vaporized material carries out solid particles which tend to shield the target to a greater extent.

Other effects that may appear in the process of cutting are beam shape distortion, excessive heat spreading, and off-centering of the beam.

All laser cutting techniques use a reactive or an inert gas delivered to the interaction region to cancel the effect of the erosion plume. The gas jet increases the cutting rate as it removes the vaporized particles, plasma, and molten material from the hole. In laser gas-assist cutting of dielectric materials, for example, the gas jet assists in disintegrating surface layers after the surface reaches its kindling temperature. The gas jet also protects the optical system against the blowoff material.

The width of the kerf in organic compounds can be smaller than the focal spot diameter because the gas evolved from the damaged target mixes up with the gas issuing from the nozzle to form a very hot gas jet which cuts through the material.

In the gas-assist cutting of the materials based on phenolformaldehyde and epoxy resins, such as cloth-base and glass-cloth-base laminates, the disintegration products form a viscous, sintered mass which is difficult to remove from the kerf by the gas jet. In this case, the laser beam must deliver more energy needed to evaporate secondary products of disintegration [109].

The values of specific energy Q_c for cutting some materials [110], which are given below, indicate that laminated plastics require much more energy for cutting than other nonmetals.

Material	Q_c, kJ/g	Material	Q_c, kJ/g
Cloth-base laminate...	50	Oil-gasoline-resistant rubber...	2.5
Glass-cloth-base laminate...	47	Gas-tight rubber...	2.1
Lining plastic...	2	Asbestos cement...	28
Pine...	0.9	Asbestos sheets...	20
Oak...	5.4	Glass ceramic...	25
Plywood...	5.4	Ceramics...	30
Cardboard...	0.8	Window glass...	31
Rigid-vinyl plastic...	1.8	Quartz glass...	45
Acrylic plastic...	2.0	Composites...	80

The production rate and the quality of cuts depend considerably on the combination of laser beam parameters and the parameters of the gas flow system. A major problem involved in the gas-assist technique, particularly when cutting thick materials, is the problem of removal of disintegration products. The thing is that it is difficult to provide a proportional increase in the cutting gas parameters at a large power of the laser beam [111].

One of the factors that limits the use of lasers for cutting thick materials is that the laser produces a narrow kerf, although it may appear at first glance that a narrow kerf is preferable in every respect. However, a small kerf width involves difficulties in implementing the process. Indeed, for the kerf to be narrow, the flow nozzle must be small in diameter. But if the length of a freely expanding jet, over which it has to smooth out its parameters, is equal to several diameters of the nozzle, the flow expands within a small length and the zone of cut takes on a tapered shape. Besides, the nozzle of a small diameter can afford only a limited gas flow rate that is insufficient to remove the molten material and disintegration products completely.

One cannot but agree with the authors of [111], who predict that further advances in laser cutting of thick materials can be expected to come only after finding the ways of correlating the laser beam parameters to those of the gas feeding system. At present much thought is being given to designs of gas dynamic systems. A large number of works have recently appeared, which propose improved versions of the systems and suggest the ways of selecting the gas dynamic parameters. However, much effort still remains to be spent to establish properly the regularities of the physico-chemical processes in gas-assist laser cutting of nonmetals.

Multikilowatt CW CO_2 lasers are promising tools for use in evaporating nonmetallic materials at high speeds. Repetitively pulsed CO_2 lasers can cut materials at a smaller size of the heat affected zone and produce kerfs with smooth edges.

In the text that follows we present the results of studies on the features of laser beam-material interaction, which help estimate the energy balance of the laser cutting process, analyze the dynamic characteristics of gas jets, and establish the regularities of various processes.

8.2. Energy Characteristics

In a first approximation, the energy characteristics of the laser cutting process can be defined from the analysis of a temperature field. The laser power density threshold q_{th}, i.e., a minimum power density required to heat the target surface to a kindling (decomposition) temperature T_0, can be found from the following formula for a fast-moving heat source:

$$q_{\mathrm{th}} = \exp\sqrt{\pi/8}\,\frac{T_0 K}{1-R}\sqrt{v_{\mathrm{c}}/kr_{\mathrm{l}}}\,, \tag{8.1}$$

where K is the thermal conductivity of a material; R is the reflectivity; k is the thermal diffusivity; v_c is the cutting speed; and r_l is the laser beam radius.

Expression (8.1) can be used to estimate q_{th} for cutting thin films of lavsan and polypropylene and also paper and cardboard, with T_0 taken equal to the vaporization temperature of the material.

In cutting sheet materials, a major portion of energy is expended on decomposing the material. Under quasistationary conditions of decomposition, the speed v_0 of travel of the decomposition front into the material bulk can be estimated from the expression

$$v_0 = \frac{q_{\mathrm{h}}}{\rho\left(cT_{\mathrm{v}} + L_{\mathrm{m}} + L_{\mathrm{v}}\right)}, \tag{8.2}$$

where q_h is the effective power density of the heat source; ρ is the material density; c is the heat capacity; T_v is the vaporization temperature; L_m is the latent heat of melting; and L_v is the latent heat of vaporization.

In expression (8.2) we can disregard L_m since L_v is much greater than L_m. The effective dwell time required to decompose the material is

$$t_{\mathrm{d}} = 2r_{\mathrm{l}}/v_{\mathrm{c}} - 8k/v_0^2. \tag{8.3}$$

The depth of cut under steady-state conditions is

$$h_c = v_0 t_d. \tag{8.4}$$

In expression (8.3) the last term can be neglected in practical calculations because $2r_1/v_c$ is much greater than $8\,k/v_0^2$. Assuming that the entire laser power P is spent on heating the material to its vaporization point, a maximum depth of cut can be given by

$$h_{max} = \frac{2P}{\pi r_1 \rho v_c (cT_v + L_v)}. \tag{8.5}$$

From equation (8.5) it follows that the penetration depth linearly varies with power. Experimental results suggest that this relation is valid for relatively thin materials. A further increase in P results in slower penetration since the blowoff material absorbs a large quantity of laser power (Fig. 8.1).

The specific energy Q_c of cutting is the energy input required to disintegrate a unit mass of the substance, which is the characteristic of a material and is independent of the conditions of cutting. The energy E needed to destroy a mass m of the substance in grams per second can be found from the formula

$$E = mc\,T_0, \tag{8.6}$$

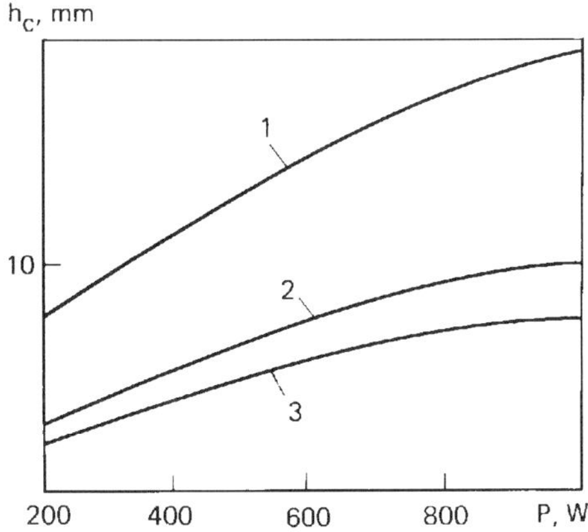

Fig. 8.1. Depth of cut versus CO_2 laser power at a cutting speed of 16 mm/s for paper laminate with polyester-acrylate binder (curve *1*), paper-based laminate with phenol-formaldehyde binder (curve *2*), and glass-cloth-base laminate with epoxy binder (curve *3*)

whence

$$Q_c = E/m = cT_0. \tag{8.7}$$

The depth of cut in dielectric materials depends on the focal length f (Fig. 8.2). The focal length as well as the aperture size determines the angle of convergence of the beam behind the optical system. The reflection of rays by the kerf walls and the waveguide character of beam propagation within the cavity are of little importance since the reflectivity of a dielectric is small. The focused beam should retain a high power density as it penetrates deeper into the material, for which reason the angle of beam convergence should be as small as possible. It should be kept in mind, however, that with an increase in the focal length above the optimal value, the focused spot diameter increases and the power density decreases.

Figure 8.3 illustrates the experimental curves of depth h_c of cut as a function of the angle of beam convergence β. As is seen, the penetration depth cut with a ring-shaped beam inherent in an unstable-resonator CO_2 laser is more strongly dependent on the angle of convergence (curve *2*). A stable-resonator laser that shapes a Gaussian beam or a circular beam with the power uniformly distributed over the spot provides a greater depth of cut under the same focusing conditions (curve *1*).

The kerf profile depends on the location of the focal point relative to the target surface. Figure 8.4 illustrates how the kerf width and the depth of cut vary with the location of the focal point above and below the surface.

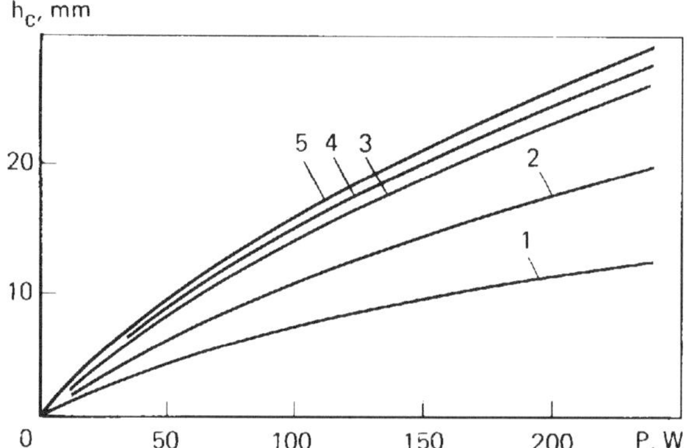

Fig. 8.2. Depth of cut in a dielectric material versus CO_2 laser power at lens focal lengths of 50, 100, 150, 200, and 250 mm (curves *1*, *2*, *3*, *4*, and *5*, respectively)

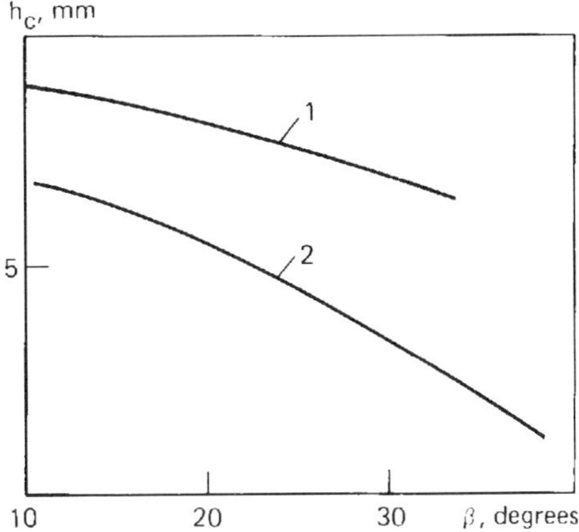

Fig. 8.3. The effect of the angle of the CO_2 laser beam convergence on the depth of cut in the glass-cloth-base laminate at P = 1.5 kW and v_c = 10 mm/s

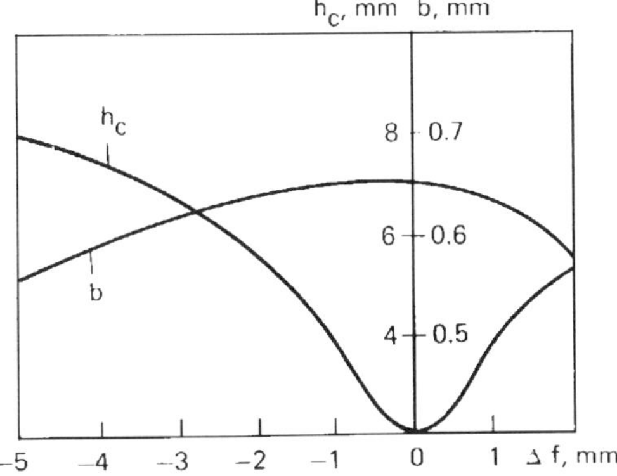

Fig. 8.4. Kerf width b and depth h_c of cut in the glass-cloth-base laminate versus focal point depth Δf relative to the target surface cut with a 1.6 = kW CO_2 laser at a speed of 11 mm/s

The kerf geometry also depends on laser polarization because the material reflectivity varies with the angle of incidence of the beam and the polarization vector position relative to the plane of incidence. The beam polarized linearly in the plane parallel to the plane of incidence delivers an increased energy to the front wall of the cavity. When the beam is polarized linearly in the plane normal to the plane of incidence, the laser beam largely couples to the sides of the kerf, so that the kerf width and the HAZ size grow and the cutting speed decreases. A maximum cutting speed depends on relative orientation of the polarization vector and the vector of the cutting speed.

The effect of relative orientation of these vectors can be different and may lead to sloping of the sides of a kerf. Light polarization has a stronger effect on the kerf profile in cutting the materials, such as semiconductors, that exhibit a rather high reflectivity at a given wavelength of laser radiation. For materials, for instance dielectrics, that display a low reflectivity in the entire near infrared region, the effect of CO_2 laser polarization is insignificant.

8.3. Gas Dynamic Parameters

The molten material and the products of disintegration are blown off the hole with a gas jet that is commonly coaxial with the laser beam. The columnar flow of gas issuing from the nozzle throat passes through the kerf and removes the decomposition products, therewith cooling the kerf walls and reducing the HAZ size.

Most of the nonmetals require a low specific energy for cutting. Material decomposition in laser cutting occurs mainly through the mechanism of vaporization, with the vaporized material removed by a gas flow. For the total removal of disintegration particles, the gas energy E_g, i.e., the work done by the gas flow, must be equal to the kinetic energy of disintegration particles. For a particle of mass m blown off at a velocity v, the kinetic energy is

$$E_p = mv^2/2. \tag{8.8}$$

The initial scatter velocity v_0 is equal to about the sonic velocity or even the supersonic velocity. The particles moving toward the lens in the direction opposite to that of the gas flow most heavily shield the target surface. The gas flow entraps the particles and carries them away from the interaction region.

To reduce the dwell time of particles in the interaction region and their shielding effect, the gas flow velocity must be equal to several units of the Mach number M at a definite value of velocity ratio ν and adiabatic exponent γ for the gas:

$$M^2 = \frac{2\nu^2/(\gamma+1)}{1-\nu^2(\gamma-1)/(\gamma+1)}. \tag{8.9}$$

The velocity ratio is a function of both the initial pressure P_0 of a gas at the nozzle inlet and the longitudinal section of the nozzle which can be of the convergent, divergent, and composite types. The profile of the nozzle is chosen with consideration for an increase in the gas volume and gas velocity in accordance with the equation

$$\frac{dS_n}{S_n} = \frac{v_s^2 - v_g^2}{\gamma v_g^2} dP_g, \tag{8.10}$$

where S_n is the cross-sectional area of the nozzle; v_s is the sound velocity; v_g is the gas velocity; and P_g is the pressure.

In equation (8.10) the increment dS_n at $dP < 0$ is opposite in sign to the term $v_s^2 - v_g^2$. This means that the nozzle must be convergent if the gas velocity at the nozzle inlet is lower than the corresponding sound velocity, otherwise it must be divergent. From equation (8.10) it follows that a convergent nozzle cannot ensure a gas velocity exceeding the corresponding sound velocity.

The gas flow reaches the critical velocity at a definite ratio of the atmospheric pressure P_a to the critical pressure P_{cr} of the gas inside the nozzle:

$$\frac{P_a}{P_{cr}} = \left(\frac{2}{\gamma + 1}\right)^{\gamma/(\gamma-1)}. \tag{8.11}$$

As seen from the above, the pressure ratio is a function of the adiabatic exponent which is 1.3, 1.6 and 1.4 for polyatomic, monoatomic, and diatomic gases, respectively. Using these value of γ, we can readily estimate the pressure ratio for the basic gases used in cutting. In particular, the gas outflow reaches the critical velocity at the 0.19 MPa critical pressure of gas inside the nozzle.

Let us consider the gas dynamic parameters of a gas-jet-assisted cutting process using a convergent nozzle (Fig. 8.5). Three versions of the process are possible depending on the initial gas pressure P_0-to-the environment pressure P_a ratio.

If the gas pressure within the nozzle is below the critical value, the gas totally expands so that P_0 drops to P_a, in which case the gas velocity is lower than the local sound velocity, $\nu_o < 1$.

If the gas pressure is equal to the critical, the pressure also drops from P_0 to P_a and the gas flows at the critical velocity equal to the sound velocity, $\nu_0 = 1$.

If the gas pressure is above the critical, the gas does not expand completely and flows out into the environment at a pressure that is higher than the ambient

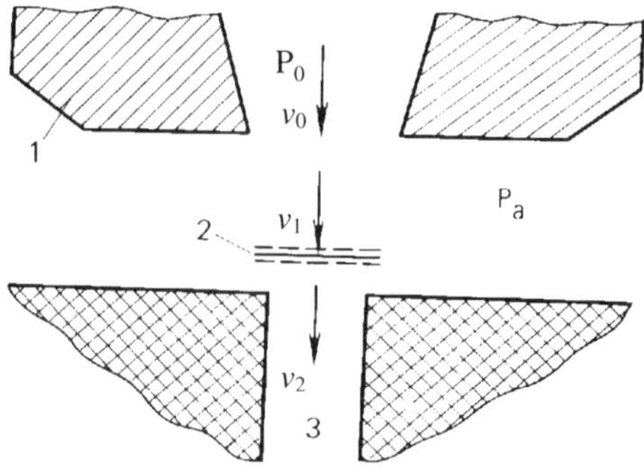

Fig. 8.5. The outflow of gas from a convergent nozzle to the interaction region: (*1*) nozzle; (*2*) shock wave; (*3*) kerf

pressure. The gas velocity at the nozzle throat is equal to the critical one, and the excess pressure causes the gas flow to accelerate behind the nozzle throat. The velocity ratio can be found from the expression of the form

$$\nu^2 = \frac{\gamma+1}{\gamma-1}\left[1-\left(\frac{P_a}{P_0}\right)^{(\gamma-1)/\gamma}\right]. \tag{8.12}$$

When $P_0 > P_a\left[2/(\gamma+1)\right]^{(\gamma-1)/\gamma}$, the velocity ratio ν is larger than unity, and so a supersonic flow of gas falls onto the target surface.

In cutting nonmetals, the width of the kerf produced is much smaller than the nozzle exit diameter. Therefore, the gas jet guided to the cut encounters an obstacle and generates a shock wave if the following condition is met:

$$\nu_1\nu_2 = 1, \tag{8.13}$$

where ν_1 and ν_2 are the velocity ratios before and after the buildup of the shock wave, respectively.

Condition (8.13) implies that the speed of the flow at the inlet to the kerf after shock wave generation decreases to a greater extent at a higher speed of the incoming flow. The expressions for the gas velocity at the point of inlet into the kerf assume the following form depending on the initial pressure P_0:

$$\text{at } P_0 < P_a\left(\frac{2}{\gamma+1}\right)^{(\gamma-1)/\gamma}$$

$$\nu_2 = \nu_1 = \left\{\frac{\gamma+1}{\gamma-1}\left[1-\left(\frac{P_a}{P_0}\right)^{(\gamma-1)/\gamma}\right]\right\}, \tag{8.14}$$

$$\text{at } P_0 = P_a\left(\frac{2}{\gamma+1}\right)^{(\gamma-1)/\gamma}$$

$$\nu_2 = \nu_1 = 1, \tag{8.15}$$

$$\text{at } P_0 > P_a\left(\frac{2}{\gamma+1}\right)^{(\gamma-1)/\gamma}$$

$$\nu_2 = \nu_1^{-1} = \left\{\frac{\gamma+1}{\gamma-1}\left[1-\left(\frac{P_a}{P_0}\right)^{(\gamma-1)/\gamma}\right]\right\}. \tag{8.16}$$

From the above relations it follows that the gas velocity at the inlet into the kerf cannot exceed the sound velocity at any gas pressure inside the nozzle. For gases such as nitrogen and air at an adiabatic exponent of 1.4, the velocity of gas in the kerf attains the critical value at $P_0 = 0.19$ MPa. If the kerf tapers downward, the velocity of gas gradually decreases. In the straight-sided kerf, the velocity corresponds to the critical at the inlet into the kerf. In the kerf diverging downward, the gas flow may acquire a velocity exceeding the critical one.

Figure 8.6 compares the coefficient of the velocity (curve *1*) estimated from equation (8.12) for a free jet issuing from the nozzle related to that of the velocity (curve *2*) estimated from (8.14) through (8.16) for a jet entering the kerf, therewith allowing for the shock wave generation. The curves are seen to coincide within the subsonic region at $P_0 < P_{cr} = 0.19$ MPa. At pressures above the critical, the relationships of viscosity coefficients illustrated by curves *1* and *2* change drastically.

To estimate the dynamic effect of the gas jet on decomposition products more accurately, account should be taken of both the pressure of the gas flow and its velocity. The total effect can be estimated from the kinetic energy of

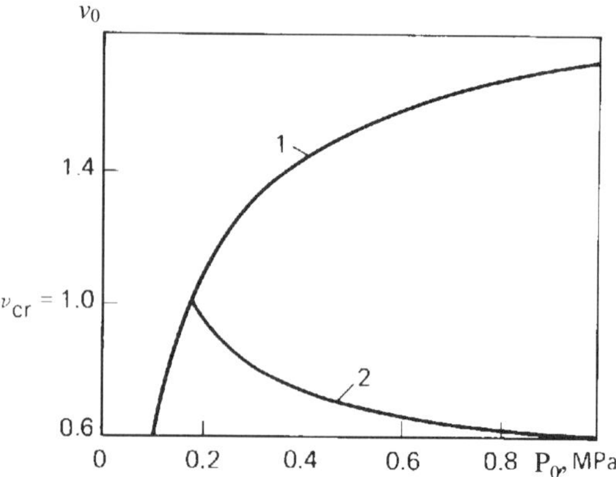

Fig. 8.6. Velocity ratio at an adiabatic exponent of 1.4 versus pressure within the nozzle

the jet, called the velocity head G which assumes the form

$$G = \frac{\gamma}{\gamma + 1} P_0 \nu_2^2 \left(1 - \frac{\gamma - 1}{\gamma + 1} \nu_2^2 \right)^{1/(\gamma - 1)} . \tag{8.17}$$

In calculating G from expression (8.17) for actual conditions of cutting, we should replace P_0 by the pressure P_f of the flow incident on the interaction region and retarded at the level of the target surface. The pressure P_f is lower than P_0 because the flow at the nozzle exit sharply expands and partially loses its energy for shock wave generation.

With a decrease in the kerf width, the gas viscosity manifests itself. At a kerf width less than 0.2 mm, the gas dynamic drag of the kerf is so large that the jet is unable to remove the disintegration products through the kerf.

The velocity head and the gas velocity in the kerf can reach maximum values under the conditions of gas outflow close to the critical. An optimum pressure P_0 for air and nitrogen is about 0.2 MPa. The pressure exerts a jump-like effect on the velocity head due to shock wave generation and the critical value of flow velocity in the kerf. At $P_0 = P_{cr}$, the velocity head reaches the highest value and ensures a maximum penetration depth.

The above regularity is only valid if the kerf width is sufficient for the effective removal of the vaporized material. If the kerf width drops below the critical value that also depends on the material thickness, the gas jet cannot flow through the kerf and a subsequent increase in the pressure causes a gradual

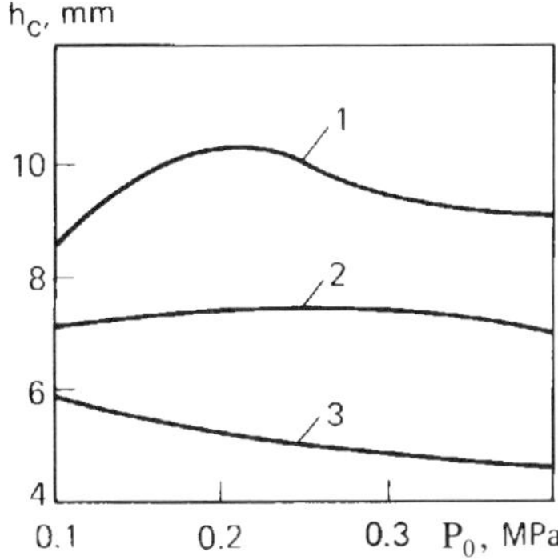

Fig. 8.7. Depth of cut in the glass-cloth-base laminate versus gas pressure at cutting speeds of 11, 17, and 20 mm/s (curves *1*, *2*, and *3*, respectively)

decrease in the depth of cut. The gas jet compacts, as it were, the particles at the surface, and a denser erosion plume near the focal plane absorbs the laser energy more heavily, thus reducing the penetration depth.

In Fig. 8.7 are shown experimental plots of the depth of cut against the initial pressure P_0 at different cutting speeds for fabric glass laminate. The depth of cut is maximum at v_c = 11 mm/s (curve *1*) and P_0 = 0.2 MPa. At v_c = 20 mm/s, an increase in the pressure due to a reduced kerf width causes the depth of cut to decrease in a monotonic fashion.

The kind of gas used in cutting nonmetals has no pronounced effect on the cutting speed and penetration depth as it has in cutting metals. For some organic materials, however, the choice of an adequate gas is quite important. The use of oxygen or air impairs the kerf quality because of the edge charring and a larger HAZ size. Neutral gases such as argon, helium or nitrogen can ensure a better kerf. The use of a gas which can react with decomposition products of a polymer provides a means for controlling the cake formation process.

The kinetic characteristics of a gas jet depend on the nozzle exit diameter D_n, the height H of the nozzle throat above the target surface, and pressure distribution over the cross section of the exit. It is possible to select an optimal diameter D which ensures the highest cutting speed at definite values of H and P_0. An increase in D_n at the same value of P_0 decreases the gas pressure along the jet axis and changes the jet area where the pressure exceeds a minimum pressure $P_{g\ min}$ requited for cutting under specified conditions (Fig. 8.8). As is evident from the graph, an increase in the diameter from D_{n1} to

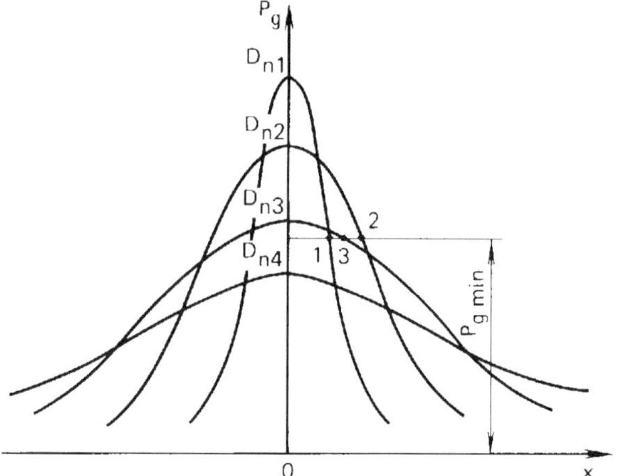

Fig. 8.8. Distribution of gas pressure in cross sections of the nozzle exits of diameters $D_{n1} < D_{n2} < D_{n3} < D_{n4}$

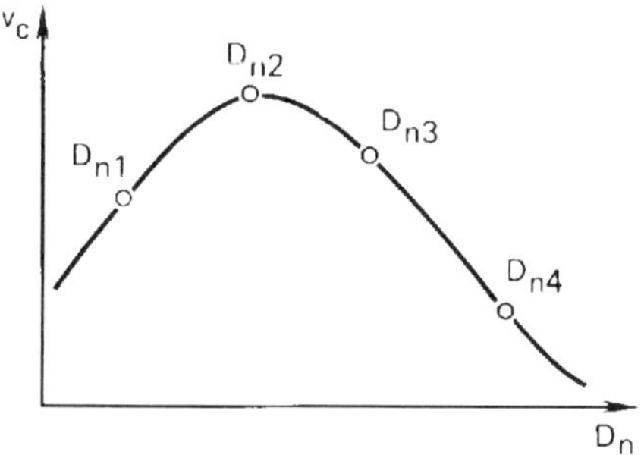

Fig. 8.9. Cutting speed versus diameter of the nozzle exit

D_{n2} increases the jet area, where P_g is greater than $P_{g\ min}$ (points *1* and *2*). A further increase in the diameter from D_{n2} to D_{n3} decreases this area (points *2* and *3*). That is why the plot of the cutting speed versus diameter D_n displays an extremum (Fig. 8.9).

The design of a nozzle depends considerably on the beam-focusing optics used. In cutting with 2 kW high-power lasers, preference should be given to metallic mirrors over focusing lenses because mirrors show an increased radiation stability and ensure a longer service life of optical elements. The beam-focusing mirror systems should preferably operate with a nozzle in which the central hole serves to provide a passage for the beam and the gas ducts located concentrically around the nozzle periphery deliver the gas so that individual jets directed at an angle to the optical axis form a single flow falling on the interaction region.

8.4. Process Parameters

This section describes some general regularities and specific features of the process of laser cutting of nonmetallic materials.

Laminated plastics. As an example, we consider the features of laser cutting of the glass-cloth-base laminate and the paper-based laminate widely used in industry, particularly in electrical engineering. In cutting these materials, carbon dioxide delivered to the interaction region through a convergent nozzle can enhance the process and reduce the charring zone.

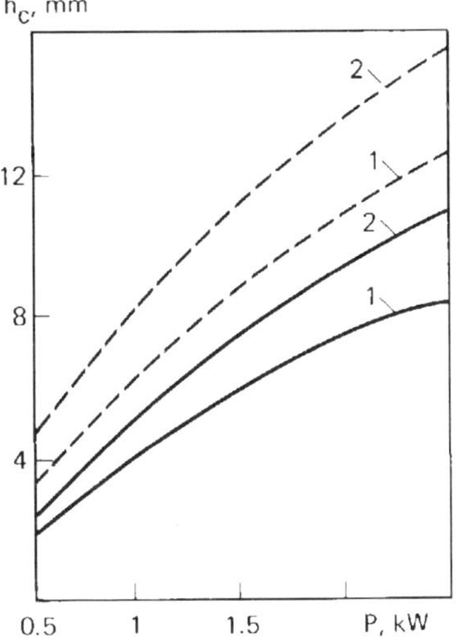

Fig. 8.10. Depth of cut in the glass-cloth-base laminate (curves *1*) and paper-based laminate (curves *2*) versus laser power at $v_c = 10$ mm/s and $f = 160$ mm. Solid and dash curves stand for cutting with unstable-resonator and stable-resonator lasers, respectively.

Experiments reveal that a circular beam with power uniformly distributed in its cross section, or a Gaussian beam, can provide a larger depth of cut than a hollow-ring beam (Fig. 8.10).

With an increase in the laser power from 0.3 to 2.5 kW and power density from 10^5 to 10^6 W/cm^2, the penetration depth grows (Fig. 8.11). The power required to cut the glass-cloth-base laminate (curve *1*) is higher than that for cutting the paper-based laminate with a phenol-formaldehyde binder (curve *2*) and a polyester-acrylate binder (curve *3*). This is because the fillers of these plastics require different values of specific energy for their decomposition. In particular, the specific energy for decomposing glass amounts to about 30 MJ/kg as against 1 MJ/kg required to decompose paper. The paper-based laminate with the thermoplastic binder (polyester-acrylate) requires less energy

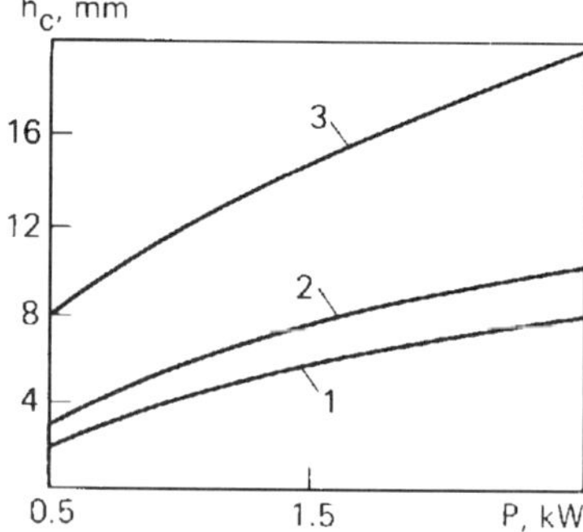

Fig. 8.11. Depth of cut versus laser power at v_c = 16.6 mm/s

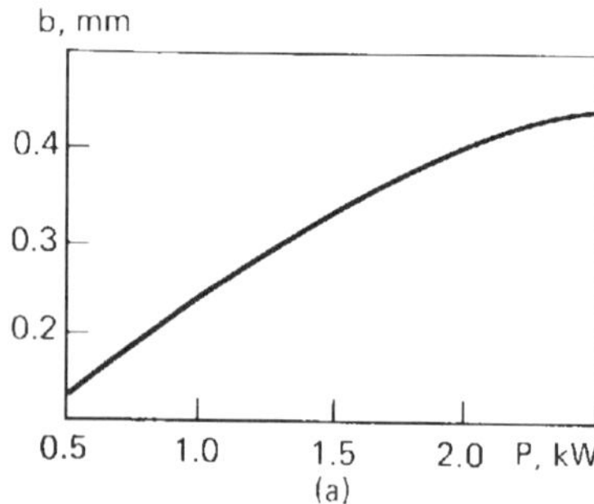

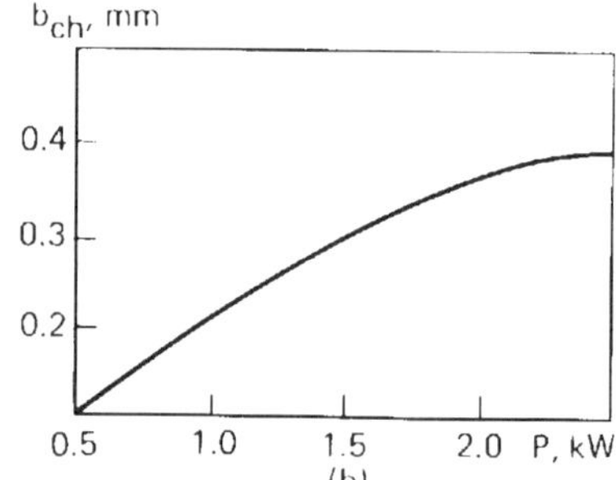

Fig. 8.12. Kerf width (*a*) and charring zone width (*b*) versus laser power for a glass-cloth-base laminate

than the paper-based laminate with thermosetting binder (phenol-formaldehyde) because polyester-acrylate completely transfers to a gaseous state, thereby reducing the amount of aerosol in the process of thermal decomposition.

Both the kerf width b and the charring zone width b_{ch} increase with laser power in cutting the glass-cloth-base laminate (Fig. 8.12). The same is also true for the paper-based laminate with a thermosetting binder. The kerf width in the paper-based laminate with a thermoplastic binder changes in the same manner, but the kerf is free of the zone of charring.

Increasing the speed of cutting with constant laser power decreases the energy per unit cut length (heat input) and, hence, the depth of cut (Fig. 8.13).

The kerf profile, particularly the kerf width at the bottom of the cut, varies strongly with the cutting speed. The kerf width at the top of the cut depends

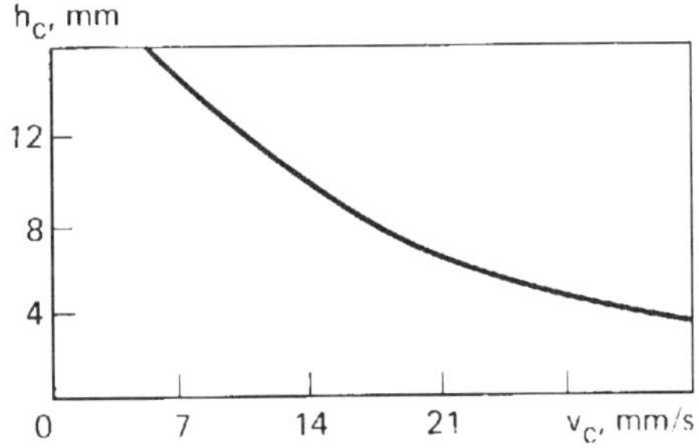

Fig. 8.13. Depth of cut in the glass-cloth-base laminate versus cutting speed at $P = 2.0$ kW

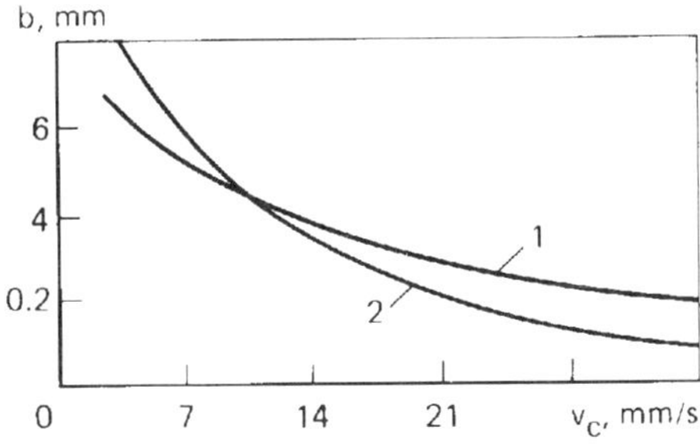

Fig. 8.14. Width at the top (curve *1*) and bottom (curve *2*) of the kerf cut in 3-mm thick glass-cloth-base laminate at $P = 1.5$ kW as a function of cutting speed

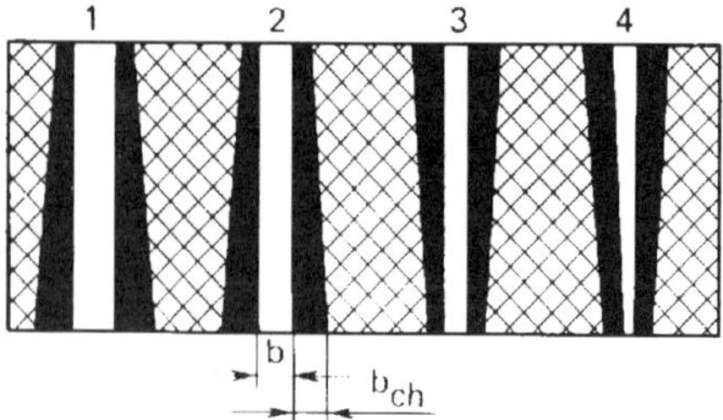

Fig. 8.15. Cross sections of kerfs in the 5-mm glass-cloth-base laminate cut with a 2-kW laser beam at speeds of 6.6, 16.6, 25, and 33 mm/s (kerf profiles *1*, *2*, *3*, and *4*, respectively)

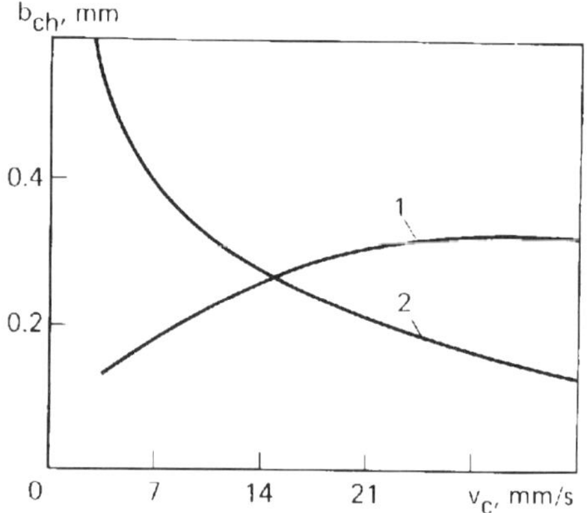

Fig. 8.16. Scorched layer width in the upper portion (curve *1*) and lower portion (curve *2*) of the kerf in the 3-mm glass-cloth-base laminate cut at $P = 1.5$ kW as a function of cutting speed

on the focal spot diameter. The kerf width at the bottom decreases faster than that at the surface with an increase in the cutting speed (Fig. 8.14). Figure 8.15 illustrates how the kerf profiles change with the cutting speed. The angle of slope of kerf sides at the 160-mm focal length f of the lens does not exceed 3 or 4°.

The zone of charring in the cross section of the kerf diverges at higher cutting speeds and contracts at lower ones. But in distinction from the kerf, the scorched layer widens in the upper portion of the cut and contracts in the lower with increasing cutting speed (Fig. 8.16).

At cutting speeds less than 5 mm/s, the zone of charring increases appreciably because a larger amount of energy coupled to the target intensifies the process of decomposition, which transfers to an uncontrolled process of thermal-oxidative decomposition. For this reason plastic laminates should not be cut at these speeds.

In cutting an organic material, a coolant such as water should be brought to the cutting region to reduce ignition of the material and its charring. Water must be added to an external layer of the gas jet around its periphery, therewith avoiding excess water supply since water vapors absorb laser radiation.

Quartz glass. This material can readily be cut with a low-power laser to provide a highly efficient kerf. The interaction zone is practically free of the molten metal pool because the silicon oxide vaporizes at a temperature close to the melting point of glass. The beam only fuses off the kerf sides which becomes smooth after vitrification.

A 100-W power laser is sufficient to cut thin quartz glass plates. A 200-W laser beam focused at the surface to a spot about 0.3 mm in diameter can cut 4-mm quartz glass with a kerf width of 1 mm. In some cases, molten quartz can fill up the kerf. Increasing the beam power, reducing the cutting speed or blowing off the molten material with a gas jet can eliminate the kerf clogging.

Window glass. The process of laser cutting of window glass is more complex [112]. If the beam does not cause the material to sublime completely, a metal pool forms in the kerf and cracks appear at a distance of 2 or 3 mm from the fused edges under the action of stresses. To safeguard the glass against damage in the process of cutting, the specimens should be preheated in the furnace to a temperature of 873 K. Preheating to a higher temperature may induce appreciable strains in the specimen.

When the laser beam interacts with glass, a plume of sublimation products arises above the surface, which should be blown off by the gas jet to prevent damage to focusing optics. The plume includes sodium and calcium components and also silicon oxide particles.

The jet of air directed to the region of laser cutting removes the molten material from the kerf and thus increases the speed of cutting and the quality of the kerf. The speed of cutting a 4-mm glass can reach 23 mm/s at a laser power of 0.4 kW. The kerf edges at the top of the cut are clean and smooth. The edges at the bottom show traces of leaked-in sintered glass that cracks and spalls quite readily.

Textile. Laser cloth cutters possess much practical promise for textile industry. The available laser cutters can cut cloths in a single ply up to 3 mm thick with a kerf width of 0.2 to 0.5 mm without scorching.

In multiple-ply cutting of capron, nylon, and other synthetic materials, an approach is taken to separate the plys with thin paper interlayers or to wet

Table 8.1. Parameters of CO_2 Laser Cutting

Material	δ, mm	P, W	v_c, mm/s	Material	δ, mm	P, W	v_c, mm/s
Quartz	3.2	500	12.3	Glass-cloth-			
Glass	3.2	5000	76.1	base laminate	8.0	2500	16.6
Plywood	6.4	850	90.1	Cloth-base			
Cardboard	19.4	200	1.6	laminate	5.0	800	12.5
Ceramics	6.5	850	10.0	Textile	0.45	500	666.6
Plexiglass	10.0	900	58.3	Nylon	0.76	200	101.6
Asbestos	5.0	500	0.83	Leather	3.20	200	10.5
Cement				Rubber	2.0	100	31.7

the ply surfaces to prevent the ply cut edges from sticking together as a result of fusion. The cutting process ensures a good quality of the kerf sides, fuses off the surface of the cut and thus keeps the cloth edges from untwisting. Table 8.1 lists the parameters of cutting of some nonmetals with a CO_2 laser.

There are also other techniques of separating nonmetallic materials with lasers, for example, wire insulation stripping, scribing, and controlled fracturing. The CO_2 lasers emitting at a wavelength of 10.6 μm are used for the removal of insulation from conductors because plastics absorb heavily CO_2 laser radiation while metals reflect it effectively. The guiding system revolves the laser beam about the wire to provide a circular cut through the insulation layer and then directs the beam along the wire to produce a longitudinal slit for stripping the conductor easily. Wires up to 20 mm in diameter can be stripped with a CW CO_2 laser of about 50-W power.

Controlled fracturing relies on laser heating with a tightly focused laser beam along the requisite path with the aim to generate thermal stresses. When the stresses exceed the ultimate strength of the material, the fracture develops along the path of the beam travel on the target surface. Laser fracturing is used for splitting brittle nonmetals such as ceramics, silicon, and glass. The processes involved in controlled fracture are given a detailed treatment in [112].

9 Cutting of Metals

9.1. General

Industry widely uses mechanical methods for dividing metals. A large variety of general-purpose and special-purpose machine tools are available for cutting sheet and shaped blanks. Machine cutting has a number of advantages but calls for costly cutting tools and presents difficulties in cutting metals along complex curvilinear paths [113].

There are also such methods of cutting as electrochemical, electrophysical, and physico-chemical machining. Oxyacetylene flame cutting, plasma arc cutting, and other physico-chemical methods provide a higher production rate than mechanical methods but do not afford a desired accuracy of the cut edges and surface finish, and the stock cut requires subsequent machining. Electrical-discharge cutting yields a narrow and a high-quality kerf but offers a low output.

Laser cutting is a promising technique of cutting by melting and vaporization with subsequent removal of the molten material and disintegration products from the kerf.

The laser beam can deliver a high power to a small spot that is sufficient to cut any low-machinability material. As noted earlier, the laser can produce cuts with a narrow kerf width and a small HAZ at a high accuracy and increased speed. Since there is no tool contact with a workpiece surface, readily deformable and flexible parts can be cut with a laser. Easy control of the laser beam offers the possibility of guiding the beam around complicated contours to cut out flat and shaped parts.

Laser cutting is one of the first techniques of material laser processing, which made its appearance back in the early 1970s. In the years since then, laser cutting has received much development effort. Gas-assist laser cutters ranging in powers from a few tens of watts to a few kilowatts are now available. The laser beam heats, melts, and vaporizes the material along the line of cut and the jet of gas removes the products of disintegration. A gas used in metal cutting, such as oxygen or air, removes the molten material and contributes to the cutting speed owing to the heat energy of an exothermal reaction between the metal and oxygen.

Industrial setups for cutting metals use solid-state and gas lasers of continuous wave and repetitively pulsed types. The areas of application of laser cutters continue to expand, although capital investments involved in the construction of laser cutter setups is still rather high. Laser cutting can be a reasonable choice and cost effective in the fields where the use of traditional techniques is either time consuming or is impractical at all.

There is a large number of works on gas-assist laser cutting of metals, which deal with both theoretical and practical aspects [64, 110, 113, 114]. In this chapter we consider the results of investigations on laser cutting of metals carried out at the Bauman Moscow State Technical University.

9.2. Physical Processes

Apart from the common features of laser energy reflection and absorption and heat propagation, laser cutting displays specific features. In the laser-metal tool zone the metal heats up to the melting temperature and the melt region so formed begins to propagate into the bulk of the solid. If the laser radiation coupled to the target is high in intensity, the temperature rises to the boiling point, which leads to vaporization of the metal. The vaporization rate varies with temperature in an exponential fashion and reaches a maximum value at a steady vaporization temperature when the rate of the melting wave is equal to that of the vaporization wave.

The amount of melt, the steady temperature, and the rates of melting and vaporization may vary with the laser power density delivered to the target. Consequently, it is possible to control the process of target damage by changing the power density and beam dwell time.

The intensity of the process of heating to vaporization depends appreciably on the absorptivity of a metal, which varies with the surface temperature, lasing wavelength, laser polarization, and the angle of incidence of radiation on the target surface. The energy coupled to the target depends on the parameters of a vapor-gas plasma induced during the interaction of the beam.

So, laser radiation incident on a metallic target can give rise to two mechanisms of cutting, melting and vaporization. The kerf separates out the target over its entire thickness and moves at a definite speed in the direction of the cut.

The use of the mechanism of vaporization alone for cutting metals involves difficulties in view of a fairly high specific energy consumption.

The assist gas used to remove the melt out of the hole helps in reducing the energy consumption. The gas overcomes the viscous-capillary force of the liquid melt and moves it away along the cut length.

Gas-assist cutting displays both a steady character of metal disintegration, with the liquid bath distributed along the length of the kerf, and an unsteady character of disintegration, in which case the molten metal is periodically removed from the kerf.

The mechanism of steady disintegration sets in when in each cross section of the kerf the rate of melting in the cut direction is equal to the rate of removal of the molten metal.

The unsteady disintegration of the metal along the line of cut occurs in the following manner. After a successive removal of molten metal from the kerf, the beam incident on the uncovered lower portion of the cut forms a metal pool again and melts down the upper front portion of the kerf as it moves relative to the target. The gas jet is yet unable to force the molten metal to flow away. After a certain time, the volume of the liquid reaches the critical value and is removed from the kerf. The beam now initiates the next cycle of target damage, attended by the formation of specific ridges on the sides of the kerf.

In the gas-assist process of cutting steels and a number of alloys oxygen is used, which provides an evolution of an additional amount of heat as a result of the exothermal chemical reaction. The oxide film formed on the metal surface increases the absorptivity of the metal and has a noticeable effect on the melt flow since the oxide viscosity is much higher than that of the molten metal.

Cutting with a repetitively pulsed laser can be done at a lower average power. When the surface temperature of the metal pool formed by a successive pulse is below the boiling point of the metal, the molten metal moves along the cut under the action of gas dynamic forces. When the successive pulse causes the surface temperature of the pool to exceed the boiling temperature of the metal, the recoil momentum of vapors comes into play and speeds up the removal of the molten metal. The metal pool can be formed and removed both during the time of the pulse action and in the interval between two consecutive pulses.

Gas-assist cutting by melting requires lesser average power than cutting by vaporization and is therefore preferable.

The use of repetitively pulsed lasers opens up new fields of use of laser cutting of metals. With properly selected parameters of the process, it is possible to control the quantity of the melt, to ensure a high quality of the kerf sides, and to increase the net energy efficiency of the process.

The pattern of metal disintegration in the process of cutting is rather intricate. In an effort to optimize the conditions of laser cutting, much theoretical work has been done to gain a deeper insight into the mechanism of laser cutting. There is a variety of models for the analysis of laser beam-solid interaction. For example, works [115, 116] examine thermal phenomena using the models of surface and volume heat sources created by the laser beam in the tool zone. These models enable us to estimate the kerf width and the cutting speed on the assumption of a stationary mechanism of target damage. Where the process uses a reactive gas such as oxygen, theoretical models apply for considering the amount of heat evolved in the course of the exothermal reaction between metal and oxygen. On thc basis of these models we can derive approximate expressions for estimating the finish of kerf sides. But these models do not allow us to evaluate the kerf profile and the amount of flash. For this, a more complex model is needed to determine the distribution of the molten metal in the kerf by solving simultaneously the thermal and the hydrodynamic problem for the metal pool.

The work [115] presents a mathematical model based on a unidimensional heat equation, hydrodynamic equation, and the equation of oxidation, the solution of which helps in verifying the thermal, hydrodynamic, and chemical processes under steady-state conditions and also in revealing the joint effect of the processes. The surface temperature of the melt region is found to vary with the gas flow rate. The model also allows for the effect of the exothermal reaction on thermal processes. However, the model relies on a rather simplified pattern of gas flow in the tool zone and is unsuitable for the analysis of the unsteady process of cutting at a low rate.

The models evolved for the analysis of the hydrodynamic flow of the melt allow us to obtain analytical expressions for estimating the performance parameters, although they consider the process of kerf shaping rather approximately and do not analyze the flow of gas in the tool zone.

Continuous laser cutting shows considerable promise. In cutting along a curvilinear path, however, the speed of travel of the beam along sharply curved path sections has to be reduced markedly. A decrease in the cutting speed distorts the kerf profile and increases the roughness of the kerf sides and width.

Cutting along a curvilinear path with a repetitively pulsed laser can provide a kerf of a higher quality since the parameters of the process are easier to control. The theoretical analysis of repetitively pulsed laser cutting commonly relies on the mechanism of vaporization, with the cut being viewed as a row of successive holes overlapping one another. This approach can provide only approximate results because it disregards the liquid phase that has a considerable effect on the quality of kerf sides.

The text below briefly describes the theoretical models worked out for the analysis of continuous laser cutting and repetitively pulsed laser cutting.

9.3. Continuous Laser Cutting

The theoretical model based on modern statements of thermal, hydrodynamic, and chemical processes enables us to analyze the effect of laser parameters, focusing system parameters, and gas-assist flow rate on the cutting speed and the quality of cut.

The theoretical studies of oxygen-assisted CW laser cutting reveal that both the steady and the unsteady mechanism of decomposition promote the cutting process. The steady mechanism entails the formation of a metal pool of definite dimensions, the depth of which does not vary with time.

Since the throughput and the kerf quality depend on the volume of the molten metal in the kerf, the design model considers the formation of the pool and melt flow in the kerf. The melt layer on the front wall of the cut varies with the absorbed laser energy, the heat of the exothermal reaction, the gas dynamic effect, and the thickness of the metal and its properties.

The roughness of the kerf sides, kerf profile, and quantity of flash depend on the thickness of the melt layer on the kerf sides. The model includes the expressions for estimating the melt layer thickness which is a function of the parameters of the laser, focusing optics, and gas-assist feeding system and also thermal and chemical properties of the metal and metal thickness.

As mentioned earlier, the unsteady mechanism of decomposition leads to the periodic formation and removal of the melt, which takes place when the rate of metal disintegration in the direction of cut is higher than the cutting speed. The quality of the cut depends on the thickness of the damaged layer in the cross section, which varies with the cutting speed and, hence, the time of formation of the metal pool and the time of its removal from the kerf.

The rate of disintegration v_{dis} varies with properties of a metal, but is approximately the same for most steels. It is thus possible to single out the ranges of CW laser cutting speeds within $v < v_{dis}$ and $v = d_{dis}$ for the unsteady and the steady mechanism of disintegration, respectively.

The studies of gas-assist cutting of steels indicate that a thin oxide film is formed on the melt surface whose absorptivity is close or equal to the absorptivity of oxides at lasing wavelengths of 10.6 μm and 1.06 μm. At actual angles of incidence of laser radiation on the target surface, the absorptivity of oxides is 2 to 5 times that of the unoxidized metal. Oxides have approximately the same capacity of absorbing radiation emitted both at 10.6 μm and 1.06 μm. The share of heat due to the exothermal reaction reaches 30 percent under steady conditions of the target damage. An inert gas used in the process of steel cutting does not contribute to the cutting speed because the absorptivity of the front wall of the cut decreases noticeably with an increase in the rate of flow of the molten metal in the kerf.

Theoretical investigations point out that there is a minimum cutting speed below which it is difficult or impossible to produce a quality kerf. In cutting along a curvilinear path, a path section of a small curvature should be cut at a maximum possible speed. In path sections with large curvature, the quality of kerf cut with the same parameters is poor. In the range of cutting speed less than 10 mm/s, which is typical of the unsteady character of disintegration, the requisite quality of cut is difficult to obtain whatever the parameters of the laser beam, focusing system, or assist-gas feeding system.

9.4. Repetitively Pulsed Laser Cutting

In order to optimize the parameters of oxygen-assist cutting with a repetitively pulsed laser, we have carried out the theoretical studies of the process assuming that the laser beam cuts through a metallic target at an average speed. The pulse energy coupled to the target and the heat of exothermal reaction form an oxide-covered metal pool which is removed by the gas jet both during the interaction of a pulse and in the interval between the successive pulses. The absorptivity of kerf sides varies with the formation of the oxide film.

The pulses of two shapes have been used in the analysis of the process: a rectangular pulse and a pulse with a linearly increasing leading edge and an exponentially decaying trailing edge. The latter pulse shape is typical of most solid-state lasers and gas TEA lasers.

In oxygen-assist cutting of steels, oxygen contributes little to the heat source created in the tool zone, i.e., the heat due to the exothermal reaction is insignificant if the bath temperature does not exceed the boiling temperature of the metal.

The estimates of the performance parameters have been made using the expressions derived for the calculation of the figures of merit of the gas-assist cutting of thin-sheet steels.

It is found that the kerf width is little dependent on the thermal constants of the metal and is generally independent of the metal thickness and the cutting speed. What strongly affects the kerf surface finish is the pulse repetition rate which need be increased with the cutting speed to reduce the roughness of the kerf walls.

The results of the analysis indicate that a repetitively pulsed laser is suitable for cutting sheet steels at a speed of about 20 mm/s. To ensure a quality kerf in the steady-state conditions of material damage, cutting should be done at a pulse repetition rate f_p of about 400 Hz, pulse duration t_p of 0.5 to 2.5 ms, pulse energy E_r of 0.5 to 5 J, and pulse period-to-pulse duration ratio of more than 2. The throughput and the quality of cut are readily controllable by adjusting f_p alone.

For the gas-assist process of cutting sheet steels along a curvilinear path to be efficient and provide a requisite quality of cut, the laser cutting setup should be able to operate both in the continuous wave mode and in the repetitively pulsed mode.

10 Surface Hardening

Laser surface hardening, or heat treatment, is the process of laser heating of a metal surface to a depth required for hardening and subsequent self-quenching of the surface layer at a high rate of cooling by heat conduction into the bulk metal.

In distinction from the conventional processes of induction hardening, electric hardening, or metal-bath hardening, laser hardening involves heating of the target surface to a shallow depth. The process of laser hardening entails specific structural transformations because of very high rates of heating and cooling.

This chapter describes some results of studies relating to laser surface hardening of various alloys.

10.1. The Mechanism of Iron-Carbon Alloy Hardening

Laser hardening of steels leads to the formation of an austenitic structure during heating and subsequent transformation of austenite into martensite at the stage of cooling [117, 118]. In surface hardening through solid phase transformations, the stage of heating determines the structural changes at the stage of quenching. On heating iron alloys to a critical point Ac_1, pearlite begins to convert to austenite at constant temperature if heating occurs at a slow rate.

Since the laser beam heats up the metal at a high rate, the pattern of austenite formation is different (Fig. 10.1). The heat energy delivered to the metal exceeds the amount of energy required to rearrange the crystal lattice at a definite finite speed. For this reason, the transformation occurs in the temperature range between the starting point Ac_{1st} and the finishing point Ac_{1f} rather than at constant temperature, i.e., in region *1* of higher temperature [119].

Because of a high heating rate, the diffusion process of conversion of excessive body-centered cubic ferrite F to face-centered austenite A may not terminate on line *GS* of the constitution diagram but shifts to region *2* of high temperatures. In a similar way, the fusion boundary between cementite C and austenite may shift within region *3*. The process of diffusion redistribution of

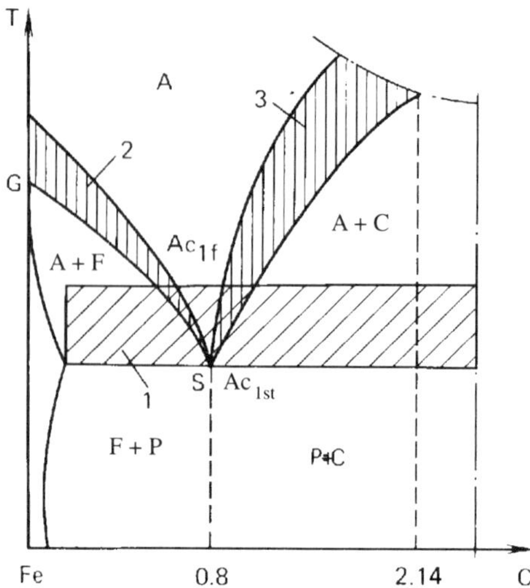

Fig. 10.1. Portion of the constitution diagram of Fe-Fe_3C for illustration of structural transformations on high-rate heating

carbon in austenite, or diffusion annealing, shifts to the region of still higher temperatures. The laser-hardened structure depends on the degree of completion of the austenization process that is governed by the heating rate, heating temperature, interaction time, and the initial structure. At a high heating temperature and a sufficiently long interaction time, the process may yield homogeneous austenite. A decrease in both the heating temperature and the interaction time slows down the homogenizing process and results in rather inhomogeneous austenite, especially as regards carbon distribution. The degree of inhomogeneity of the laser-hardened structure is the lower the higher the degree of dispersion of the initial structure.

An increase in the heating rate aids in grain refinement, but the process of austenite formation is complicated by the effects of restoration of the grain shape and size. In the process of heating and even during cooling, austenite grains grow, although the carbide particles that withstand high temperatures inhibit the grain growth. Under these conditions the size of the austenite grain depends on the relationship between the heating temperature and the interaction time. Since the dwell time of the laser beam is short, the grain has no time to grow large. However, it is desirable to heat the metal up to its melting point in order to increase the depth of hardening. But in this case, the austenite grains cannot be too fine. Fine grains only form at the boundary of the laser-heated zone where the temperature is not high.

As noted above, a high cooling rate ensures self-quenching and hardening of the surface layer. To obtain martensite in iron-carbon alloys within the temperature range of from 673 to 873 K, where austenite is the least stable, the cooling rate should exceed the critical values that range from 50 to 200 K/s for most iron alloys. The cooling rate in transformation hardening is very high. An increase in the cooling rate does not change the composition of phases and structures such as martensite, cementite (carbides), and retained austenite. But high cooling rates can be responsible for a highly nonuniform structure due to inhomogeneous austenite and can lead to structural defects because of the enhanced precipitation hardening and retarded processes of recovery and recrystallization.

In the course of cooling, the structure portions get finer and the density of dislocations and the stresses in the metal grow. Laser hardening yields a finer martensite than that formed by the conventional techniques. The microhardness of laser-treated steels can be by 2000 MPa higher than that of steels hardened by the conventional methods.

As mentioned above, surface hardening with melting (liquid phase quenching) affords heat treatment to a larger depth. The grain size decreases at a high cooling rate, but not in a unique fashion because the crystal growth rate and the nucleation rate vary in a complex manner with overcooling, i.e., the difference in liquidus and actual crystallization temperatures.

At cooling rates of 10^3 to 10^6 K/s, dendritic segregation takes place. This process enriches the grain regions solidified at the start and at the end of crystallization in high-melting and low-melting elements, respectively. The cooling rates above 10^6 K/s cause the formation of a flat crystallization layer which retards the distribution of elements in the liquid phase, so that the dendritic segregation slows down sharply.

With an increase in the cooling rate, the concentration of a dissolved element in the solid solution grows in comparison with the equilibrium concentration and leads to the formation of a metastable (supersaturated) solution. At super-high cooling rates, the crystallization centers have no time to grow and the metal solidifies into a glassy mass with an amorphous structure that displays a certain short-range order in the arrangement of atoms.

10.2. Structural Changes in Laser-Hardened Iron-Carbon Alloys

The laser beam heats up the metal target in depth to different temperatures, so that the laser-treated zone has a layer structure and consists mainly of three layers.

The first layer is the hardened region formed from the melt, which, generally, has a columnar structure with crystals stretched in the direction of heat removal. The structure mainly consists of martensite, carbides being commonly dissolved. In optimal conditions of hardening, decarburization does not occur and the layer is free of craters and slag inclusions. In solid phase transformation hardening, the structure does not contain the first layer. The second layer is the hardening region formed from the solid phase, with its lower boundary determined by the temperature of heating up to the critical point Ac_1.

This layer is the region of both full and incomplete hardening. The structure is inhomogeneous in depth. The upper portion consists of martensite and retained austenite formed from homogeneous austenite at the stage of cooling. The lower portion contains, along with martensite, the elements of the initial structure, namely, ferrite in hypoeutectoid steel and cementite in hypereutectoid steel.

The third layer is the transition structure region formed in the metal after its heating to a temperature below the critical point Ac_1. In steel initially hardened or tempered, the layer of troostite or sorbite is formed in the process of laser treatment, which has a reduced microhardness. This laminated structure is typical of steels heat-treated with CW and pulsed lasers. The text which follows briefly describes the structure of some laser-hardened steels.

Hypoeutectoid carbon steels. In low-carbon steels containing less than 0.3 percent carbon, the first liquid phase hardened layer consists of fine-needled martensite, the microhardness of which is comparatively low, from 5000 to 6000 MPa.

The solid-phase transformation region exhibits a more inhomogeneous structure. The upper portion of the region treated with a pulsed laser or a CW laser at high speeds may consist of martensite with a hardness of up to 6000 MPa, which is formed in small areas of the former pearlite. At a lower speed of CW laser treatment, the layer displays homogeneous plate-like martensite whose hardness ranges from 4300 to 5000 MPa. The lower portion of the layer includes coarse grains of ferrite. In general, laser hardening of low-carbon steels offers little promise for industrial applications.

In laser-treated medium-carbon steels containing up to 0.6 percent carbon, the martensite hardness increases appreciably. In normalized steel treated with a CW CO_2 laser, the first liquid-phase hardened layer exhibits fine-grain plate-like martensite with a hardness of 7000 to 8500 MPa. The second solid-phase hardened layer consists of the upper portion of the martensitic structure whose hardness is equal to that of the first layer and the lower portion of an inhomogeneous structure. The inhomogeneity of the components in the lower portion changes in depth in the following sequence: martensite-troostite; martensite and troostite; troostite-ferrite; and ferrite at the boundary with the original structure.

In normalized or annealed steels laser-hardened at a high rate, the region of homogeneous martensite is absent and the troostite-ferrite network around martensite may extend up to the surface, thus reducing the hardness. For this reason, heat treatment must be carried out at low rates, less than 15 mm/s, with the beam dithered, where possible, to increase the region of homogeneous martensite. Laser hardening of steels preliminarily subjected to high-temperature tempering produces the same effect.

Eutectoid and hypereutectoid carbon steels. In these steels laser-hardened with melting, the first layer consists of fine-grain martensite and retained austenite whose content can be as high as 45 percent. Since undissolved cementite is absent, both martensite and austenite are quite rich in carbon. As a result, in steels containig 1.0 to 1.2 percent C the hardness of martensite increases and can reach 12 000 to 13 000 MPa.

In the second layer, the upper portion contains dissolved carbides and has the solid solution saturated with carbon, which leads to an increased content of retained austenite. The lower portion contains undissolved carbides and has a much lower content of retained austenite, which results in a maximum hardness of the structure. This necessitates hardening of hypereutectoid steels at highest speeds to impart the hardened layer structure with undissolved carbides.

Alloyed steels. The alloying elements of these steels exert diverse effects on the structure of the laser-treated zone. The microhardness of a low-carbon alloy steel after laser hardening is greater than that of a laser-hardened unalloyed steel with the same carbon content.

The hardness of laser-treated medium-carbon steels containing even a small amount of alloying elements increases quite appreciably. For example, the microhardness of chromium steel after CW laser hardening with maximum melting reaches 11 500 MPa, which is much higher than that of laser-hardened carbon structural steel.

In high-carbon steels, small additions of alloying elements render the laser-hardened zone more inhomogeneous because of a decreased diffusivity of carbon and increased stability of carbides. In laser hardening of chromium steels, the liquid-phase hardened layer displays a fine-grain structure of high-carbon martensite and retained austenite. A high rate of laser hardening preserves carbides in the surface layer. The alloying elements add to the hardness of martensite in the layer formed from the melt, which can be as high as 12 000 MPa. However, this layer generally exhibits microcracks, and therefore it is inadvisable to use laser hardening with melting for these steels.

The solid-phase hardened layer in laser-hardened chromium steels is inhomogeneous. Because of the incomplete process of austenization at the heating stage, the lower portion of the layer contains martensite, undissolved cementite, and retained austenite. In annealed steels, this portion of the layer has a low hardness ranging from 2700 to 3700 MPa. The middle portion heat-treated to higher temperatures exhibits high-alloy martensite and undissolved carbides, which impart the structure a hardness of up to 12 000 MPa. The upper portion of the layer contains dissolved carbides. Because this portion heats up to a high temperature, austenite homogenizes completely. After its cooling, the metal of this portion transforms into fine-needled martensite and contains an increased amount of retained austenite of a relaitvely low microhardness of 700 MPa.

Varying the parameters of heat treatment offers a means for control of the individual regions of the laser-treated zone. A small interaction time ensures a partial dissolution of carbides that is sufficient to saturate martensite and simultaneously preclude an increase in the amount of retained austenite. Transformation hardening with a pulsed laser or a CW laser at a high rate can provide such a structure.

High-alloy tool steels. Since these steels display a low mobility of carbon, it is difficult to carry out laser hardening at an optimal degree of austenization, i.e., with minimum dissolution of the carbide phase and sufficient saturation of a solid solution. At a low laser power, austenite becomes saturated at the heating stage, and so low-carbon martensite and retained austenite appear at the stage of quenching. In the process of surface hardening at a high laser power, the dissolution of carbides results in supersaturated austenite which is responsible for a large amount of retained austenite in the hardened layer.

So, for each grade of high-alloy steels, it is necessary to specify a narrow range of heat treatment conditions so as to ensure the formation of martensite with a sufficient amount of carbon and preclude the dissolution of carbides to a certain extent at the heating stage. Both transformation hardening and surface hardening with minimum melting can provide this type of structure.

In laser hardening of die and high-speed steels, the hardened layer can acquire a high hardness of 9000 to 10 500 MPa. This layer hardened with a pulsed laser or a CW laser consists of martensite, carbides, and a small amount of retained austenite. An optimal process of hardening with melting can be set up by properly selecting the pulse length or using the energy of a pulse that is 2 or 3 joules lower than the critical value. A CW laser can ensure a high hardness in the process of heat treatment with minimum melting.

Cast irons. In the process of heat treatment of these alloys with melting, graphite dissolves in the melt to yield the structure of chilled iron at a high cooling rate, which tends to equalize the concentration of silicon over the bulk of the hardened layer. The structure contains very fine dendrites or cellular austenite, the interdendritic spacings being filled with a two-phase component, namely, ledeburite which almost totally consists of cementite. This points to the quasi-eutectic solidification of the metal. A large amount of cementite increases the hardness of this layer.

The hardness of the liquid-phase hardened layer varies with the grades of irons and reaches 6400 to 10 000 MPa for nodular iron, 7400 to 9000 MPa for gray iron, and 6000 to 8000 for malleable iron.

The upper portion of this layer can sometimes have a low hardness if graphite dissolves incompletely or floats up from the lower portion.

The boundary between the liquid-phase region and the solid-phase one is irregular because of the saturation of the metal matrix near the graphite inclusions with carbon and a decrease in the melting temperature, as is evident from the constitution diagram of Fe–Fe_3C.

The degree of saturation with carbon varies with distance from the graphite inclusions and results in different compositions arranged in the following sequence: cementite; lamellar ledeburite; ledeburite and austenite; homogeneous austenite; and austenite-martensite of the acicular type. The hardness of these compositions is different and reaches 6400 to 6700 MPa for austenite and austenite-martensite. The hardness of cementite and ledeburite ranges from 10 000 to 12 000 MPa. The lower portion of the laser-treated zone contains a small amount of graphite and the structure here consists of martensite and retained austenite.

10.3. Surface Hardening of Titanium and Zirconium Alloys

The text below presents the results of laser hardening of alloys with melting. The parameters of surface hardening of titanium alloys with a 2-kW CW CO_2 laser are given in Table 10.1.

Laser heating with melting produces the zone of melting on the surface and the heat affected zone of the solid phase below the melt layer. In laser hardening in the atmosphere of air or nitrogen, the first layer, i.e., the liquid-phase hardened layer, has a dendritic structure with needle-like inclusions and, sometimes, middle-size grains of the beta phase. As seen from Table 10.1, the microhardness of the first layer varies in a wide range and decreases with an increase in the content of alloying elements.

Hardening in helium yields a liquid-phase hardened layer that consists of the network of alpha-phase plates oriented in the beta-phase grains and elongated in the direction of heat removal. The surface of the layer hardened in helium is white and that of the layer hardened in air is dark.

The diffractograph pictures of specimens hardened in air display the intense lines of the oxide TiO_2 and nitride TiN, which attests that these compounds form a surface film of increased absorptivity. That is why heat treatment in air ensures a larger melt depth than in helium, as is evident from Table 10.1.

Table 10.1. Parameters of Hardening of Ti Alloys with a 2-kW CO_2 Laser

Alloy	Rate, mm/s	Medium	Melt depth, mm	Microhardness, MPa		
				Initial	First layer	Second layer
Grade 1	14	Air	0.77	3280-3660	7700-165000	3960-4660
	14	Helium	0.43		5570-7520	3660-4660
Grade 2	4.2	Air	1.0	3400-3810	5320-10100	6440-6770/3280-5080
	14.0	Air	0.89		7520-13100	3660-5570
	140	Helium	0.3		5080-7520	4120-5320
Grade 3	4.2	Air	1.02	3530-3960	6440-8900	4860-5080/2960-4290
	14.0	Air	0.85		5080-6440	3400-4290
	14.0	Helium	0.4		4570-5830	3280-4290

Note. The numerator and denominator denote hardness of the acicular region and the two-phase region, respectively.

The first hardened layer in all alloy grades contains titanium nitrides. These nitrides, the martensitic α'-phase, and nitrogen and oxygen that saturate the melt add to the hardness of the surface layer.

The second layer of the heat affected zone consists of a few sublayers primarily because the heat source heats them to different temperatures. At the boundary between this layer and the first one the structure is acicular with the needles oriented within the grains of the beta phase. Below this acicular region there lies a sublayer of partially dissolved particles of the beta phase. This region probably heats up to a temperature of polymorphic transformation and can contain α'-, α''-, and β-phases. At the boundary of the original structure there lies a coarse-grained sublayer heat-treated at a temperature below the critical point. This region consists of course particles of the alpha and the beta phase. Its structure changes from acicular martensite to the composition of the original structure.

Laser hardening of zirconium alloys with melting yields hardened layers arranged in a similar pattern. Heat treatment with a pulsed laser in air is conducive to an intensive removal of the molten metal from the interaction region and results in an irregular surface of the hardened layer. A CW CO_2 laser provides a smoother surface.

The liquid phase solidifies into the martensitic α'-phase of the needle-like structure which is the supersaturated solid solution of the alloying element in hexagonal closed-packed zirconium. After its hardening with a CW CO_2 laser in air, Zi alloy with one percent Nb content reaches a hardness of 4000 to 6000 MPa, which, however, is lower than the hardness of Ti alloys because it contains a smaller amount of nitrides.

Laser hardening of the same alloy in a vacuum yields the hardened layer of a much lower hardness of 2200 to 2500 MPa on account of a strong effect of gases dissolved in the melt. The second layer hardened with a CW laser is a solid-phase transformation layer. At the stage of heating, the alpha phase in the upper portion of the layer transforms to the beta phase which then cools down rapidly to yield the martensitic α'–phase of the acicular structure whose hardness ranges from 2000 to 2800 MPa. The lower portion exhibits a transition zone heated to a temperature below the critical point, which contains somewhat coarser grains and has a hardness of 1600 to 2000 MPa that is equivalent to the hardness of the original structure.

10.4. Surface Hardening of Aluminum and Copper Alloys

The laser-treated zone in an aluminum alloy heat-treated with melting consists of only one layer resulting from melt quenching, the solid-phase hardened layer being quite neglible. Table 10.2 illustrates the hardness of aluminum alloys before and after the laser heat treatment.

The original structures of Al-Si alloy (grade A1) and duralimins (grades D1 and D2) consist of the alpha solid solution with various secondary phases. Artificial or natural aging imparts these alloys a maximum strength through the formation of pre-precipitation Guinier-Preston zones or secondary phases in the metastable state. Laser hardening appreciably refines the grains of the alpha solid solution, but the secondary metastable phases are absent. As seen from Table 10.1, the hardness of the liquid-phase laser-hardened layer in pretreated alloys of the above grades is somewhat lower than that of the original structure.

The results of laser hardening of silumins are quite different. The original structures of hypoeutectoid alloys of the Al-Si system (grades A2 and A3) and of the Al-Si-Cu-Mg system (grades A4, A5, A7, and A8) consist of the original grains of the alpha solid solution and the alpha-silicon eutectic. In hypereutectoid alloy grade A6, the primary crystals are silicon grains. The structures of these alloys may also include various intermediate phases in the stable or metastable state.

Table 10.2. Hardness of Al Alloys

Composition	Hardness, MPa				
	A1	D1	D2	A2	A3
Initial:					
primary-grain	1110-111180	1220-1860	1310-1650	700-920	810-900
eutectic	1660-1790	—	—	1010-1460	790-970
Laser-treated	940-1790	1030-1650	840-1070	970-1790	950-1220

Composition	Hardness, MPa				
	A4	A5	A6	A7	A8
Initial:					
primary-grain	970-1140	1000-1330	7000-12 720	1030-1140	790-970
eutectic	1720-1950	1070-1890	970-2200	—	1180-1780
Laser-treated	1460-2570	2350-2650	940-1790	1790-2030	1220-2570

Laser hardening changes considerably the structure of silumins as it suppresses the crystallization and growth of the primary crystals of the alpha phase or silicon. The hardened layer displays a quasieutectic structure noted for a high dispersity of phases. The hardness of the layer exceeds somewhat the hardness of the eutectic in the original structure. Laser heating changes the morphology of phases. The acute-angled phases in the eutectic attain the globular shape, which improves the mechanical properties.

One more factor that adds to the hardness of laser treated silumins is that the dissolved intermediate phases and silicon supersaturate the solid solution [120]. What attests to a high degree of supersaturation is the increase in the hardness of laser-treated silumins during aging. At an aging temperature of 423 K, the hardness of the liquid-phase laser-hardened layer in alloy grade A5 increases for 17 hours from 2460 to 3050 MPa, whereas the hardness of the layer not treated with the laser beam increases moderately.

The results of laser hardening of copper alloys with melting depend on the original structure of each alloy. Ni-Cr-Si bronze has a dispersion-hardened structure without the eutectic, the hardness of which varies from 1520 to 1860 MPa. At the stage of laser heating, the metastable intermediate phases of the bronze dissolve, but at the stage of quenching the liquid phase transforms into the coarse-grain structure of the alpha-phase solid solution. The hardness of the surface layer does not increase and reaches 1460 to 1900 MPa.

The hardness of the first layer decreases in depth. In the solid-phase hardened layer, the hardness sharply drops to 1030 MPa, which points to a possible dissolution of the hardening phases.

The Fe + 30% Cu alloy has an original structure that consists primary of copper carbides with a hardness of 1820 ± 140 MPa and an alpha-phase-copper eutectoid whose hardness amounts to 2340 ± 110 MPa. The first laser-treated layer exhibits a finely dispersed quasi-eutectic whose hardness rises to 4000 MPa and 4300 MPa after its heat treatment with a CW laser and a pulsed laser, respectively.

In alloys that solidify in a wide temperature range, a high cooling rate can entail an interdendritic segregation and a change in the phase composition. For example, in the laser-treated bronze whose original structure consists of the alpha-phase solid solution with a hardness of 860-1070 MPa, the surface layer includes the networks of the second phase along with the alpha-phase grains. The interdendritic segregation apparently causes the two-phase region on the constitution diagram to shift to the region of small concentrations, thus forming intermediate beta and delta phases. As a result, the hardness of the alloy treated with a CW laser and a pulsed laser grows to 1650 MPa and to 1720 MPa, respectively.

10.5. Properties of Laser-Hardened Alloys

Laser surface hardening is an effective tool for increasing the surface hardness, structure dispersity, load-carrying capacity, and wear resistance of parts designed to operate in friction.

Table 10.3 displays the results of the comparison tests for wear resistance of 10 × 10 × 100-mm carbon-structural-steel specimens subjected to normalizing, annealing, furnace hardening, and laser heat treatment. The test unit was designed to perform dry friction tests in air at a slow speed of reciprocating motion of the pressure sensitive pin from chromium ball-bearing steel subjected to hardening and low-temperature tempering.

The 1-kW CO_2 laser beam was moved along the track on the target surface to heat treat the steel to a depth of about 1 mm without melting at a speed of 25 mm/s. The wear-out rate w was found from the formula $w = S/ln_c$, where S is the mean area of the wear track, determined from the roughness indicator readings of eight tracks, and l is the wear track length.

The comparison of test results shows that laser hardening ensures the lowest friction coefficient, reduces the run-in period to merely two or three cycles, and results in a small number of sound radiation pulses that cover a narrow range. The reason is that the laser-hardened specimen acquires a homogeneous structure.

Table 10.3. The Results of Wear Resistance Tests on Carbon Structural Steel

Type of treatment	n_c	n_p	n_r	C_f	S, mm^2	$w \times 10^3$, mm^2/m
Laser hardening	310	20-70	2-3	0.39	14.3	0.96
Normalizing	160	190-420	15	0.44	38.7	5.04
Annealing	150	4-120	25	0.42	46.8	6.5
Furnace hardening	180	—	20	0.68	72.0	8.4

Designations: n_c is the number of test cycles; n_p is the number of acoustic radiation pulses per cycle; n_r is the number of test cycles over the running-in period; and C_f is friction coefficient.

The sliding friction resistance of irons and aluminum alloys hardened with a CW laser increases noticeably. In laser-treated irons, graphite retained in the hardened region promotes an additional friction resistance. Steels and some other alloys offer an increased friction resistance in alkaline and acid media.

Tests for rolling friction resistance were made with the aid of a carriage moved on rollers and pressed against the specimen with a force of 130 N. The track length of the carriage was 200 mm and the number of motions executed in a minute was 25. The test procedure involved both grain iron and chromium steel specimens heat-treated without melting by a moving CO_2 laser beam.

After 120 000 motions of the carriage on the surface of an iron specimen, there appears a 80-μm thick layer of damaged surface with a network of cracks obviously initiated by graphite inclusions. Transformation hardening thus imparts to the iron an unsatisfactory resistance to rolling friction. The wear resistance of laser-hardened chromium-steel specimens is found to be much higher than that of the specimens hardened and tempered by conventional techniques. The mating parts run in well and their surface roughness decreases. So, laser-induced transformation hardening of this steel intended for work in rolling friction can be a rather promising process.

There is a large number of works concerned with wear resistance of laser-hardened metals and alloys [111, 122-126]. The information presented in these works and in a number of other papers has formed the basis for establishing the regularities of laser hardening.

Heat treatment with a pulsed laser finds practical uses for hardening tools [127-129]. The side surfaces of blanking dies retain the laser-hardened layers after multiple sharpening. The durability of dies increases two to five times. Industrial set-ups are available for pulsed laser hardening of parting-off and straight-turning tools, reamers, taps, drills, cutters, broaches, knives, and other cutting tools.

A stationary CW laser beam incident on a workpiece moving at constant speed provides a higher throughput. The beam forms a heat treat track 1 to 10 mm wide. The depth of hardening in steels and irons treated without melting reaches 2.0 mm. Liquid phase quenching offers a greater depth, but then the surface quality becomes poorer. Continuous laser hardening can assure a more uniform hardness along the length of the track. The heat-treated tracks can be run with or without overlapping. Dithering the beam perpendicular to the track at a rate of 200 hertz and above increases the track width up to 20 mm and provides uniform hardening of the target surface.

The main goal of continuous laser hardening is to impart to the surface of a workpiece a high wear resistance, primarily the resistance to rolling and sliding friction [130]. Laser hardening can often increase the heat stability of an alloy structure, which generally depends on the degree of structure metastability and the diffusive mobility of atoms. In carbon-bearing alloys, the crystal lattice distortion and dislocation density prove to be higher after laser hardening than after conventional hardening. This will cause a more intense process of disintegration of laser-hardened martensite into ferrite-cementite mixture and, hence, a faster decrease in hardness during tempering. On the other hand, the process of heating of laser-hardened iron and steels to 300 °C can transform retained austenite into martensite, in which case the hardness changes in a more complex fashion.

Let us consider how the Vickers hardness of Cr steel specimens hardened with a CW CO_2 laser and by the conventional technique changes in the process of heating to 473 K (curves *1* and *3*) and to 573 K (curves *2* and *4*) at a holding time of up to 24 hours (Fig. 10.2). As is evident from curves *1* and *3*, during heating the hardness of the steel treated by the conventional method decreases more rapidly than the laser-hardened steel since the latter contains a large amount of retained austenite. In the process of heating, retained austenite rapidly converts to tempered martensite, so the hardness decreases slower.

This effect shows up more vividly during heating at 573 K. After a holding time of up to 4 hours, the hardness of the laser-treated steel drops markedly (curve *2*). During the subsequent heating over the course of 8 hours at the same temperature, the hardness grows appreciably because the retained austenite intensively transforms into tempered martensite. At the next stage of holding, the hardness smoothly decreases. The hardness of the laser-treated steel held for a definite length of time at 473 K and 573 K is found to be 1000 to 2000 MPa higher than that of the steel hardened in a conventional way.

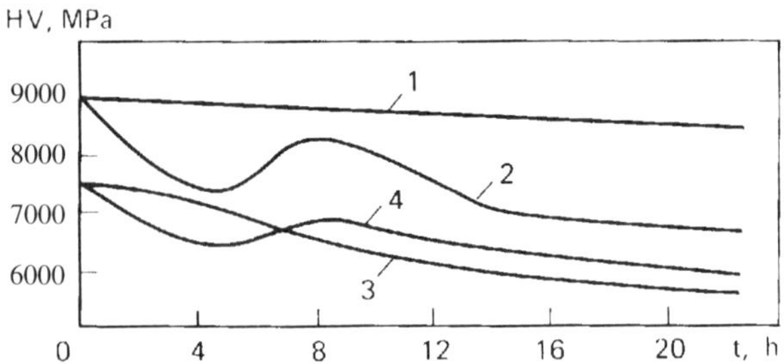

Fig. 10.2. Hardness of Cr steel heat-treated with a CW CO_2 laser and by the conventional technique as a function of time of holding at 473 K (curves *1* and *3*, respectively) and at 573 K (curves *2* and *4*, respectively)

Irons hardened with melting by a CW laser beam behave in a similar manner. In the course of heating for one or two hours, the hardness of the first layer heated to 300-350°C rises and then drops sharply. The hardness of the second layer that contains a much smaller amount of retained austenite gradually declines with increasing temperature.

Because of a high supersaturation of solid solutions, particularly in the liquid-phase hardened layer, the process of heating can lead to an increased segregation and the formation of intermediate phases, which increases the hardness. Figure 10.3 illustrates variations in the hardness of Al-Si-Cu-Mg alloy treated with a CW CO_2 laser during long-time aging at different temperatures. Heat aging at 423 K tends to increase the hardness in a steady fashion. At temperatures of 473 and 523 K, the hardness begins to grow slightly for about an hour at the initial stage and then gradually declines. At 548 K, the hardness decreases steadily. In all cases, the hardness of the laser-treated region is higher than the original hardness of the alloy hardened and aged by the conventional technique. This attests that laser hardening supersaturates the solid solution to a higher level than the conventional technique.

The parts intended to carry intermittent loads must have high endurance strength. The fatigue resistance of hardened parts depends on the surface roughness, structural defects, residual stresses, grain size, structural elements, and other factors. The combined effect of the above factors is rather diverse, and so recourse should be made to experimental studies for estimating the fatigue resistance.

Multiple-cycle tests on smooth cylindrical specimens subjected to cantilever bending reveal that the fatigue resistance of pulse laser-hardened specimens decreases by about 40 percent. Continuous-laser transformation hardening of steels increases the fatigue resistance from 200-300 MPa to 520 MPa. As a

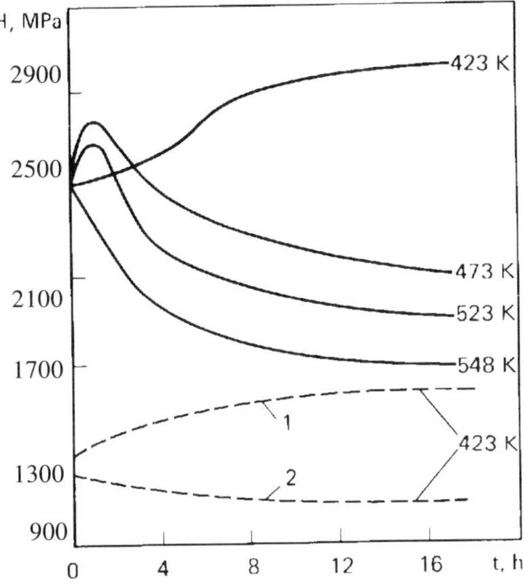

Fig. 10.3. Hardness of aluminum alloy treated with a CW CO_2 laser (solid curves) and by the conventional technique (dashed curves) as a function of time of aging. Curves *1* and *2* identify the eutectic and solid solutions, respectively.

rule the fatigue resistance of parts subjected to liquid phase quenching decreases.

The corrosion stability of alloys generally grows as the phase composition becomes more homogeneous. Therefore, liquid-phase laser quenching and laser-induced amorphization of alloys promote the corrosion resistance. The tests for the corrosion resistance of gray iron hardened by a CW laser beam with melting indicate that the flow of critical current through a specimen in the H_2SO_4 solution is 5 to 10 times lower and the process of anode oxidation slows down. The corrosion current through laser-hardened silumins in the solutions of H_2SO_4, NaCl, and KOH decreases noticeably.

If the process of laser hardening of corrosion-resistant steels dissolves chromium carbides and uniformly distributes chromium in the solid solution, then the intercrystalline corrosion resistance of steels grows.

In laser hardening of carbon steels the structure can become more inhomogeneous, original ferrite or carbides can be preserved, and retained austenite can appear, with the result that the corrosion resistance of steels can become insufficient. Thus, the corrosion resistance of laser-hardened alloys varies with the type of alloy and heat treatment conditions.

Surface hardening with pulsed lasers is suitable for heat treatment of cutting tools, dies, and wearing parts. However, since the energy coupled from the pulsed laser to the target is comparatively small, the process can provide a shallow depth of hardening at a low throughput and does not ensure uniform properties of the hardened layer.

Continuous laser hardening is largely free of the above shortcomings. Examples of industrial applications of CW CO_2 lasers confirm a high efficiency of these devices [131–133]. CW lasers can be used to best advantage where it is difficult to rely on the conventional techniques if the parts are subject to buckling, have difficult-to-reach areas, and are large in size. Laser hardening will gain still wider acceptance with the development of more reliable heat treating set-ups equipped with lasers of 1 kW power and over and automated control systems.

11 Surface Alloying and Cladding

There are a number of conventional techniques for metal surface coating to improve corrosion resistance and strength while retaining the properties of the base metal [134]. Last years have seen ever increasing application of lasers for glazing, amorphization, shock hardening, alloying and cladding, or facehardening.

This chapter describes some experimental results of studies on laser alloying and cladding.

11.1. Alloying with Nonmetallic Elements

Laser alloying of metals with carbon, silicon, and boron involves spreading the powders or slurries containing the alloying elements over the surface and melting the deposited material to produce an alloyed layer on the surface. Laser alloying in a liquid or a gaseous medium sometimes finds application too [135]. An average thickness of the alloyed layer produced with a pulsed laser and a continuous laser reaches 0.3 or 0.4 mm and 0.3 to 1.0 mm, respectively. In distinction from alloying, cladding requires a minimum fusion of the base metal that is enough to produce a bond between the alloy material and the substrate.

Carburizing. The carburizers used for the purpose include graphite and carbon black in the form of solutions in acetone, alcohol, and other solvents. In use are also the solutions in various varnishes. The powder-solvent slurries can contain active agents such as borax and ammonium chloride.

Gas carburizing is the hardening process in which carbon-containing gases are held at a pressure of 10 MPa to enhance the diffusion saturation of the surface layer with carbon. Liquid carburizing is the process of surface hardening in a carbon-containing liquid such as hexane, toluene, carbon tetrachloride, and petroleum oil (Fig. 11.1). Laser beam *1* incident on the surface of liquid *4* vaporizes hole *2* through which it penetrates deeper into the liquid and falls on the surface of workpiece *5*. Vapor-gas hole *2* ends up above the target surface in dome-like cavity *3* filled with carbon vapors.

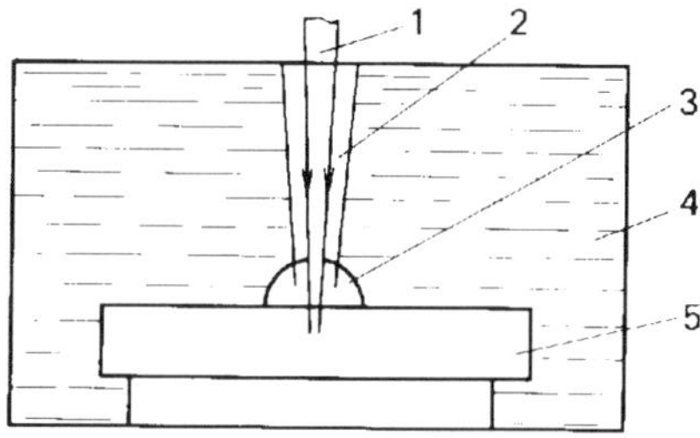

Fig. 11.1. Laser-assisted liquid carburizing: (1) laser beam; (2) vapor-gas hole; (3) gas cavity; (4) liquid; (5) piece of work

The composition of the alloy layer and its hardness vary with the amount of carbon diffused into the metal. A small amount of carbon diffused into a low-carbon steel yields a structure composed of martensite and retained austenite, the hardness of which can be as high as 9000 MPa owing to a high saturation of martensite. A large amount of carbon introduced into the surface layer may lead to an increased content of austenite, in which case the hardness will vary over a wide range from 4500 to 9000 MPa. At a still higher content of carbon alloyed, for example, into a high-carbon steel, the surface layer displays carbides apart from martensite and austenite and its hardness can rise from 9000 to 14 000 MPa.

Nitriding. Liquid nitriding uses slurries based on ammonium salts or carbamides to form hard nitrides in the laser-treated layer, which increase the hardness and wear resistance. In gas nitriding, the pressure of nitrogen should be held at about 9 MPa. Gas nitriding in which the jet of nitrogen is brought directly to the laser-melted region is a more efficient process. In this process titanium, zirconium, hafnium, and their alloys can be readily nitrided. The solidified melt of a titanium alloy forms an alpha-phase layer containing titanium nitrides whose hardness is as high as 17 000-20 000 MPa.

The structure of the nitrided zone in steels includes nitrous martensite, retained austenite, and iron nitrides. Nitrous martensite can withstand high temperatures, thus adding heat stability to steels.

Siliconizing. The process of surface impregnation uses dry silicon powders or silica gel suspensions applied to a metallic surface and melted with a laser beam to form a surface alloy layer of high heat stability and increased corrosion resistance and wear resistance.

In siliconizing steels to a silicon content of up to 0.1 percent, the structure of the solidified melt transforms to alpha iron, apparently martensite. A further increase in the silicon content yields, apart from alpha iron, silicides Fe_3Si,

Fe_5Si_3, FeSi, and $FeSi_2$, which increase the hardness of the laser-treated layer from 8000 to 15 000 MPa.

Borating. Prior to laser melting, the surface is given the coat of an alloy material applied by different methods such as plasma spraying and spraying or spreading the powders of boron, boron carbide, boric anhydride, borax, ferroboron with binders, and water suspensions of these powders.

If the borated layer contains a small amount of boron, the structure consists of alpha iron and the borated eutectic with hardness ranging from 6000 to 12 000 MPa. At a higher content of boron, the structure includes borides such as Fe_3B, Fe_2B, and FeB, which increase the hardness to 21 000 MPa.

The structure becomes hard and wear-resistant because the borated layer does not contain retained austenite. The resistance to abrasive wear increases with the content of FeB. The Fe_3B and Fe_2B borides enhance the toughness of the layer. The structure retains a high hardness even at a temperature of up to 600 °C if it contains Fe_2B and FeB.

11.2. Alloying with Metallic Elements

Laser alloying is a fusion process in which a pulsed or a CW laser melts the alloy material applied to the surface. A convenient method is spraying powder over the surface and simultaneously melting it by a CW laser beam.

The structure of the laser-alloyed layer contains supersaturated solid solutions and sometimes intermetallic compounds. Refractory and carbide-stabilizing elements melted into iron-carbon alloys increase the hardness of alloyed regions.

Chromium alloyed into iron and steel adds the corrosion resistance to the metal and increases the impact strength and wear resistance. Alloying a low-carbon Ni-Mo steel with chromium and carbon gives a 1.25-mm layer of high heat stability, whose Rockwell C hardness falls in the range of up to 55.

The alloying elements and compounds commonly used to provide laser-alloyed layers on the surface of aluminum alloys include Fe, Ni, Ti, ferrovanadium, Ni-Cr-B-Si, Co-Cu, and other systems. The alloy layer has the structure of the solid solution of an alloying agent in aluminum with such intermetallic compounds as $FeAl_3$, VAl_3, $TiAl_3$, CoAl, $NiAl_3$. These compounds acquire a nearly globular shape, thereby improving the properties of the laser-alloyed surface. The hardness of the alloyed region increases to 3000 MPa and that of intermetallic clusters reaches 8000 to 10 000 MPa.

In Fig. 11.2 are shown the plots of the Vickers hardness of laser-hardened and laser-alloyed surfaces of Al alloy specimens against the time of holding

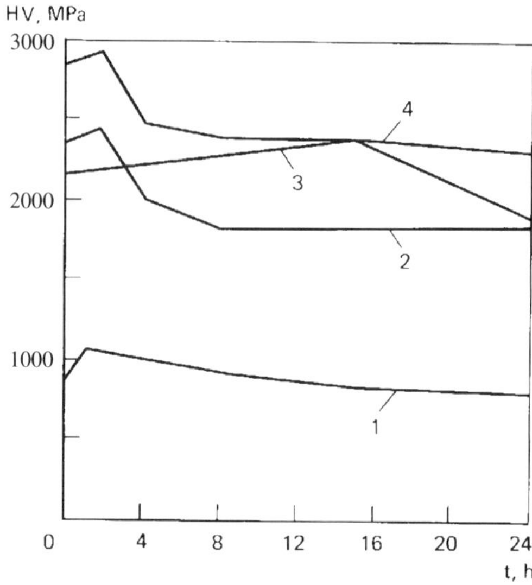

Fig. 11.2. Hardness versus the time of holding at 250 °C for All alloy: laser-hardened (curve *1*) and laser-treated to saturate its surface layer with Fe (curve *2*), Ni (curve *3*), and Ni-Cr-B-Si (curve *4*)

at a temperature of 250 °C. The hardness of the laser-alloyed layer is 1000 to 1500 MPa higher than that of the laser-hardened layer. The layer alloyed with the Ni-Cr-B-Si compound has the highest hardness.

Saturating the surface of a titanium alloy with chromium which acts as a beta stabilizer yields a beta-phase structure of enhanced wear resistance, whose hardness increases from 2800 to 6400 MPa.

Alloying carbides into refractory metals such as TiC, NbC, VC, TaC, and WC provides a high-quality alloy layer. For carbides to be fused into a laser-heat-treated region, the powder of a refractory metal and graphite can be placed on the surface and melted by a laser beam. This process can increase the hardness of the surface layers of steels to 10 000 MPa, which, however, is much lower than the hardness of carbides that dissolve in the alloyed region.

The conditions of alloying with a continuous laser should be chosen so that carbides can be dissolved incompletely in the molten metal. In this case, the surface layer acquires hardness that is equivalent to that of the original carbide, thus appreciably increasing the wear resistance. Carbon carbides melted onto the surface of steel raise the hardness of the alloy layer to 17 000 MPa.

The process of melting the borated powders of Ni, Co, Fe, and Cr on the metal surfaces imparts valuable properties to the alloy layer. The MoS_2

composition fused into the surface layers of carbon steels, chromium steels, and titanium alloys can increase the wear resistance of surfaces two-to-five times.

11.3. Fusion of Powder-Binder Coating Mixtures

The quality of laser melting of coating mixtures largely depends on the compositions and types of binders used in conventional processes of hardfacing [137, 138] and ion laser-induced hardfacing processes [139-141]. These are sizing agents, self-curing plastics, fats, oils, and paintwork materials classified into four groups in accordance with the similarity of their properties (Table 11.1).

Powder mixtures using group 1 binders adhere well to the substrate. The laser beam incident on the coatings of these mixtures initiates an intensive decomposition and burning of the binders with the formation of solid disintegration products ejected at a high speed from the heat affected zone. The particles of alloying elements are also removed, which decreases the dimensions of the bead bonded to the surface. The decomposed particles of group 1 binders periodically shield the target surface so that the surface layer melting depth along the heat treat track differs. This drawback can be overcome by adjusting the parameters of laser radiation.

The molten zone contains about 70 percent of the base metal, the contents of alloying elements being small, for example, 11 to 17% Ni and 3 to 7% Cr. The X-ray microscopic analysis reveals that the built-up layer has uniform distribution of alloying elements in depth (Fig. 11.3).

The dimensions of the fusion zone and the content of alloying elements of the zone depend appreciably on the specific heat energy per unit area:

$$E_h = P\eta_n / v_m d_f, \tag{11.1}$$

where P is the laser power, η is the net efficiency; v_m is the melting rate; and d_f is the focused spot diameter.

A higher specific heat energy E_h involves an increase in the molten zone size and reduces both the content of alloying elements in the layer and the layer hardness because of a high saturation of the melt with the base metal.

Group 2 binders of higher heat stability decompose slowly, thus yielding a much smaller amount of disintegration products, but provide a poorer adhesion of the coating mixture to the substrate. As a result, the built-up layer of the bead has quite irregular dimensions. The bead shape can be improved by raising the laser power or reducing the melting rate. However, the depth

Table 11.1. Results of Laser Melting of Powder-Binder Mixtures

Group	Binder	Dilution coefficient	Melting depth, mm	Melt region width, mm	Hardness, MPa	Dry residue, %	Notes
1	Synthetic resin adhesive, epoxy resin, self-curing plastics, solution of borax in acetone,solution of rosin in alcohol or oil	0.6-0.8	0.14-0.34	0.14-0.38	3800-5000	10-80	Intensive soot blowoff Irregular bead
	Isopropyl alcohol	0.1	0.9	2.23	3660-7940	0	
2	Water glass	0.1	1.0	2.1	5830-6770	50	Irregular bead
	Silicate glue	0.05	1.2	2.18	—		Irregular bead
	Dextrin glue	0.1	0.6-1.4	1.1-2.5	6440-10 100		Rolls on bead
3	All-purpose adhesive	0.1	1.1-1.28	2.0-2.8	5400-9400	6-10	Uniform bead
	Nitrocellulose lacquer	0.05-0.15	1.25-1.55	2.4-3.3	5400-10 000		Uniform bead
	with boric acid	0.1	1.0-1.26	1.72-2.97	4660-7940		Porosity
	with borax		0.9-1.19	1.50-3.02	5400-9400		Porosity
4	Nitrocellulose lacquer with flux	—	1.48-1.58	2.14-2.8	6700-10 100	15-20	Porosity
	with graphite	0.6-0.7	0.12-0.41	0.17-0.45	3800-5400		Porosity
	with ceramic				3600-5000		Strong gas plume

of melting of the base metal then grows, which increases the dilution coefficient

$$m = \frac{S_o}{S_o + S_d} 100, \tag{11.2}$$

where S_o is the area of the melted base metal and S_d is the area of the deposited metal.

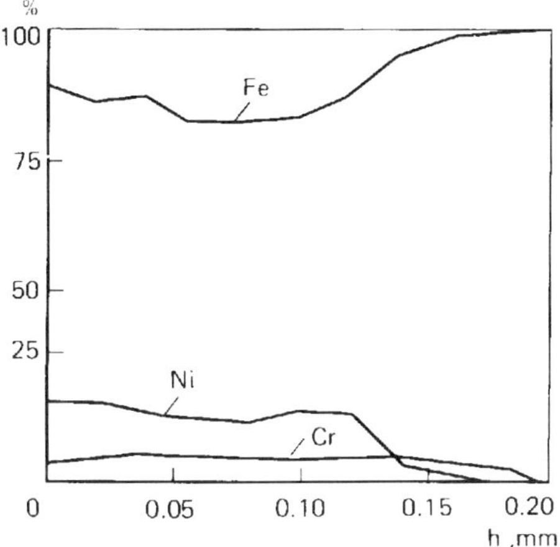

Fig. 11.3. Content of alloying elements versus melting depth in laser melting of the slurry with synthetic resin adhesive

Heat-resistant binders are apparently more adequate for alloying than for cladding.

Fusion mixtures using groups 3 and 4 binders based on nitrocellulose lacquer afford a better quality of the bead with regular dimensions along its length, minimum fusion of the base metal, small coefficient m, and reduced porosity. The text below describes the experimental results of cladding using the slurries with nitrocellulose lacquers.

Work [142] presents the theoretical relations derived from the solution of a model problem on the laser-induced melting of a porous medium. These relations are useful for establishing tentative parameters of laser cladding, which certainly need experimental verification.

Proceeding from the experimental results of heat power density q_h versus laser-induced melting rate v_m, we can divide the laser treatment parameters into two regions separated by a straight line that represents q_h as a linear function of v_m (Fig. 11.4). The parameters of the lower region do not provide a uniform adhesion of the melt to the substrate and lead to the drop-by-drop build-up of the layer with irregular dimensions of the bead. The parameters within the upper region ensure a uniform bead.

The quality of a built-up layer depends heavily on the adhesive strength and the difference in chemical compositions between the coating and the alloy metal. An increased fusion of the substrate offers a stronger built-up layer-to-base metal bond, but results in a greater dilution of the base metal in the

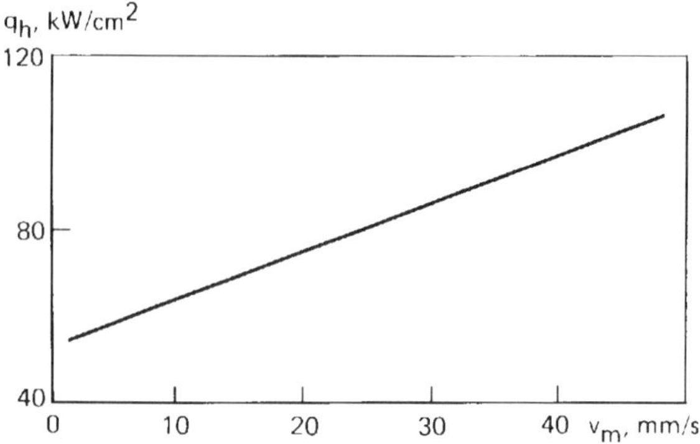

Fig. 11.4. Heat power density as a function of melting rate

melt, which impairs the properties of the built-up metal. For this reason, the dilution coefficient should be held up within 5 to 15 percent, which varies with the laser power density and the melting rate.

A large number of experiments on laser cladding by fusion of 1-mm thick coating mixtures confirm that in a first approximation the dilution coefficient m varies in a linear fashion with the laser power density q in the focused spot at the target surface (Fig. 11.5). In this case, account is taken of the relation between the heat power density q_h and the melting rate v_m, as illustrated in Fig. 11.4.

From the analysis of plots of Figs 11.4 and 11.5 we can conclude that the process can provide a quality bead at a minimum melting depth of the base

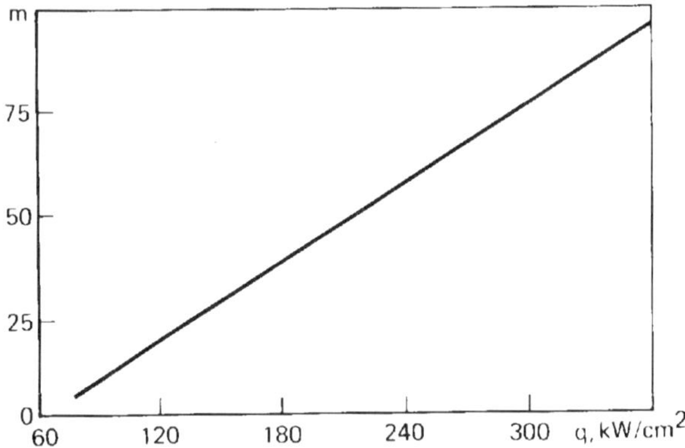

Fig. 11.5. Dilution coefficient as a function of laser power density

metal in a narrow range of values of P, v_m, and d_f, with E_h varying from 60 to 70 J/mm^2. This regularity holds for an increased thickness of the mixture layer.

A higher laser power density cannot intensify the process of laser cladding since the dilution coefficient m then sharply grows because the heat source promotes deeper penetration.

Under optimal conditions of melting the coating mixture with the nitrocellulose lacquer, the process can secure acceptable dimensions of beads. For example, at a thickness of the slurry coating of 1.5 mm, the bead width is 2.4 to 3.3 mm and the bead height approaches the thickness of the slurry layer and varies from 1.25 to 1.55 mm.

The content of the nitrocellulose lacquer of the powder mixture can be varied with the conditions of preparation of the slurry and the method of its spreading over the surface. At a content of this lacquer less than five percent, the mixture crumbles, while at a content of more than 30 percent the deposited layer gets fluid. The optimal content ranges from 10 to 20 percent.

The built-in layer has a high hardness that can reach about 10 000 MPa. It is evident from the microscopic analysis that the content of alloying elements of the hardfaced layer differs little from the original content of these elements in the powder: 65 to 75% Ni, 12 to 14% Cr, 4 to 8% Fe, and up to 1.7% Si (Fig. 11.6). The content of iron increases in comparison with the original content only at the fusion boundary.

In the course of melting the mixture, the binder of the slurry layer at the edges of the built-up bead disintegrates, thus uncovering the base metal, because the laser power around the periphery of the focused spot is not sufficient to melt up the alloying elements. Since the binder no longer provides bonds between the powder particles at the bead edges, the surface tension forces attract these particles wetted with the melt to the bead being built up. For this reason,

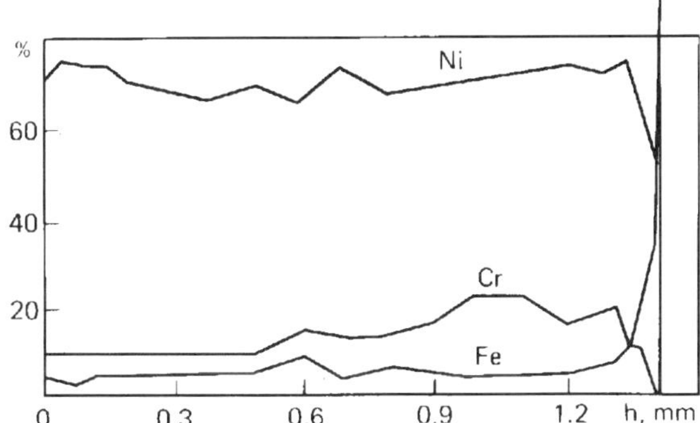

Fig. 11.6. Percentage of alloying elements versus melting depth in laser melting of the slurry with nitrocellulose lacquer

the adjacent bead cannot be run until after the slurry layer is restored.The beam scanned normal to the direction of travel of the target can increase the bead width, but does not eliminate the mixture decomposition at the bead edges.

Reactive fluxing agents added to the binders to increase the melt fluidity and to reduce the angle of contact are able to decrease the decomposition of the mixture layer near the bead. It was found that adding boric acid, borax, and flux composed of alkali metal chlorides to the nitrocellulose lacquer decreases the uncovered region width, but tends to increase appreciably the bead metal porosity. Heat-resistant agents such as graphite and ceramics also reduce the width of the uncovered region, but promote the metal porosity too.

Thus, the process of melting the slurries with the nitrocellulose lacquer can be used to deposit single-pass beads particularly in difficult-to-reach areas, and also to alloy the surfaces of metals.

11.4. Cladding with Plasma Spraying

Laser melting of the coatings sprayed onto the metal surfaces is an efficient process which increases the strength of the adhesion between the base metal and the alloy layer. In this section we consider the results of experimental studies on laser melting of coatings applied by the method of plasma spraying [134].

The experiments were performed on low-carbon steel specimens in the form of 50-mm diameter tubes with 3-mm thick walls. An industrial plasma spraying set-up was used to deposit coatings 1 mm thick at a porosity of 8 to 10 percent under the following conditions: current, 400 A; arc voltage, 40 V; flow rate of plasma-generating 90% Ar + 10% N_2 gas mixture, 1.2×10^{-3} m^3/s; rate of powder spraying, 0.83×10^{-3} kg /s; distance to a target, 120 mm; angular velocity of a specimen, 6.28 s^{-1}; and linear speed of a specimen, 4 mm/s.

The power of the laser beam required to melt the spray coatings was 2 to 3.5 kW, the focused beam diameter d_f was 0.5 to 1.6 mm, and the specific heat energy E_h was 25 to 400 J/mm^2.

Figure 11.7 illustrates the experimental results. The volume of the molten metal tends to increase somewhat with the specific energy since both the melting depth h and the molten zone width grow. A specific energy of less than 180 J/mm^2 causes the deposited layer to melt only partially, and so the solidified metal exhibits rounded pores formed by the gas evolved from the lower portion of the layer.

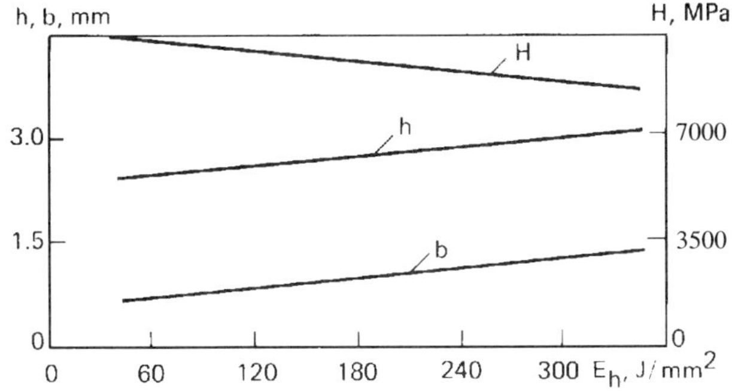

Fig. 11.7. Depth, width, and mean hardness of the laser-treated region versus specific heat energy

At a specific energy of 180 to 330 J/mm^2 the heat source melts through the layer and fuses off the substrate metal to a depth of 0.25 mm. The solidified melt does not generally contain pores. A further increase in E_h above 330 J/mm^2 favors a greater depth of melting of the substrate, so that the dilution of the cladding by the base material grows and results in poorer properties of the clad metal.

The laser-treated layers have a highly dispersed structure and are free from oxide inclusions and pores. The hardness of the laser-treated region is slightly lower than that of the plasma-sprayed layer. But this comparison relies on the values of the sprayed layer hardness measured selectively beyond the phase boundaries and pores, whereas the hardness of the laser-treated region is taken to be the value averaged over the entire volume of the clad metal [143].

As is evident from the plot of Fig. 11.7, the hardness of the laser-treated coating slightly declines with an increase in E_h, but commonly does not drop below 8000 MPa. The X-ray spectrum analysis shows that the percentage of alloying agents in the treated layer differs little from that in the powder and reaches 60 to 74% Ni, 11 to 17% Cr, 1.5 to 2.5% Si, and about 4% Fe.

Laser melting ensures uniform distribution of alloying elements throughout the molten layer, excepting the layer portion at the fusion boundary. The liquid metal then rapidly cools down and retains the alloying elements.

Experiments were performed to investigate the process of cladding with partial overlapping of heat treat tracks. For this, a tubular specimen was revolved and simultaneously moved in the longitudinal direction to melt the cylindrical surface along the helical path. At the initial stage of laser heating, the melting depth slightly exceeded the steady value until the process was brought up to the quasistationary state.

In laser treatment with partial overlapping of adjacent heat-treat tracks, the laser beam causes the overlapped band to melt twice. This band differs little in hardness from the single-pass treated region since the process of remelting leads to insignificant phase transformations and does not commonly change the structure and composition of the remelted metal. Melting the substrate to a depth of 0.25 mm provides a good bond between the alloy layer and the substrate surface. Also, a short time of laser interaction precludes the process of active diffusion.

Thus, the laser cladding as a two-stage process of plasma spraying and laser melting of the plasma arc coatings can provide high-quality alloy layers on the substrates. This process looks promising for improving the properties of spray coatings applied to heavy-load-bearing parts.

11.5. Cladding with Gas-Powder Spraying

In this process of hardfacing, the gas jet sprays the powder through an orifice onto the surface and the laser beam simultaneously melts the cladding material which quickly solidifies to yield a hardfaced layer [144]. The powder particles that get into the region of laser radiation heat up to the melting temperature for about 1 ms and melt up completely on the substrate surface for about 10 ms. In actual conditions, a powder particle covers a path no larger than 10 mm at a speed of 10 to 30 m/s in the region of the laser beam until it falls on the substrate.

The parameters of gas-powder spray hardfacing include: the laser power P; melting (cladding) rate v_m that is equivalent to the speed of travel of a specimen; focused spot diameter d_f; mass flow rate G_p of the powder sprayed on the surface; distance l from the nozzle throat to the target surface; and angle of incidence of the nozzle α relative to the beam axis (Fig. 11.8).

The experiments were carried out to verify the effects of individual parameters on the process of cladding with a CW CO_2 laser which provided hardfaced coatings on low-carbon steel specimens whose flat surfaces were cleaned of scale on a grinder. The beam was focused with a 400-mm focal length lens. The diameter of the focused beam spot was varied by adjusting the distance Δf between the focal point and the substrate surface (Table 11.2).

An increase in the laser power P from 1.5 to 3.5 kW causes the volume of the melt to grow, with the result that the width b and the height h of the built-up bead become larger (Fig. 11.9). The dilution coefficient m rises from 0.02 to 0.11 since the melting depth increases with the laser power.

Increasing the melting rate from 8.3 to 50 mm/s at the same laser power reduces the width and the height of a bead by one half and three fourths,

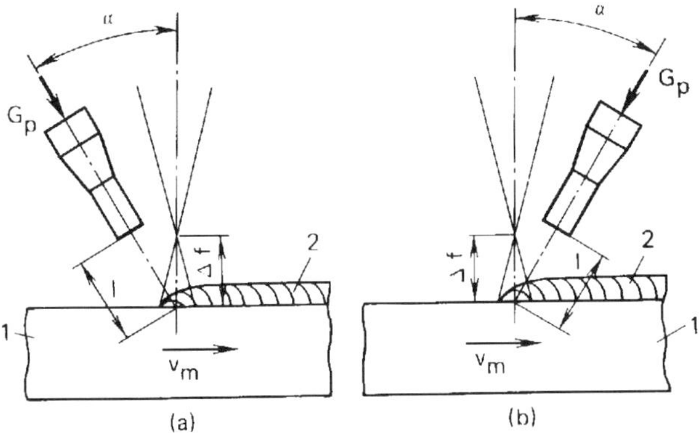

Fig. 11.8. Building up the alloy layer z by laser melting of the powder sprayed in the direction of motion of the specimen (a) and in the opposite direction (b): (1) specimen; (2) deposited bead

respectively, because the specific heat energy and mass flow rate of the powder tend to decrease at a higher heating rate (Fig. 11.9*a*).

Table 11.2. Parameters of Laser Cladding

Parameter	P, kW	v_m, mm/s	Δf, mm	$C_p \times 10^3$, kg/s	l, mm	α, deg
Value	1.5-3.5	8.3-50	15-60	0.4-1.2	5-35	10-70
Control step	0.5	8.3	15	0.1-0.2	5	10

The degree of beam defocusing has different effects on the geometry of a bead. With an increase in Δf at a constant power P, the laser power density decreases and results in a reduced bulk of the molten powder and in a lower height of the hardfaced bead (Fig. 11.9*c*). The bead width first grows with the hot spot size and then decreases apparently because the power density around the spot periphery heavily diminishes. The dilution coefficient m generally declines with an increase in Δf.

At a higher mass flow rate and a constant laser power, the height and width of the bead grow and the dilution coefficient goes down from 0.14 to 0.06.

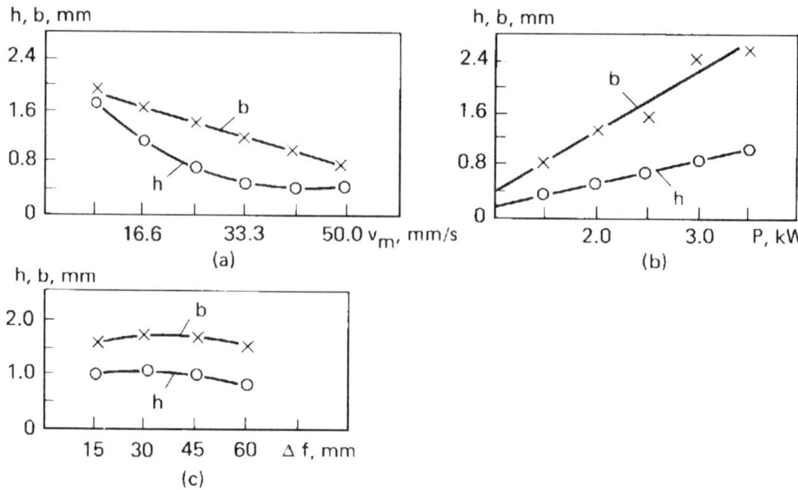

Fig. 11.9. Height and width of the bead built up by laser melting of the powder sprayed at $G_p = 0.845$ g/s, $l = 15$ mm, and $\alpha = 45°$ versus operating parameters: (a) $P = 2.5$ kW and $\Delta f = 30$ mm; (b) $v_m = 25$ mm/s and $\Delta f = 45$ mm; (c) $P = 2.5$ kW and $v_m = 16.6$ mm/s

As the angle of inclination of the nozzle α climbs up to 45°, the bead height increases under the action of the dynamic pressure of the gas-powder jet striking the molten metal. With a further increase in the angle α, a much smaller amount of powder gets into the melt and the built-up layer becomes thinner. The bead becomes somewhat wider with a larger value of α, whereas the coefficient m rises from 0.05 to 0.47.

In comparison with the processes of laser melting of powder slurries and spray coatings, laser cladding with gas-powder spraying requires a specific energy ranging from 30 to 50 J/mm^2.

The direction of the gas-powder jet relative to the direction of a specimen travel has a noticeable effect on the bead shaping.

Figure 11.10 illustrates the versions of powder spraying in the direction of motion of a specimen at an angle $\omega = 0$ of the nozzle to the axis of the specimen motion, in the opposite direction to the specimen at $\omega = 180°$, and at angles $0 < \omega < 180°$.

Spraying the powder in the direction of travel of a specimen ensures a stable built-up of the bead whose height and width vary insignificantly in the range from 10 to 15 percent. The gas-powder jet exerts pressure on the molten metal and forces it toward the solidified metal.

In spraying the powder in the direction opposite to that of the specimen motion, the gas jet pushes back the molten metal which spreads over the surface to a certain extent and increases the melt area, so that the height and width

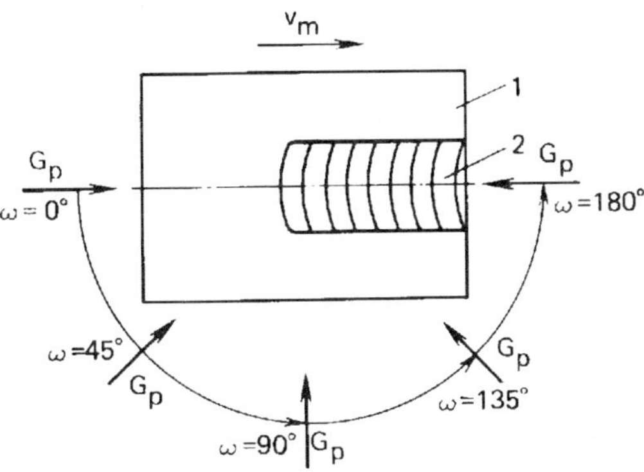

Fig. 11.10. Versions of spraying the powder onto specimen *1* to build up bead *2*

of the bead become unstable at $\omega = 180°$ and vary along the bead length within 50 to 60 percent.

With an increase in the angle ω from 0° to 90°, the jet displaces the liquid metal from the bead in the direction normal to the bead axis. This distorts the geometry of the bead, which is particularly the case when the angle ω increases from 45° to 135°.

If the laser beam axis passes through the center of the area of the gas powder sprayed on the target, the bead surface gets rough because the powder particles fall on the rear portion of the metal that has not yet solidified completely. The bead remelted with the beam in the second pass without applying the powder assumes a smooth surface.

The process of cladding with the laser beam split into two beams, the powder being sprayed in the direction of the specimen motion onto the first hot spot, yields a uniform bead noted for a smooth surface.

An important characteristic of gas-powder spray cladding is the powder utilization factor U defined as the ratio of the mass of the built-up metal to the mass of the powder consumed in the process. This factor increases with the laser power and decreases at a higher melting rate. Variations in U with Δf display an extremum. The plots of the powder utilization factor against the parameters of laser hardfacing with the powder sprayed in the direction of the specimen motion are shown in Fig. 11.11. The utilization factor increases with the mass flow rate G_p (Fig. 11.11*a*) and sharply decreases with increasing distance l (Fig. 11.11*b*). An increase in the angle of incidence α causes U to grow, which reaches the maximal values at α of 40 to 50 ° (Fig. 11.11*c*).

The utilization factor is much higher at $\omega = 180°$ than at $\omega = 0°$ because the melt area in the former case is larger.

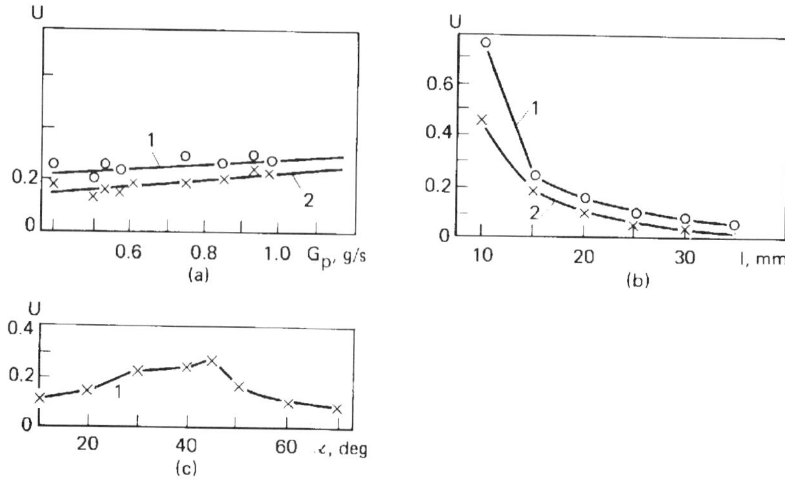

Fig. 11.11. Powder utilization factor U versus G_p, l, and α at $P = 2.5$ kW and $\Delta f = 30$ mm: (a) $l = 15$ mm, $\alpha = 45°$, $v_m = 16.6$ mm/s (curve *1*), and $v_m = 25.0$ mm/s (curve *2*); (b) $G_p = 0.845$ g/s, $\alpha = 45°$, $v_m = 8.3$ mm/s (curve *1*), and $v_m = 16.6$ mm/s (curve *2*); (c) $G_p = 0.845$ g/s, $l = 15$ mm, and $v_m = 16.6$ mm/s (curve *1*)

Proceeding from experimental results, we have worked out a statistical model of the process under consideration and derived regression equations for defining the dimensions of beads and the powder utilization factor [144]. These equations can also be used to solve an inverse problem, namely, to determine the optimal process parameters that afford the bead of requisite dimensions at an appropriate utilization factor U.

In the process of cladding of metals with Ni–Cr–B–Si alloy coatings, the optimal values of the parameters can be chosen to vary in the following ranges: $P = 1$ to 3 kW; $v_m = 16.7$ to 33.3 mm/s; $\Delta f = 30$ to 45 mm at $f = 400$ mm; $G_p = 0.6$ to 0.9 g/s; $l = 15$ to 20 mm; and $\alpha = 30°$ to 35°. The height h and width b of a bead can vary from 0.5 to 2.0 mm and from 1.2 to 2.6 mm, respectively. The factor U can range from 0.4 to 0.7. The dilution coefficient at the above values of the parameters is equal to or smaller than 0.15.

For the beads to be built up at m of 0.05 to 0.15, the powder should be sprayed in the direction of the target motion. As noted above, powder spraying in the opposite direction provides a somewhat irregular bead, but results in a lower heat input and shallower depth of melting of the substrate surface [71].

The gases used to eject the powder are air, nitrogen, helium, argon, and carbon dioxide. The type of gas is of little importance for the process of melting of self-fluxing powders. If the coatings are built up of nonself-fluxing powders, inert gases should be used to protect the molten metal against oxidation. The optimal size of the powder particle should be within 40 to 160 μm. The finer particles are liable to clotting and the coarser ones become embedded in the nozzle throat. The process of gas-powder spray hardfacing suffers from incomplete utilization of the powder, calls for a complex powder spraying device, and needs an arrangement for collecting the particles not used in the process.

The process that applies powder mixtures, primarily the powders of self-fluxing alloys based on nickel and cobalt, holds much practical promise. These powders prepared with addition of tungsten, titanium, and boron carbides and also ferrovanadium and ferrotitanium enable us to build up layers that can feature special properties [145].

Laser cladding offers a means for adjusting the dwell time of the liquid phase and for cooling the liquid metal at a high rate. The built-up layer exhibits a better structure than the layer deposited by other techniques using the same powders.

The comparison has been made between the structures of built-up layers formed in the processes of induction melting and in laser melting of an alloy mixture. The induction melting process yields structures consisting of coarse, complex-shaped, primary carbides of Cr_7C_3 and of Cr_3C_2 carbides of the rod-like shape, the hardness of which can range between 15 000 and 23 000 MPa. The structure can also contain other carbides, including finely dispersed borides. Besides, the layer includes white grains of the gamma solid solution of the cellular or dendritic structure with a hardness of 3780 to 5250 MPa and a gray fine-grain structure component with a hardness of 6340 to 8200 MPa, which is the binary eutectic composed of the gamma phase and nickel boride Ni_3B.

The structure and the phase composition of the laser-hardfaced layer are quite different. At a rate of laser cladding exceeding 10 mm/s, the process provides the structure without primary carbides, which includes cellular grains of the gamma solid solution heavily supersaturated with alloying elements and the gamma-Ni_3B eutectic with metastable phases representing one of the forms of carbides and borides. The alloy layer has a hardness of 7900 to 8900 MPa that varies more smoothly in depth.

Laser melting of an iron powder gives the structure of chilled iron composed of the fine-grain austenite-cementite mixture with martensite partially transformed from austenite. The hardness reaches 5250 to 8900 MPa. The process of laser cladding refines the structure of coating, dissolves brittle carbide phases, and forms supersaturated solid solutions, thus enhancing the wear resistance of the built-up layer.

The tests for sliding friction show that the laser clad coatings of the Ni–Cr–B–Si alloy have 3 to 5 times the wear resistance of the plasma arc coatings melted with a gas burner and 10 times the wear resistance of the coatings produced from the same cladding alloy by induction melting. The run-in period for laser clad coatings is found to be much shorter.

The laser clad deposit has a high strength of adhesion equal to the strength of the substrate metal or built-up metal and is 3 to 5 times the adhesive strength of spray coatings.

Laser cladding with gas-powder spraying is a promising process for increasing the strength of high-load-bearing portions of parts, such as seal edges of the valves of ICE valve gear, mounting surfaces of the parts of gas- and water-distribution mechanisms, and various pieces of equipment. Laser cladding can serve as a useful tool in repair and renewal operations for coating the worn-out parts.

REFERENCES

1. Basov, N.G., and Danilychev, V.A. *Industrial High-Power Lasers. Science and Mankind.* Moscow, Znanie, 1985, pp. 261–278 (in Russian).
2. Prokhorov, A.M. (ed.). *Laser Handbook.* Moscow, Sov. Radio, 1978, vol. 1, p. 504 (in Russian).
3. Prokhorov, A.M. (ed.). *Laser Handbook.* Moscow, Sov. Radio, 1978, vol. 2, p. 400 (in Russian).
4. Stelmakh, M.F. (ed.). *Industrial Lasers.* Moscow, Energiya, 1975, p. 216 (in Russian).
5. Veyko, V.P. *Laser Treatment of Film Components.* St. Petersburg, Mashinostroenie, 1986, p. 248 (in Russian).
6. Rykalin, N.N., Uglov, A.A., Zuev, I.V., and Kokora, A.N. *Laser and Electron Beam Material Processing Handbook.* Moscow, Mir Publishers, 1988, p. 591 (in Russian).
7. Kovalenko, V.S. *Techniques of Laser Material Processing.* Kiev, Visshaya Shkola, 1985, p. 88 (in Russian).
8. Rykalin, N.N., Uglov, A.A., and Kokora, A.N. *Laser Material Processing.* Moscow, Mashinostroenie, 1975, p. 296 (in Russian).
9. Proceedings of the All-Union Conference, *Industrial Laser Applications.* Moscow, Nauka, 1986, p. 216 (in Russian).
10. Mikaelyan, A.L., Ter-Mikaelyan M.L., and Turkov, Yu.G. *Solid-State Lasers.* Moscow, Sov. Radio, 1967, p. 384 (in Russian).
11. Abilsiitov, G.A., Velikhov, E.P., Golubev, V.S. et al. *High-Power Gas-Discharge Lasers and Their Applications.* Moscow, Nauka, 1984, p. 106 (in Russian).
12. Prokhorov, A.M., and Shcherbakov, I.A. Chromium-doped garnet lasers. *Izv. Akad. Nauk,* 1987, vol. 51, No. 8, pp. 1341–1353 (in Russian).
13. Danilov, A.A., Nikolskiy, M.Yu., and Shcherbakov, I.A. On thermal and optical generation processes in solid-state lasers. *Izv. Akad. Nauk,* 1987, vol. 51, No. 8, pp. 1431–1440 (in Russian).
14. Basov, N.G., Kabaev, N.K., Danilychev, V.A., et al. A CW CO electroionization closed-cycle laser. *Kvant. Elektronika,* 1979, vol. 6, No. 4, pp. 772–776 (in Russian).
15. Averin, A.B., Basov, N.G., Glotov, E.P., et al. Industrial general-purpose electroionization-type CO_2–CO lasers. *Izv. Akad. Nauk,* 1983, vol. 47, No. 8, pp. 1519–1526 (in Russian).
16. Generalov, N.A., Kosynkin, V.D., et al. A steady-state nonself-maintained electrodeless discharge in the closed-cycle laser, *Fizika Plazmy,* 1980, vol. 6, No. 5, pp. 1152–1161 (in Russian).
17. Generalov, N.A., Zimakov, V.P., and Kosynkin, V.D., An industrial composite-type fast gas transport laser. *Kvant. Elektronika,* 1982, vol. 9, No. 8, pp. 1549–1558 (in Russian).

18. Kozlov, G.I., and Kuznetsov, V.A. A multibeam gas-discharge CW CO_2 laser. *Kvant. Elektronika*, 1986, vol. 12, No. 3, pp. 553–558 (in Russian).
19. Antyukhov, V.V., Bondarenko, A.I., Glova, A.F., et al. An AC-excited high-power multibeam CO_2 laser. *Kvant. Elektronika*, 1981, vol. 8, No. 10, pp. 2234–2237 (in Russian).
20. Kolesnikov, V.Yu., Orlov, B.V., Polskiy, Yu.E., et al. A CO_2-laser discharge chamber. *Kvant. Elektronika*, 1984, vol. 11, No. 5, pp. 957–961 (in Russian).
21. Aleynikov, V.A., Bibikova, V.V., Lysogorov, O.S., et al. A compact CO_2 closed-cycle laser. *Elektron. Promyshlen.*, 1981, Series 5–6, pp. 71–75 (in Russian).
22. Kosyrev, F.K., Kosyreva, N.P., and Lunev, E.I. An experimental laser set-up. *Avtomat. Svarka*, 1976, No. 9, pp. 72–73 (in Russian).
23. Abilsiitov, G.A., Artamonov, A.V., Velikhov, E.P., et al. An industrial 10-kW CO_2 laser. *Kvant. Elektronika*, 1980, vol. 7, No. 11, pp. 2467–2473 (in Russian).
24. Ivanchenko, A.I., Krasheninnikov, V.V., and Ponomarenko, A.G. On CO_2 lasers development. Institute of Theoretical Mechanics, Academy of Sciences, Novosibirsk, 1986, No. 6, p. 34 (in Russian).
25. Levin, G.I. An industrial high-power CO_2 laser. *Kvant. Elektronika*, 1983, No. 12, pp. 2493–2496 (in Russian).
26. Baranov, V.Yu., Velikhov, E.P., Kazakov, S.A., et al. The method of multiphoton molecular dissociation for isotope separation by a high-power CO_2 laser beam. *Kvant. Elektronika*, 1979, vol. 6, No. 4, pp. 811–823 (in Russian).
27. Vedenov, A.A., Drobyazko, S.V., and Korzinkin, M.M. Characteristics of a periodically pulsed closed-cycle CO_2 laser. *Kvant. Elektronika*, 1980, vol. 7, No. 6, pp. 1186–1191 (in Russian).
28. O'Shea, D.C., Callen, W.R., and Rhodes, W.T., *Introduction to Lasers and Applications*, Reading, MA: Addision-Wesley Publishing Co., 1978.
29. Garashchuk, V.P. Requirements for a CO_2 laser for welding set-ups. *J. Avtomat. Svarka*, 1980, No. 2, pp. 49–52 (in Russian).
30. Tenkins, J.A., and White, W.E. *Fundamentals of Optics*. New York, McGraw-Hill Book Co., 1976.
31. Grigoryants, A.G., and Fromm, V.A. Optimizing the focused laser beam for welding. *Akad. Nauk*, 1984, No. 5, p. 56 (in Russian).
32. Klimkov, Yu.I. *Design of Optoelectron Devices with Lasers*. Moscow, Sov. Radio, 1978, p. 262 (in Russian).
33. Churilovskiy, V.N. *The Theory of Third-Order Chromatic Aberrations*. St. Petersburg, Mashinostroenie, 1968, p. 366 (in Russian).
34. Rayzer, Yu.P., *Basic Physics of Gas-Discharge Processes*. Moscow, Nauka, 1980, p. 416 (in Russian).
35. Vedenov, A.A., and Gladush, G.G. *Physics of Laser Material Processing*. Moscow, Energoatomizdat, 1985, p. 207 (in Russian).
36. Shestakov, N.V., Shanin, O.I., Shanin, Yu.I., et al. A Controlled Focusing Optical Welding Head for Laser Welding. *Automatic Welding*, 1992, No. 6, pp. 47–49 (in Russian).
37. Vedenov, A.A., Gladush, G.G., and Yavokhin, A.N. On the theory of steady-state optical breakdown in gases near a metallic surface. *Kvant. Elektronika*, 1981, vol. 8, No. 7, pp. 1485–1490 (in Russian).

38. Kovalev, A.S., and Popov, A.M. On gas breakdown by CO_2 laser radiation near a metallic surface without intensive vaporization. *Zh. Teor. Fiz.*, 1981, vol. 51, No. 1, pp. 73–77 (in Russian).
39. Ready, J.F. *Industrial Applications of Lasers.* New York, Academic Press, 1978.
40. Kozlov, G.I. A plasma generator with gas flow. *Zh. Teor. Fiz.*, 1978,vol. 4, No. 10, pp. 586–589 (in Russian).
41. Askaryan, G.A. Self-focusing effect. *Uspekhi Fiz. Nauk*, 1973, vol. 3, No. 2, pp. 249–260 (in Russian).
42. Grigoryants, A.G., Moryashchev, S.F., and Fromm, V.A. Gaseous atmosphere and its effect on weld penetration. *Izv.Vuzov*, 1980, No. 5, pp. 109–112 (in Russian).
43. Marushchenko, V.V., Grigoryants, A.G., and Ivanov, V.V. Gas flow and its effect on the melt depth in CO_2 laser welding of structural materials. *Avtomat. Svarka*, 1983, No. 12, pp. 38–44 (in Russian).
44. Ginzburg, V.A. *Electromagnetic Wave Propagation in Plasma*, Moscow, Fizmatgiz. 1960, p. 314 (in Russian).
45. Gladkov, E.A., Ivanov, V.V., and Tulubenskiy, M.G. Diagnosis of shaping a laser-welded bead using a flat two-fold probe. *Svarochn. Proizvodstvo*, 1985, No. 3, pp. 40–42 (in Russian).
46. Mirkin, L.I. Physical concepts of laser material processing. Moscow, Izdatelstvo MGU, 1975, p. 383 (in Russian).
47. Krylov, K.I., Prokopenko, V.T., and Mitrofanov, A.S. *Laser Applications in Machine Building and Instrument Making.* St. Petersburg, Mashinostroenie, 1978, p. 336 (in Russian).
48. Anisimov, S.I., Imas, Ya.M., Romanov, G.S., et al. A high-power laser beam and its effect on metals. *Izv. Akad. Nauk*, 1987, vol. 51, No. 8, pp. 1431–1440 (in Russian).
49. Andriyakhin, V.M., Grigoryants, A.G., Mayorov,V.S., et al. Carbon steel hardening with a CW CO_2 laser using absorbent coatings. *Izv. Vuzov*, 1983, No. 8, pp. 121–126 (in Russian).
50. Arata Y. and Miyamoto, I. Laser welding. *Technocrat*, 1978, vol. 11, No. 5, pp. 33–42.
51. Sepold, G., Bödecker, V., and Juptner, W. Intensitatsabhabhangige Schmelzbad-geometrie beim Schweissen mit Laser Stahlen, Laser-77 Opto-Electronics. *Proc. of Conference*, JPV Business Press, 1977, pp. 109–112.
52. Rykalin, N.N. *Calculation of Thermal Processes in Welding.* Moscow, Mashgiz, 1951, p. 296 (in Russian).
53. Megaw, J.H. and Kaye, A.S. Multikilowatt laser processing. Laser-77 Opto-Electronics. *Proc. of Conference,* JPC Business Press, 1977, pp. 291–296.
54. Schawlow, A.L. Laser Interactions with Materials. Laser-77 Opto-Electronics, *Proc. of Conference,* JPC Business Press, 1977, pp. 109–112.
55. Velichko, O.A., Molchan, I.V., and Moravskiy, V.E. Continuous laser welding today. *Avtomat. Svarka*, 1977, No. 5, pp. 44–50 (in Russian).
56. Carslow, H.S., and Jaeger, J.C. *Conduction of Heat in Solids.* New York, Oxford University, 1959.

57. Swift-Hock, D.T., and Gick, A.E.F. Penetration welding with lasers. *Welding Journal*, 1973, vol. 52,. No. 11, pp. 492–499.
58. Willgoss, R.A., Megaw, J.H., and Clark, J.N. Laser welding of steels for power plants. *Optics and Laser Technology*, 1979, No. 2, pp. 73–78.
59. Breinan, E.M. and Banas, C.M. High power laser welding. Advanced Welding Technology. *Proc. of 11th Intern. Symp. of JWS*. Osaka, 1975, pp. 1–6.
60. Banas, C.M. High power laser welding. *Optical Engineering*, 1978, vol. 17, No. 3, pp. 210–216.
61. Mazumder, J.M. and Steen, W.M. Welding of Ti-6Al-4V by continuous wave CO_2 laser. *Metal Construction*, 1980, vol. 12, No. 9, pp. 423–427.
62. Olshanskiy, N.A. (ed.). *Handbook of Welding in Machine Building* (in 4 volumes). Moscow, Mashinostroenie, 1978, vol. 1, p. 504 (in Russian).
63. Ivanov, V.V., Baykov, V.V., Grigoryants, A.G., et al. Increasing the melting efficiency in laser welding with dynamic beam focusing. *Welding Technology*, 1984, No. 5, pp. 9–11 (in Russian).
64. Duley, W.W. *Laser Processing and Analysis of Materials*. New York and London, Plenum Press, 1983.
65. Rykalin, N.N. Uglov, A.A., and Smurov, I.Yu. Three-dimensional problems relating to laser heating of metals. *Fizika i Khimiya Obrabotki Materialov*, 1979, No. 2, pp. 3–8 (in Russian).
66. Ainsworth, N.S. Laser Cutting. *Paper Technology and Ind.*, 1978, vol. 19, No. 7, pp. 220–225.
67. Moy, Y.P. Utilization et perspectives des faiscerux de haute puissene dans le traul des metaux, *Soudage et Techniques Connexes*, 1977, vol. 31, No. 5, pp. 197–206.
68. Norenkov, I.P. *Introduction to Automated Design of Devices and Systems*. Moscow, Vysshaya Shkola, 1986 (in Russian).
69. Korshunov, Yu.M. *Mathematical Basis of Cybernetics*. Moscow, Energiya, 1980 (in Russian).
70. Mazumder, J. and Steen, W.M., Heat transfer model for CW laser material processing. *Journal of Applied Physics*, 1980, vol. 51, No. 2, pp. 941–947.
71. Arkhipov, V.E. The Effect of the Ways of Powder Feed on the Process of Laser Surfacing. *Welding Engineering*, 1992, No. 2, pp. 33–35 (in Russian).
72. Ermakov, S.M. *The Monte Carlo Method and Related Problems*. Moscow, Nauka, 1971, p. 327 (in Russian).
73. Warren, R.E., and Sparks, M. Laser heating of a slab having temperature-dependent surface absorptance. *Journal of Applied Physics*, 1979, vol. 50, No. 11, pp. 7952–7957.
74. Tribelskiy, M.I. On the liquid phase surface in laser melting of strongly absorbing materials. *Kvant. Elektronika*, vol. 5, 1978, No. 4, pp. 804–812 (in Russian).
75. Kaydalov, A.A. and Nazarenko, O.K. Liquid metal front motion in electron beam welding. *Avtomat. Svarka*, 1974, No. 12, pp. 60–61 (in Russian).
76. Leskov, G.I., and Nesterenkov, V.M. Electron beam welding of metals. Thermal and hydrodynamic processes in the vapor-gas cavity. *Avtomat. Svarka*, 1978, No. 6, pp. 23–26 (in Russian).
77. Paton, B.E., Leskov, G.I., and Zhivaga, L.I. Bead shaping in electron beam welding. *Avtomat. Svarka*, No. 3, pp. 1–5, 1976 (in Russian).

78. Zuev I.V., Selishchev, S.V., and Skobelkin, V.I. Free oscillations initiated by high-power-density energy sources, *Dokl. Akad. Nauk*, 1980, vol. 254, No. 6, p. 1326 (in Russian).
79. Rykalin, N.N., Zuev, I.V., and Uglov, A.A. *Basics of Electron Beam Material Processing*. Moscow, Mashinostroenie, 1978, p. 239 (in Russian).
80. Bashenko, V.V., Lopota, V.A., Mutkevich, E.A. et al. *Use of High-Power-Density Energy Sources for Increasing Welding Efficiency*. St. Petersburg, LDNTP, 1980 (in Russian).
81. Arata, Y., and Miyamoto, I. Metal heating by laser beam. *Welding Construction*, 1978, vol. 57, No. 8, pp. 1–43.
82. Bashenko, V.V., Gorny, S.G., Lopota, A.A., et al. *On Bead Shaping in CW CO_2 Laser Welding*. St. Petersburg, LDNTP, 1983 (in Russian).
83. Poveshchenko, Yu.A., and Popov, Yu.P. Program package for solving heat source problems. Inst. of Applied Mathematics, 1978, No. 65, p. 24 (in Russian).
84. Grigoryants, A.G., Zakharov, A.V., Kuznetsov, O.A., et al. Mathematical modeling of thermal processes in laser deep-penetration welding. Inst. of Applied Mathematics, 1985, No. 14, p. 26 (in Russian).
85. Vedenov, A.A., *Physics of Electric-Discharge CO_2 Lasers*. Moscow, Energoizdat, 1982, p. 111 (in Russian).
86. Samarskiy, A.A. *Difference Schemas Theory*. Moscow, Nauka, 1977, p. 614 (in Russian).
87. Samarskiy, A.A. *Difference Schemas of Gas Dynamics*. Moscow, Nauka, 1980, p. 351 (in Russian).
88. Vinokurov, V.A., and Grigoryants, A.G. *The Theory of Welding Stresses and Strains*. Moscow, Mashinostroenie, 1984, p. 279 (in Russian).
89. Nikolaev, G.A. *Welded Constructions*. Moscow, Mashgiz, 1962, p. 552 (in Russian).
90. Makhnenko, V.I. *Analysis of Kinetics of Welding Stresses and Strains*. Kiev, Naukova Dumka, 1976, p. 320 (in Russian).
91. Grigoryants, A.G. A method for analysis of welding stresses and strains, *Izv. Vuzov*, 1978, No. 5, pp. 146–150 (in Russian).
92. Kasatkin, B.S., Kudrin, A.B., Lobanov, L.M., et al. *Experimental Methods of Investigation of Stresses and Strains*. Kiev, Naukova Dumka, 1981, p. 584 (in Russian).
93. Vasil'ev, D.M. and Trofimov, V.V. X-ray measurement of macrostresses. *Zavod. Laborat.*, 1984, No. 2 (in Russian).
94. Prokhorov, N.N. *Physical Processes in Metal Welding*. Moscow, Metallurgiya, 1976, vol. 2, p. 599 (in Russian).
95. Vinokurov, V.A. (ed.). *Welding Handbook*, vol. 3. Moscow, Machinostroenie, 1979, p. 568 (in Russian).
96. Gavrilyuk, V.S. and Shcheglov, M.E. A set-up for verifying weld bead solidification. *Avtomat. Svarka*, 1982, No. 8, pp. 70–71 (in Russian).
97. Fedorov, V.G., Grigoryants, A.G., Popov, I.F., et al. Laser-welded bead structure, *Izv. Vuzov*, 1979, No. 2, pp. 122–125 (in Russian).
98. Gavrilyuk, V.S., Grigoryants, A.G., Ivanov, V.V., et al. Laser weld crystallization. *Avtomat. Svarka*, 1983, No. 6, pp. 27–29 (in Russian).
99. Makarov, E.L. *Cold Cracks Developed in Welding Alloy Steels*. Moscow, Mashinostroenie, 1981, p. 248 (in Russian).

100. Uglov, A.A. The Present State of and Promises for the Laser Production Process. *Physics and Chemistry of Materials Processing*, 1992, No. 4, pp. 32–42 (in Russian).
101. Nikolaev, G.A. and Grigoryants, A.G. Laser material processing in machine building. *Izv. Akad. Nauk*, 1983, vol. 47, No. 8, pp. 1458–1467 (in Russian).
102. Bashenko, V.V., and Lopota, V.A. Welding by an inclined laser beam. *Svaroch. Proizvodstvo*, 1981, No. 7, pp. 19–21 (in Russian).
103. Bashenko, V.V., Kulikov, N.V., and Surkov, A.V. The effect of metal surface state on laser weld penetration. *Svaroch. Proizvodstvo,* 1984, No. 5, pp. 16–18 (in Russian).
104. Gurevich, S.M. *Nonferrous Metal Welding Handbook*. Kiev, Naukova Dumka, 1981, p. 375 (in Russian).
105. Mayfield, J. Hornets fly on composite wings. *American Machinist*, 1978, No. 11, pp. 107–110.
106. Desforges, C.D. Laser applications. *Engineering*, 1977, No. 10, pp. 1–8.
107. Strizhev, Yu.N. Laser set-ups for cutting wood materials. *Plity i Fanera*, 1983, p. 32 (in Russian).
108. Sobol, E.N. Material decomposition by a laser beam. *Zh.Tekhnich. Fiziki*, 1982, vol. 52, No. 8, pp. 1697–1699 (in Russian).
109. Grigoryants, A.G., and Sokolov, A.A. Energy aspects of CW CO_2 laser cutting of laminated plastics. *Svaroch. Proizvodstvo*, 1986, No. 1, pp. 32–3?? (in Russian).
110. Babenko, V.P., and Tychinskiy, V.P. Gas-Assist Laser Cutting. St. Petersburg, LDNTP,1976, p. 34 (in Russian).
111. Banas, C.M. and Webb, R. Macromaterials processing. *Proc. of IEEE*, 1982, vol. 70, No. 6, p. 533.
112. Machulka, G.A. *Laser Glass Processing*. Moscow, Sov. Radio, 1979, p. 136 (in Russian).
113. Gornyi, S.G., Emelchenkov, I.R., et al. An Investigation of the Process of Gas-Laser Metal Cutting. *Welding Engineering*, 1992, No. 3, pp. 31–32 (in Russian).
114. Kovalenko, V.S., Romonenko, V.V., and Oleshchuk, L.M. *Laser Cutting with Low Waste*. Kiev, Tekhnika, 1987, p. 112 (in Russian).
115. Tikhomirov, A.V. Prospects of Laser Cutting. *Izv. Akad. Nauk*, 1983, vol. 47, No. 8, pp. 1481–1486 (in Russian).
116. Steen, W.M. and Kamalu, J.N. *Laser Cutting*. Imperial College of Science and Technology, London, 1985, SW7, 2BP. NK, p. 111.
117. Safonov, A.N., and Grigoryants, A.G. *Laser Processing in Machine Building*. Moscow, Mashinostroenie, 1986, p. 47 (in Russian).
118. Kovalenko, V.S., Verkhoturov, A.D., Golovko, L.F., et al. *Laser and Electroerosion Material Hardening*. Moscow, Nauka, 1986, p. 276 (in Russian).
119. Kim, E.I., Grigoryants, A.G., Safonov, A.N., et. al. Analysis of kinetics of steel austenization by a CW laser beam. *Inzhenerno-Fizich. Zhurnal*, 1987, No. 3, pp. 444–449 (in Russian).
120. Volgin, V.I. The effect of laser alloying on the hardness of an aluminum alloy surface layer. *Poverkhnost'*, 1983, No. 1, pp. 125–128 (in Russian).
121. Denisov, V.I., Kuznetsov, A.A., Sarychev, G.A., et al. Abrasive Wear Test Set-

ups with Ultrasonic Wave Recording. *Radiation Experiment Technique*, Moscow, Energoizdat, 1982, pp. 73–78 (in Russian).
122. Kovalenko, V.S., Golovko, L.F., Merkulov, G.V., et al. (ed.). *Strengthening of Parts by Laser Beam.* Kiev, Tekhnika, 1981, p. 132 (in Russian).
123. Andriyakhin, V.M. and Chekanova, N.T. The effect of a multikilowatt CO_2 laser beam on the structure of irons. *Poverkhnost'*, 1983, No. 3, pp. 129–137 (in Russian).
124. Velikikh, V.S., Goncharenko, V.P., Kartavtsev, V.S., et al. Wear and heat resistance of laser-treated tool steels. *Tekhnologia i Organizats. Proizvodstva*, 1978, No. 4, pp. 52–53 (in Russian).
125. Andriyakhin, V.M., Vasilev, V.A., Sedunov, V.K., et al. Laser hardening of cylinder liners, *Metalloved. i Termich. Obrabotka Metallov*, 1982, No. 9, pp. 41-43 (in Russian).
126. Kraposhin, V.S. Laser treatment of metallic surfaces. *Poverkhnost'*, 1982, No. 3, pp. 1–12 (in Russian).
127. Krishtal, M.A., Zhukov, P.A., and Kokora, A.N. *Laser-Treated Alloy Structure.* Moscow, Metallurgiya, 1973, p. 192 (in Russian).
128. Kovalenko, V.S. Pulsed Laser Material Processing. Kiev, Visshaya Shkola, 1977, p. 144 (in Russian).
129. Ready, J.F. Material processing. An overview. *Proc. of the IEEE*, 1982, vol. 70, No. 6, p. 533.
130. Lukin, V.D., Povalyaev, V.A., and Puzhevskiy S.N. Induction hardening of seat-valve gear surfaces. *Dvigatelestroen.*, 1980, No. 11, pp. 39–43 (in Russian).
131. Arkhipov, V.E., Birger, E.M., Greachin, A.N., et al. Industrial laser applications. *Tekhnologiya Avtomobilestroen.*, 1980, No. 5, pp. 24–27 (in Russian).
132. Andriyakhin, V.M., and Fishkis, M.M. *Prospects of Laser Applications in Automative Industry.* Moscow, NIIN, 1980, p. 63 (in Russian).
133. Kalner, V.D., Volgin, V.I., and Andriyakhin, V.M. CO_2-laser hardening of aluminum alloys. *Poverkhnost'*, 1982, No. 12, pp. 131–134 (in Russian).
134. Kudinov, V.V. *Plasma Arc Coatings.* Moscow, Nauka, 1977, p. 184 (in Russian).
135. Akakov, Yu.A., and Edneral, N.V. Laser surface alloying, *Izv. Acad. Nauk*, 1983, vol. 47, No. 8, pp. 1487–1495 (in Russian).
136. Spivak, A.V. CW CO_2 laser beam-liquid interaction. *Papers of Academy of Sciences*, 1986, vol. 290, No. 5, pp. 1107–1111 (in Russian).
137. Kolomytsev, P.T. *Heat-Resistant Diffusion Coatings.* Moscow, Metallurgiya, 1979, p. 271 (in Russian).
138. Lakhtin, Yu.M., and Arzamasov, B.N. *Chemical Heat Treatment of Metals.* Moscow, Metallurgiya, 1985, p. 254 (in Russian).
139. Burakov, V.A., Baryshevskaya, E.A., and Burakova, N.M. Pulsed laser heating for iron carburizing, *Izv. Vuzov.* 1981, No. 11, pp. 28–31 (in Russian).
140. Gnanamuthu D.S. Laser surface treatment. *Optical Engineering*, 1980, No. 5, pp. 783–792.
141. Lakhtin, Yu.M., Kogan, Ya.D., and Tarasova, T.V. Investigations into laser alloying of corrosion-resistant steels. *Elektron. Obrabotka Materialov*, 1985, No. 3, p. 28 (in Russian).
142. Grigoryants, A.G., Redkoborody, Yu.N., Shibaev, V.V., et al. Analysis of

hardfacing by laser-assisted powder melting. *Izv. Vuzov.* 1973, No. 3, pp. 155–159 (in Russian).

143. Grigoryants, A.G., Sofonov, A.N., Shibaev, V.V., et al. Analysis of the process of laser treatment of Cr–B–Ni plasma-arc coatings. *Proc. of TSNITTmash,* 1982, No. 168, pp. 52–55 (in Russian).

144. Grigoryants, A.G., Safonov, A.N., and Shibaev, V.V. On bead shaping in laser-assisted gas-powder spray hardfacing. *Poroshkov. Metallurgiya,* 1984, No. 9, pp. 39–42 (in Russian).

145. Dudko, D.A., Zemzin, V.I., Prikhno, I.G., et al. *Powder Cladding Materials.* Kiev, Naukova Dumka, 1978 (in Russian).

INDEX

UNIV DUNELM

DURHAM UNIVERSITY LIBRARY
3 0104 00772564 5